Development of Alternative Green Sample Preparation Techniques

Development of Alternative Green Sample Preparation Techniques

Editor

Emanuela Gionfriddo

MDPI • Basel • Beijing • Wuhan • Barcelona • Belgrade • Manchester • Tokyo • Cluj • Tianjin

Editor
Emanuela Gionfriddo
The University of Toledo
USA

Editorial Office
MDPI
St. Alban-Anlage 66
4052 Basel, Switzerland

This is a reprint of articles from the Special Issue published online in the open access journal *Separations* (ISSN 2297-8739) (available at: https://www.mdpi.com/journal/separations/special_issues/green_sample_preparation).

For citation purposes, cite each article independently as indicated on the article page online and as indicated below:

LastName, A.A.; LastName, B.B.; LastName, C.C. Article Title. *Journal Name* **Year**, *Article Number*, Page Range.

ISBN 978-3-03943-468-8 (Hbk)
ISBN 978-3-03943-469-5 (PDF)

© 2020 by the authors. Articles in this book are Open Access and distributed under the Creative Commons Attribution (CC BY) license, which allows users to download, copy and build upon published articles, as long as the author and publisher are properly credited, which ensures maximum dissemination and a wider impact of our publications.

The book as a whole is distributed by MDPI under the terms and conditions of the Creative Commons license CC BY-NC-ND.

Contents

About the Editor vii

Preface to "Development of Alternative Green Sample Preparation Techniques" ix

Emanuela Gionfriddo
Development of Alternative Green Sample Preparation Techniques
Reprinted from: *Separations* 2020, 7, 31, doi:10.3390/separations7020031 1

Priscilla Rocío-Bautista, Iván Taima-Mancera, Jorge Pasán and Verónica Pino
Metal-Organic Frameworks in Green Analytical Chemistry
Reprinted from: *Separations* 2019, 6, 33, doi:10.3390/separations6030033 5

Eduardo Carasek, Gabrieli Bernardi, Sângela N. do Carmo and Camila M.S. Vieira
Alternative Green Extraction Phases Applied to Microextraction Techniques for Organic Compound Determination
Reprinted from: *Separations* 2019, 6, 35, doi:10.3390/separations6030035 27

Jason S. Herrington, German A. Gómez-Ríos, Colton Myers, Gary Stidsen and David S. Bell
Hunting Molecules in Complex Matrices with SPME Arrows: A Review
Reprinted from: *Separations* 2020, 7, 12, doi:10.3390/separations7010012 45

Ronald V. Emmons, Ramin Tajali and Emanuela Gionfriddo
Development, Optimization and Applications of Thin Film Solid Phase Microextraction (TF-SPME) Devices for Thermal Desorption: A Comprehensive Review
Reprinted from: *Separations* 2019, 6, 39, doi:10.3390/separations6030039 65

Atakan Arda Nalbant and Ezel Boyacı
Advancements in Non-Invasive Biological Surface Sampling and Emerging Applications
Reprinted from: *Separations* 2019, 6, 52, doi:10.3390/separations6040052 87

Katarzyna Madej and Wojciech Piekoszewski
Modern Approaches to Preparation of Body Fluids for Determination of Bioactive Compounds
Reprinted from: *Separations* 2019, 6, 53, doi:10.3390/separations6040053 113

Govind Sharma Shyam Sunder, Sandhya Adhikari, Ahmad Rohanifar, Abiral Poudel and Jon R. Kirchhoff
Evolution of Environmentally Friendly Strategies for Metal Extraction
Reprinted from: *Separations* 2020, 7, 4, doi:10.3390/separations7010004 131

Andre Cunha Paiva, Daniel Simões Oliveira and Leandro Wang Hantao
A Bottom-Up Approach for Data Mining in Bioaromatization of Beers Using Flow-Modulated Comprehensive Two-Dimensional Gas Chromatography/Mass Spectrometry
Reprinted from: *Separations* 2019, 6, 46, doi:10.3390/separations6040046 159

Alessandra Honjo Ide and José Manuel Florêncio Nogueira
Determination of Hydrophilic UV Filters in Real Matrices Using New-Generation Bar Adsorptive Microextraction Devices
Reprinted from: *Separations* 2019, 6, 45, doi:10.3390/separations6040045 175

Anh V. Le, Sophie E. Parks, Minh H. Nguyen and Paul D. Roach
Optimised Extraction of Trypsin Inhibitors from Defatted Gac (*Momordica cochinchinensis* Spreng) Seeds for Production of a Trypsin Inhibitor-Enriched Freeze Dried Powder
Reprinted from: *Separations* **2019**, *6*, 8, doi:10.3390/separations6010008 . . . . **187**

Marc André Althoff, Andreas Bertsch and Manfred Metzulat
Automation of μ-SPE (Smart-SPE) and Liquid-Liquid Extraction Applied for the Analysis of Chemical Warfare Agents
Reprinted from: *Separations* **2019**, *6*, 49, doi:10.3390/separations6040049 . . . . **203**

About the Editor

Emanuela Gionfriddo is an Assistant Professor of Analytical Chemistry at the Department of Chemistry and Biochemistry of The University of Toledo (OH, USA). Research work in Dr. Gionfriddo's laboratory focuses on the development of advanced analytical separation tools for the analysis of complex biological and environmental matrices, with emphasis on alternative green sample preparation methodologies. She received her B.Sc. (2008) and M.Sc. (2010) in Chemistry and her Ph.D. in Analytical Chemistry (2013) at the University of Calabria (Italy). She joined Prof. Pawliszyn's group at the University of Waterloo (Ontario, Canada) in 2014 as Post-Doctoral Fellow and manager of the Gas-Chromatography section of the Industrially Focused Analytical Research Laboratory (InFAReL), and within three years became a Research Associate. Dr. Gionfriddo is one of the founding members of the Dr. Nina McClelland Laboratory for Water Chemistry and Environmental Analysis at The University of Toledo.

Preface to "Development of Alternative Green Sample Preparation Techniques"

Sample preparation has been, for many years, an overlooked field in separation science. However, in the last three decades, significant progress has been made in the development of techniques that enable the efficient extraction and preconcentration of targeted analytes from a broad range of samples. One of the main challenges that many research groups are tackling from diverse angles is the analysis of complex matrices—in fact, when aiming at the efficient and fast extraction of targeted molecules the analyst must also ensure the minimum coextraction of matrix components that could bias the analysis or contaminate analytical instrumentation. Moreover, considering the growing concerns for environmental protection and resources depletion, the development of new sample preparation techniques should comply, as much as possible, with the principles of Green Analytical Chemistry, promoting reusability and reducing energy consumption and the production of laboratory waste.

This Special Issue of Separations presents the state-of-the-art in the development of alterative green sample preparation techniques and collects 11 outstanding contributions that describe advances in extraction phases chemistry and geometry, evolution of automation for sample preparation and use of chemometric approaches for method development and data interpretation.

Emanuela Gionfriddo
Editor

Editorial

Development of Alternative Green Sample Preparation Techniques

Emanuela Gionfriddo

Department of Chemistry and Biochemistry, Dr. Nina McClelland Laboratory for Water Chemistry and Environmental Analysis, School of Green Chemistry and Engineering, The University of Toledo, Toledo, OH 43606, USA; Emanuela.Gionfriddo@UToledo.Edu

Received: 16 April 2020; Accepted: 8 May 2020; Published: 4 June 2020

Although chemistry disciplines are often regarded by the public as polluting sciences, in the last three decades, the concept of "Green Chemistry" has fueled the development of more sustainable and environmentally friendly chemical processes that are mainly aimed at minimizing the production of toxic laboratory waste, to maximize pollution prevention [1]. Since the establishment of the 12 principles of Green Analytical Chemistry, the analytical chemistry community is striving to apply these principles in the analytical chemistry laboratory, which redefines analytical procedures, with a drastically changed philosophy on analytical method development [2–5]. Among the various steps that constitute the analytical workflow, sample preparation and extraction showed great potential for improvement toward greener approaches, especially for complex matrices, whether for targeted or non-targeted analyses, which present many analytical challenges. Many researchers in the analytical chemistry community have subsequently embraced the challenge and focused their research efforts toward greener and faster sample preparation approaches, guaranteeing minimal consumption of organic solvents, promoting the production of reusable extraction devices, enhancement of analysis throughput through the use of automated systems, use of natural sorptive materials, etc. [6].

The Special Issue of Separations, "Development of Alternative Green Sample Preparation Techniques", aims to provide an update on recent trends in green sample preparation to readers already familiar with the topic and hopefully spark the curiosity and the attention of more analytical chemists toward the importance of this topic.

This Special Issue of Separation collates 11 impressive contributions that describe the state-of-the-art in the development of green extraction technologies, from green materials for microextraction to the development of new sampling devices geometries for enhanced extraction efficiency and analysis throughput.

Seven review articles describe important aspects of green sample preparation:

In terms of green materials for sample preparation, two interesting reviews provide insights on green synthesis of sorbents and use of biomaterials.

Veronica Pino and coworkers at the University of La Laguna (Spain) reviewed the use of Metal–Organic Frameworks (MOFs) as sorbent materials for green sample preparation. The authors survey the characteristics of these materials, giving particular emphasis to potential toxicity issues of neat MOFs and alterative synthetic routes, to ensure green approaches in their preparation [7].

As an alternative to synthetic sorbents, and to further green sample preparation techniques, the use of biopolymers, obtained from renewable natural sources, has attracted the interest of many researchers in the field of sample preparation. Eduardo Carasek and coworkers at the Universidade Federal de Santa Catarina (Brazil) described how various biosorbents can be used as extraction phases for different extraction techniques and discussed several applications to environmental, food, and biofluids analysis. Moreover, the authors describe the use of alternative environmentally-friendly extraction phases, such as Ionic Liquids and SupraMolecular Solvents [8].

Two reviews in this Special Issue discuss the development of the geometries of microextraction devices for enhanced sample throughput and extraction efficiency:

The first review article in the literature exclusively dedicated to the development and application of Arrow SolidPhase Microextraction (SPME) was written by Jason S. Herrington, German A. Gómez-Ríos and coworkers at Restek Corporation (USA). This article reports the development of a novel SPME geometry, Arrow SPME, which guarantees enhanced mechanical robustness, compared to classical SPME fibers, with improved extraction efficiency. Practical aspects of the use of Arrow SPME were described, together with several applications in environmental, food, cannabis, and forensic analysis. Moreover, novel interfaces for direct coupling of Arrow SPME to mass spectrometry and recent developments in coating materials were discussed [9].

Moreover, my research group at The University of Toledo (USA) provided a review article that focused on the development and applications of thin-film microextraction devices for thermal desorption. In this article, we provided a comprehensive discussion on the development of TF-SPME for thermal desorption. Practical tips for method development and optimization of Thin-Film-SolidPhase Microextraction–ThermalDesoptionUnit–Gas Chromatography/Mass Spectrometry (TF-SPME–TDU–GC/MS) protocols were discussed. Additional detailed outlook on the current progress of TF-SPME development and its future has also been discussed, with emphasis on its applications to environmental, food, and fragrance analysis [10].

The importance of green extraction techniques as applied to clinical/bio-analysis was reviewed in two contributions in this Special Issue:

The applicability of sampling techniques in clinical settings requires the use of non-invasive protocols that pose no harm to living systems, while providing a high degree of pre-concentration and specificity for the analytes of interest. In this context, Ezel Boyaci and coworkers at the Middle East Technical University (Turkey) wrote a review on new approaches to non-invasive biological surface sampling and discussed recent developments in non-invasive in vivo and in situ sampling methods from biological surfaces. Directions for the development of future technology and potential areas of applications, such as clinical, bioanalytical, and doping analyses, were also discussed [11].

Katarzyna Madej and Wojciech Piekoszewski from Jagiellonian University (Poland), provided an interesting overview of the most commonly used microextraction techniques for analysis of biofluids. Considering the complexity of biofluids, modern clinical and forensic toxicological analysis seeks sample preparation methods characterized by high selectivity and enrichment capability, and easy automation and miniaturization, with minimal sample size requirement. These unique features were well documented and described in this review, focusing on microextraction approaches, such as liquid-phase techniques (e.g., single-drop microextraction, SDME; dispersive liquid–liquid microextraction, DLLME; hollow-fiber liquid-phase microextraction, HF-LPME) and sorbent-based extraction techniques (solid-phase microextraction, SPME; microextraction in packed syringes, MEPS; disposable pipette tip extraction, DPX; stir bar sorption extraction, SBSE) [12].

Finally, among the review articles is an interesting contribution on the evolution of green-sample preparation strategies for metal extraction from Jon R. Kirchhoff's research group at the University of Toledo (USA). Extraction of metal analysis is a topic rarely discussed in terms of microextraction techniques and green sampling procedures. The review article thus offers readers a clearly defined viewpoint on various extraction strategies that minimize the use of organic solvents, such as the application of micro-methodology to minimize waste with reduced costs, improved safety, and the utilization of benign or reusable materials for extraction of metal ions from environmental samples by solvent- and sorbent-based extraction techniques [13].

In this Special Issue, four research papers have been collated from experts in the field of analytical chemistry:

At the University of Campinas (Brazil), Leandro W. Hantao's group developed a HeadSpace-SolidPhase Microextraction comprehensive two-dimensional gas chromatographic/Mass Spectrometry (HS-SPME-GCxGC/MS) method to establish the contribution of *Brazilian Ale 02* yeast strain to the aroma profile of beer compared with the traditional Nottingham yeast. The use of HS-SPME enabled sampling of the beer aroma and the introduction of the extracted analytes into the comprehensive two-dimensional gas chromatographic system, without the use of organic solvents. Critical in this study was also the application of a multiway principal component analysis approach for data processing to distinguish beer samples based on yeast strain [14].

José Manuel F. Nogueira and coworkers at the University of Lisbon (Portugal), developed a new generation of bar adsorptive microextraction devices (BAµE) that, combined with micro-liquid desorption, followed by high-performance liquid chromatography–diode array detection (BAµE-µLD/HPLC–DAD), enabled the determination of two very polar ultraviolet (UV) filters (2-phenylbenzimidazole-5-sulfonic acid (PBS) and 5-benzoyl-4-hydroxy-2-methoxybenzenesulfonic acid (BZ4)) in aqueous media. This contribution showed an innovative analytical cycle that includes the use of disposable devices, making BAµE a user-friendly and suitable approach for routine work and a remarkable analytical alternative for trace analysis of priority compounds in real sample matrices [15].

Another contributing work to the green extraction procedure can be seen from the work by Paul D. Roach and coworkers, as they explored the use of water (non-toxic solvent) for effective extraction of trypsin inhibitors from defatted Gac (*Momordica cochinchinensis* Spreng) seeds, to produce trypsin-inhibitor-enriched freeze-dried powder. The optimization of the extraction procedure resulted in high-quality powder in terms of its highly specific trypsin inhibitor activity (TIA) and physical properties [16].

An additional important aspect of green extraction technology is, without doubt, the ability of miniaturization and automation. Considering these factors, Marc A. Althoff and coworkers developed a Smart-SolidPhase Extraction (Smart-SPE) protocol for analysis of chemical warfare agents. The innovation demonstrated in the ability to use fully automated analytical workflows that included several steps, such as sample vortexing, liquid–liquid extraction, and µ-SPE. Such fully automated analytical workflows avoid errors prone to off-line protocols and can help to minimize health risks for lab personnel when toxic substances may be analyzed [17].

Serving as Guest Editor of this Special Issue has been a very exciting experience. I would like to express my deepest gratitude to all the authors for their brilliant contributions and invite the readers of *Separations* to take full advantage of all the important information that this Special Issue provides, hoping that many of these strategies will be broadly applied in many analytical chemistry laboratories.

Funding: This research was funded by The University of Toledo.

Conflicts of Interest: The authors declare no conflict of interest.

References

1. Anastas, P.T.; Warner, J.C. *Green Chemistry: Theory and Practice*; Oxford University Press: New York, NY, USA, 1998; Available online: https://books.google.com/books/about/Green_Chemistry.html?id=SrO8QgAACAAJ (accessed on 17 July 2019).
2. Tobiszewski, M.; Mechlińska, A.; Namieśnik, J. Green analytical chemistry—Theory and practice. *Chem. Soc. Rev.* **2010**, *39*, 2869–2878. [CrossRef]
3. Gałuszka, A.; Migaszewski, Z.; Namieśnik, J. The 12 principles of green analytical chemistry and the SIGNIFICANCE mnemonic of green analytical practices. *TrAC—Trends Anal. Chem.* **2013**, *50*, 78–84. [CrossRef]
4. Armenta, S.; Garrigues, S.; de la Guardia, M. The role of green extraction techniques in Green Analytical Chemistry. *TrAC—Trends Anal. Chem.* **2015**, *71*, 2–8. [CrossRef]
5. Armenta, S.; Garrigues, S.; de la Guardia, M. Green Analytical Chemistry. *TrAC—Trends Anal. Chem.* **2008**, *27*, 497–511. [CrossRef]
6. Armenta, S.; Garrigues, S.; Esteve-Turrillas, F.A.; de la Guardia, M. Green extraction techniques in green analytical chemistry. *TrAC—Trends Anal. Chem.* **2019**, *116*, 248–253. [CrossRef]
7. Rocío-Bautista, P.; Taima-Mancera, I.; Pasán, J.; Pino, V. Metal-organic frameworks in green analytical chemistry. *Separations* **2019**, *6*, 33. [CrossRef]

8. Carasek, E.; Bernardi, G.; Carmo, S.N.D.; Vieira, C.M.S. Alternative green extraction phases applied to microextraction techniques for organic compound determination. *Separations* **2019**, *6*, 35. [CrossRef]
9. Herrington, J.S.; Gómez-Ríos, G.A.; Myers, C.; Stidsen, G.; Bell, D.S. Hunting molecules in complex matrices with spme arrows: A review. *Separations* **2020**, *7*, 12. [CrossRef]
10. Emmons, R.V.; Tajali, R.; Gionfriddo, E. Development, optimization and applications of thin film solid phase microextraction (TF-SPME) devices for thermal desorption: A comprehensive review. *Separations* **2019**, *6*, 39. [CrossRef]
11. Nalbant, A.A.; Boyacı, E. Advancements in non-invasive biological surface sampling and emerging applications. *Separations* **2019**, *6*, 52. [CrossRef]
12. Madej, K.; Piekoszewski, W. Modern Approaches to Preparation of Body Fluids for Determination of Bioactive Compounds. *Separations* **2019**, *6*, 53. [CrossRef]
13. Sunder, G.S.S.; Adhikari, S.; Rohanifar, A.; Poudel, A.; Kirchhoff, J.R. Evolution of Environmentally Friendly Strategies for Metal Extraction. *Separations* **2020**, *7*, 4. [CrossRef]
14. Paiva, A.C.; Oliveira, D.S.; Hantao, L.W. A Bottom-Up Approach for Data Mining in Bioaromatization of Beers Using Flow-Modulated Comprehensive Two-Dimensional Gas Chromatography/Mass Spectrometry. *Separations* **2019**, *6*, 46. [CrossRef]
15. Ide, A.H.; Nogueira, J.M.F. Determination of Hydrophilic UV Filters in Real Matrices Using New-Generation Bar Adsorptive Microextraction Devices. *Separations* **2019**, *6*, 45. [CrossRef]
16. Le, A.V.; Parks, S.E.; Nguyen, M.H.; Roach, P.D. Optimised Extraction of Trypsin Inhibitors from Defatted Gac (Momordica cochinchinensis Spreng) Seeds for Production of a Trypsin Inhibitor-Enriched Freeze Dried Powder. *Separations* **2019**, *6*, 8. [CrossRef]
17. Althoff, M.A.; Bertsch, A.; Metzulat, M. Automation of μ-SPE (Smart-SPE) and Liquid-Liquid Extraction Applied for the Analysis of Chemical Warfare Agents. *Separations* **2019**, *6*, 49. [CrossRef]

© 2020 by the author. Licensee MDPI, Basel, Switzerland. This article is an open access article distributed under the terms and conditions of the Creative Commons Attribution (CC BY) license (http://creativecommons.org/licenses/by/4.0/).

Review

Metal-Organic Frameworks in Green Analytical Chemistry

Priscilla Rocío-Bautista [1,2,†], Iván Taima-Mancera [1,2,†], Jorge Pasán [2] and Verónica Pino [1,3,*]

1   Department of Chemistry, Analytical Division, University of La Laguna, 38206 Tenerife, Spain
2   X-ray and Molecular Materials Lab (MATMOL), Physics Department, University of La Laguna, 38206 Tenerife, Spain
3   University Institute of Tropical Diseases and Public Health, University of La Laguna (ULL), La Laguna, 38206 Tenerife, Spain
*   Correspondence: veropino@ull.edu.es; Tel.: +34-922-318-990
†   These authors contribute equally to this work.

Received: 31 May 2019; Accepted: 21 June 2019; Published: 27 June 2019

Abstract: Metal-organic frameworks (MOFs) are porous hybrid materials composed of metal ions and organic linkers, characterized by their crystallinity and by the highest known surface areas. MOFs structures present accessible cages, tunnels and modifiable pores, together with adequate mechanical and thermal stability. Their outstanding properties have led to their recognition as revolutionary materials in recent years. Analytical chemistry has also benefited from the potential of MOF applications. MOFs succeed as sorbent materials in extraction and microextraction procedures, as sensors, and as stationary or pseudo-stationary phases in chromatographic systems. To date, around 100 different MOFs form part of those analytical applications. This review intends to give an overview on the use of MOFs in analytical chemistry in recent years (2017–2019) within the framework of green analytical chemistry requirements, with a particular emphasis on possible toxicity issues of neat MOFs and trends to ensure green approaches in their preparation.

Keywords: metal-organic frameworks; analytical chemistry; sorbent materials; stationary phases; sensors; sample preparation; green considerations

---

1. Introduction

Metal-organic frameworks (MOFs) belong to a subclass of 3D coordination polymers, formed by metallic clusters and organic ligands through coordination bonds [1–3]. These hybrid materials are characterized by their crystallinity, ultra-low densities, permanent porosity, the presence of accessible cages, tunnels and modifiable pores, and exhibit the highest known surface areas [4,5]. Furthermore, they present adequate mechanical and thermal stability. The proper selection of certain metallic centers and specific ligands, together with a rational control of the synthetic approach, serve to design crystals with a controlled structure and topology, with the resulting MOF even being able to undergo post-modifications. This tuneability has resulted in the characterization of more than 80,000 MOFs [6]. Figure 1 schematically presents well-known MOFs structures to highlight their impressive versatility.

Given this group of interesting properties, it is not surprising that these tailorable materials have been incorporated into an impressive number of applications within different scientific fields: as photovoltaic materials [7], in gas purification and separation [8,9], and gas storage systems [9,10], in catalysis [11], in biomedicine [12], etc. Analytical chemistry is not an exception; around a hundred MOFs form part of these studies [13]. MOFs are mainly involved in analytical chemistry as sorbents in analytical sample preparation methods [14–19], as stationary phases in analytical separation techniques [13,17,18,20,21], and as sensors in spectroscopy and/or electroanalytical methods [22–24].

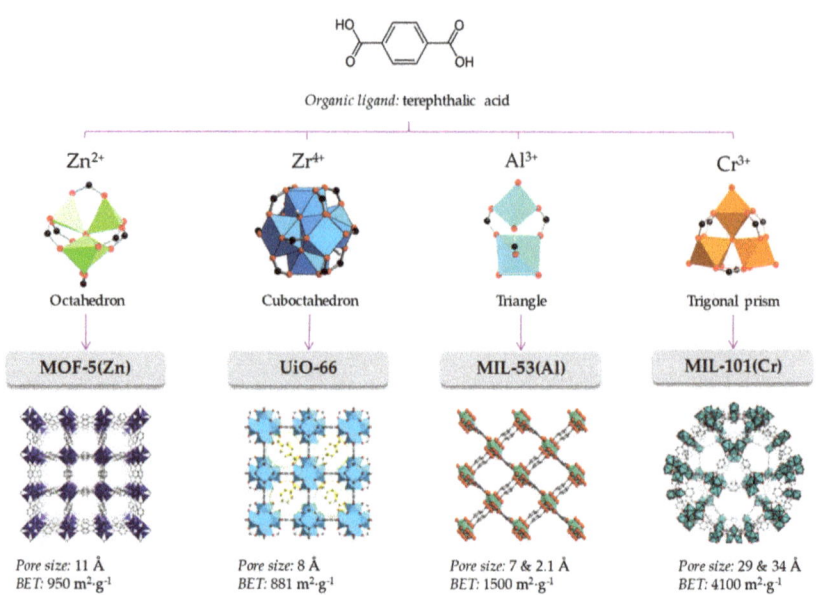

Figure 1. The node-and-connector approach to prepare MOFs. The adequate selection of the organic linker (linear in the case of terephthalic acid) and connection geometry of the metal cluster lead to the desired topology. Each framework topology has its characteristic pore size and available surface.

Current goals in analytical chemistry imply fulfilling as much as possible the requirements of green analytical chemistry (GAC) [25–27]. This way, the quality of an analytical method cannot link exclusively to its common quality features, e.g., accuracy, precision, detection limits, but it must also compile an estimation of method risks for the operator and for the environment [28,29]. GAC enforces the development of methods that ensure the elimination (or minimization of the consumption) of highly toxic chemical reagents (particularly toxic organic solvents), the minimization of gases emissions, the minimization of effluents and solid wastes discharges, the reduction of analysis times and number of analytical steps, and the incorporation of automation to the procedures [30–32]. When direct analysis is not possible due to sample complexity or difficulty in the analytical determination, miniaturization of the analytical method is the most common approach to follow GAC trends [25,32,33]. In any case, the use of novel solvents and/or sorbents, with environment-friendly characteristics is another important trend associated with the GAC approach [34,35].

Considering the rapid expansion of the analytical chemistry field of MOFs, it is important to consider that these and future analytical MOF-based methods must be incorporated into GAC. From this perspective, environment-friendly methods based on MOFs should include (partially or all) the following aspects: green design and synthesis of MOFs, evaluation of toxicity issues of MOFs, and incorporation of MOFs in GAC methods, mainly in miniaturized procedures. This review article will place particular emphasis upon trends which ensure the use of green approaches in the preparation of neat MOFs and/or in their incorporation into less-harmful analytical chemistry methods, limiting the overview to articles published in the years 2017–2019. It is also important to highlight that the attention will center upon bare MOF crystals rather than MOFs-composites.

2. Green Considerations during MOFs Preparation

2.1. MOFs Design

The design of the MOF material constitutes a critical stage, primarily because the proper choice of metal centers and ligand connectivity is responsible of generating a particular crystal structure (a framework topology) with specific characteristics of pore size and window [36,37]. Metal centers act as nodes of the framework, and are characterized by the number and direction of its connecting positions. In Figure 1, for example, the Zn_4O of MOF-5 is an octahedral node with six available connections located at the vertices of an octahedron. Ligands are characterized by their connectivity, or in other words, by how many metal centers can be linked. Thus, for example, in Figure 1, the terephthalic acid is a linear rod-like connector. The adequate combination of linker and nodes produces a particular topology; in the case of octahedral nodes with linear linkers, a primitive cubic framework is produced. When an application is pursued for a MOF, that application's requirements have to be taken into consideration. Therefore, a sorbent is required for an analytical sample preparation method, the desired MOF should be able to adsorb and release the analytes, while being stable in different solvents. If the MOF is going to be a chromatographic stationary phase, other considerations such as adequate permeability and high mechanical and/or thermal stability are additional aspects of interest.

This section will only focus on those aspects concerning the MOFs design that directly relate with the GAC considerations (assuming the obvious pursue of other important characteristics from an analytical chemistry performance). The main aspects to be considered when fabricating greener and sustainable metal-organic framework materials include: (i) Proper selection of the metal and metal source (reducing toxic metals and byproducts); and (ii) Adequate selection of the organic ligand (to ensure safer synthetic procedures and sustainable MOFs). The requirements in the MOF design and synthesis to comply for GAC issues and analytical application needs are summarized in Figure 2.

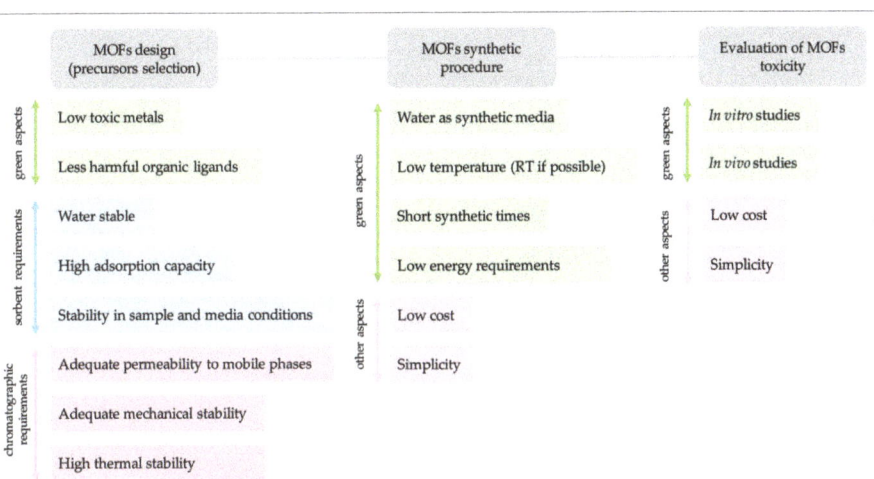

Figure 2. Considerations when preparing a MOF for analytical chemistry applications.

MOFs are chemical and physical stable materials, but in the event of degradation or decomposition, there are concerns due to the quantities of metal released to the environment. Thus, low-toxicity metal ions are preferable in the synthesis of MOFs, such alkaline earth metals (Mg or Ca), or Mn, Fe, Al, Ti and Zr [38,39]. Another consideration when selecting metals is the handling of their salts when preparing MOFs. Thus, metal nitrates and perchlorates have risks of explosion due to oxidation upon handling, while chloride may induce corrosion. Metal oxides and hydroxides only liberate water as by-products, and are therefore ideal inorganic reactants; however, these salts often exhibit insufficient solubility/reactivity [40]. Another strategy to avoid problems with the anions of the metal salts is the use of zero-valence metal precursors and recently, the syntheses of MIL-53(Al), HKUST-1 or ZIF-7 have been reported to undergo from elementary metal precursors [41].

In any case, this general statement about ideal green metals for MOF does not mean that one cannot select other metals to prepare green MOFs or MOFs for green applications. For example, MOFs-containing rare earth metals are typically included in sensing [42], because the balance between the toxicity issues of the metal, the low amount involved in the sensor, and the benefits of the resulting material, clearly shift towards the selection of those metal ions. Finally, and even more importantly, the costs associated to the metal selection are of great significance when an industry-based application is concerned.

Regarding the organic linkers, some of the carboxylic acids are readily available in low cost, such as fumaric, succinic or terephthalic. However, tailored linkers used to reach the enormous surface areas reported for MOF materials in research laboratories often require multistep synthetic reactions with intermediates and byproducts. To comply with GAC requirements, the number of steps has to be as low as possible, while ensuring the low-toxicity of intermediates and byproducts. In general, it should be an equilibrium between low-toxicity and adsorption capacity of the MOF when tailoring a specific ligand for analytical applications. Figure 3 includes a group of representative organic ligands. In any case, it is advisable the use of simple and even biodegradable ligands in MOF design such as peptides, carbohydrates, amino acids, and cyclodextrins.

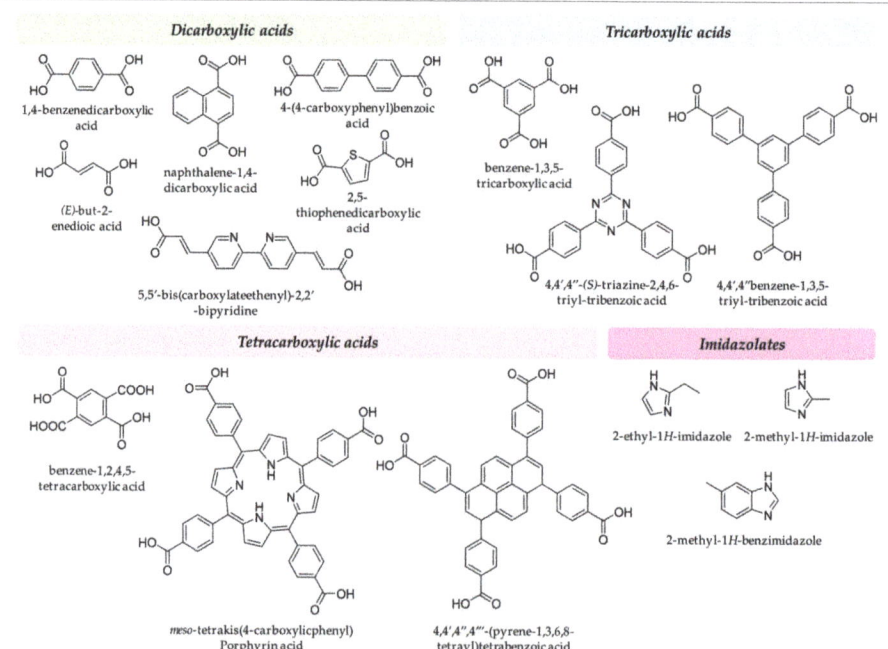

Figure 3. Representative examples of organic ligands used in MOF synthesis.

2.2. MOFs Synthesis

If there is a proper design of the MOF, thus implying adequate selection of precursors, the main parameters exerting an influence in the "green" synthesis of a MOF are: (i) type and duration of the input energy for the coordination bond formation, and (ii) solvent nature and volume [43].

Metal-organic frameworks can be synthesized following different crystallization procedures, although the most common strategy is the solvothermal method. In the solvothermal approach, the metal salt and the organic linker are mixed in a solvent (usually with a high boiling temperature). Then, this solution is introduced in an autoclave, followed by heating at an adequate temperature (between 80 and 300 °C), during a certain time (in most cases, less than 96 h). Other synthetic approaches (classified as a function of the energy input) include: microwave-assisted, electrochemical, mechanochemical, and sonochemical [44–47].

Recent advances in crystallization methods for MOF materials focus on environment-friendly alternatives to the classic solvothermal method [48]. It is worth noting that most of the reactions originally carried out by the solvothermal approach can also be performed in good yields through different, more environment-friendly, and also low-cost methods [46]. Microwave heating or ultrasound irradiation are simple, inexpensive and scalable routes towards the fabrication of MOFs. The reaction times using these techniques can be shortened, resulting in high-purity phases and high yields [49]. Mechanochemical synthesis has also appeared as a promising strategy to scale up the fabrication of MOFs. It is simple, and rapid, and although it may require some solvent addition (for example in liquid-assisted grinding, LAG), they are always in small amounts [47].

A simple change in the solvothermal reaction to make the MOF synthesis more sustainable implies the use of water instead of organic solvents. It must be noted that the most commonly used solvent, dimethylformamide, is a teratogen that readily decomposes at solvothermal temperatures into corrosive products. When using water, the main concern is the solubility of the reactants, but the use of the carboxylic acids in the form of sodium salts can help, while also precluding the formation of corrosive byproducts like HNO_3 or HCl (coming from the metal salt and the deprotonation of the acid) [50]. Other green solvents can be used to better comply with the GAC requirements, among them, ionic liquids or urea derivatives [51]. In recent years, an increasing number of hydrothermal preparations of MOFs have been accomplished at atmospheric pressure with promising results in term of yield and simplicity [52–55].

As a final consideration regarding greener synthesis of MOFs, it would be advisable to use chemometric tools during the optimization of the synthesis, such as an experimental design [56,57], rather than a one-factor-at-a-time optimization. This type of procedure would make it possible to minimize time and quantities of reagents when preparing an adequate MOF, and thus, would be in consonance with green chemistry.

2.3. Evaluation of MOFs Toxicity

The few studies dealing with the toxicity evaluation of MOFs are mostly related to in vitro cytotoxicity studies [58,59], which were usually performed to evaluate the performance of those MOFs for use as drug nano-carriers [60]. Moreover, the majority of these studies focuses on MOFs included in composites forming such carriers. In fact, most studies do not try to evaluate the MOF cytotoxicity itself, but, in general, to compare with the composite in which the MOF is included.

In the reported in vitro cytotoxicity studies, different cell lines or macrophages incubate in presence of MOFs suspensions for a certain amount of time. Afterwards, because of the appearance in the culture media of a compound coming from the cell lysis, a colorimetric detection is commonly used. The released compound is then able to react with the specific reagent of the cytotoxicity test. Among in vitro cytotoxicity tests, MTT [58,61–65], the alamarBlue® cell viability assay [54,66], and CCK-8 [67], are the most widely used with MOFs. Table 1 includes a list of cytotoxicity tests performed with neat MOFs so far, resulting in adequate cytotoxicity values [54,58–61,63,65,67–70]. These results support the fact that it results safe the use of relatively high concentrations of these MOFs aqueous suspensions. As a note, all these results must be taken only as qualitative data with MOFs, because conventional cytotoxicity studies are designed for soluble compounds (and only water suspensions can be prepared with MOFs). It is also important to mention that cell lines may also play an important role in the toxicity assays performed since for example macrophages are more robust cells than HeLa ones. An alternative approach to evaluate these MOFs suspensions should be proposed since the assays described so far are likely to be distant from what it is actually occurring in real conditions.

Table 1. Representative examples of cytotoxicity studies of MOFs.

MOF (metallic ion & ligand)	Cytotoxicity Test Type	Cell Line Type	Toxicity Value (mg·L^{-1})	Ref.
UiO-66 (Zr^{4+} & H$_2$bdc [1])	MTT [2]	MCF-7	>80 mg·L^{-1}	[61]
MIL-100(Fe) (Fe^{3+} & H$_3$btc [3]) UiO-66 (Zr^{4+} & H$_2$bdc [1]) ZIF-8 (Zn^{2+} & 2-MIm [4]) among many others	MTT [2]	HeLa & J774.2 HeLa & J774.2 HeLa & J774.2	1100 & 700 mg·L^{-1} 400 & 60 mg·L^{-1} 100 & 25 mg·L^{-1}	[58]
[Zn$_6$(L)$_3$(DMA)$_4$]·5DMA (Zn^{2+} & H$_4$L [5])	MTT [2]	Hela & HEK293	5000 mg·L^{-1}	[63]
MIL-101(Cr) (Cr^{3+} & H$_2$bdc [1])	MTT [2]	Hep-2 MCF-7	744 mg·L^{-1} 200 mg·L^{-1}	[64]
ZIF-8 (Zn^{2+} & 2-MIm [4])	MTT [2]	MDA-MB-231 MDA-MB-468	400 mg·L^{-1} 400 mg·L^{-1}	[65]
NanoMOF based on Fe (Fe^{3+} & TCPP [6])	MTT [2]	4T1	>200 mg·L^{-1}	[68]
MIL-100(Fe) (Fe^{3+} & H$_3$btc [3])	MTT [2]	HL-7702 & HepG2	80 mg·L^{-1}	[59]
MIL-88A (Fe^{3+} & fumaric acid)	MTT [2]	J774.A1	57 mg·L^{-1}	[60]
[Zn(H$_3$btc [3])(HME [7])] (DMAc [8])(H$_2$O) (Zn^{2+} & H$_3$btc [3] & HME [7])	MTT [2]	SCC-251 & HSC-4	100 mg·L^{-1}	[69]
UiO-64 (Zr^{4+} & fumaric acid)	MTT [2]	J774 & PBLs	200 mg·L^{-1}	[70]
CIM-80 (Al^{3+} & mesaconic acid)	alamarBlue®[9]	J774.1	>2000 mg·L^{-1}	[54]
ZIF-67 (Co^{2+} & 2-MIm [4]) ZIF-8 (Zn^{2+} & 2-MIm [4])	CCK-8 [10]	293T	>80 mg·L^{-1} 20 mg·L^{-1}	[67]

[1] terephthalic acid (benzene-1,4-dicarboxylic acid)
[2] 3-(4,5-dimethylthiazol-2-yl)-2,5-diphenyltetrazolium bromide dye
[3] trimesic acid (benzene-1,3,5-tricarboxylic acid)
[4] 2-methylimidazole
[5] [1,1':3',1''-terphenyl]-3,3'',5,5''-tetracarboxylic acid
[6] tetrakis(4-carboxyphenyl)porphyrin
[7] protonated melamine
[8] N,N-dimethylacetamide
[9] from resazurin (blue) to resorufin dye (fluorescent red-pink)
[10] cell-counting kit assay, from WST-8 (colorless) to WST-8 formazan dye (orange)

Tamames-Tabar et al. studied thoroughly the in vitro cytotoxicity of up to 14 MOFs. Among other interesting results, authors observed that there were no differences in the cytotoxicity of polymorphs, thus indicating a poor effect of the MOF topology into the resulting cytotoxicity [54]. The same authors also showed that, for the same type of linker, Zn-based MOFs were more cytotoxic than Zr-based MOFs, being the less cytotoxic those MOFs based on Fe [54]. From the group of 14 MOFs tested, in vivo cell penetration studies were carried out only with MIL-100(Fe) for being the best candidate as drug nano-carrier. The studies indicated an immediate cell internalization in J774 mouse macrophages, faster than in epithelial HeLa cell lines [54].

In the last two years, an increasing number of studies have been undertaken on the effects of MOFs on living cells from safety aspects while also taking into account therapeutic considerations [71]. Thus, Shen et al. demonstrated that the MOF MIL-101(Fe) functionalized with amino groups was able to induce cyto-protective autophagy in mouse embryonic fibroblast cells instead of cytotoxicity [71].

3. Analytical Methods Incorporating MOFs

Nowadays, the main uses of neat MOFs in analytical chemistry range from analytical sample preparation (as novel sorbents) to chromatography (as novel stationary or pseudo-stationary phases) and sensing (as novel materials in the sensors) [13]. In the majority of these applications, the GAC requirements [72] cover partially by incorporating the MOFs in microextraction approaches within sample preparation, by requiring quite low amounts of MOFs when preparing the chromatographic phases, or by incorporating minimum amounts of MOFs in the sensors. In few applications, it is also possible to find many GAC aspects fulfilled, that is: green MOF designed, MOF as prepared following a green synthesis and demonstrating low cytotoxicity, plus incorporation in a miniaturized method [54].

3.1. MOFs in Analytical Sample Preparation

Improvements in analytical extraction methods from a GAC perspective include the reduction of analysis times by using, for example, ultrasounds or microwaves to accelerate the extraction step, particularly when dealing with solid samples [73]. Regarding miniaturization, GAC approaches include the generic liquid-phase microextraction (when the extraction solvent amount rarely exceeds 0.5 mL) [74,75] and solid-based miniaturized extraction (when the extraction sorbent amount is below 0.5 g) [76–78] methods. Among sorbent-based miniaturized approaches, it is possible to distinguish many sub-modes. Thus, micro-solid-phase extraction (µ-SPE) is quite similar to conventional solid-phase extraction, but requiring lower sorbent amounts in devices such as micro-columns, syringe tips or bodies, mini-disks, etc. Dispersive miniaturized solid-phase extraction (µ-dSPE) utilizes the sorbent material in direct contact with the sample, followed by further separation and desorption. The magnetic-based version (m-µ-dSPE) needs a magnetic material as sorbent, thus avoiding further centrifugation and/or centrifugations steps during the method, because the sorbent separates easily from the sample with the aid of an external magnet. Solid-phase microextraction (SPME), in its more conventional version, utilizes thin fibers coated with a sorbent material, whereas stir-bar sorptive microextraction (SBSME) uses a stir bar coated with a sorbent material.

Table 2 shows representative examples of performance for different analytical sorbent-based microextraction techniques (µ-SPE, µ-dSPE, m-µ-dSPE, SPME and SBSME) incorporating MOFs as novel sorbent materials [54,79–99].

If we take into consideration the idea of avoiding the use of heavy metals when designing MOFs, it is clear that many MOFs included in Table 2 present non-toxic metallic ionic centers [54,65], such as Zn^{2+} [93], Zr^{4+} [82] or Al^{3+} [54]. Regarding the synthetic solvents in MOFs preparations, most studies employ DMF [82,90,94], avoiding in this way toxic chlorinated ones, such as dichloromethane or chloroform. Furthermore, the DMF volumes required are usually low (around 30 mL for 15 mg of MOF). It is noticeable that several MOFs used in these microextraction methods include water as solvent in their preparation (hydrothermal synthesis) [54,81]. The temperatures needed during the preparation of these crystals are usually moderate, i.e., between 80 and 150 °C [94,96], and sometimes even room temperature [93]. Regarding the synthetic times required, they are usually less than 24 h [83,88], and even down to one hour or half an hour [80,83]. Recent publications have favored the utilization of so-called bio-MOFs [90], incorporating a bioorganic ligand in its structure (i.e., adenine).

In general, quite low amounts of MOF as sorbents are always needed [54,82,84] in all microextraction procedures listed, from 2 mg [80] to 60 mg [89]. Regarding extraction times, there are reported values around 3 min or even less [85] for µ-dSPE and some of its applications in the magnetic variant [89,91].

If we focus on the analytical applications, microextraction techniques using MOFs commonly determine organic pollutants in environmental and food samples [86,88], from drugs in waters [90] to PAHs in foods [96]; and also trace metals in food and water samples [80,83].

Table 2. Representative examples of environment-friendly microextraction methods incorporating MOFs as novel sorbents in the period 2017–2019.

MOF (metallic ion & ligand) Synthetic solvent (mL)/T (°C)/time (h)	Microextraction forma/Extraction time (min)	MOF amount	Analytes (number)/Sample (amount)	LOD [1] (ng·L^{-1})/RSD [2] (%)	Analytical technique	Ref.
μ-SPE						
UiO-66 (Zr^{4+} & H_2bdc [3]) DMF [4] (40)/120/24	PP [5] cartridge (packing the MOF suspension)/40	10 mg	hormones (4)/waters (20 mL)	2.0–10/<6.5	LC [6]-MS/MS [7]	[79]
PCN-222 (Zr^{4+} & H_2TCPP [8]) DMF [4] (40)/100/1	pipette tip (packing the MOF powder)/<7	2 mg	Hg^{2+} (1)/fish (100 mL aqueous extract)	20/8	CVAAS [9]	[80]
μ-dSPE						
MIL-53(Al) (Al^{3+} & H_2bdc [3]) water (-)/210/72	MOF powder/30	8 mg	hormones (8)/water & urine (8 mL)	1.5–1000/<7.8	LC [6]-MS/MS [7]	[81]
UiO-66-NO_2 (Zr^{4+} & O_2N-H_2bdc [10]) DMF [4] (15)/150/24	MOF powder/3	20 mg	EDCs [11] (9)/waters (20 mL)	1.5–90/<14	LC [6]-DAD [12]	[82]
Al-Fu nano-flakes (Al^{3+} & fumaric acid) water (200)/90/0.5	MOF powder/20 s	30 mg	Cu^{2+} (1)/waters & food (25 mL sample or aqueous extract)	0.2 μg·L^{-1}/6.5	GFAAS [13]	[83]
CIM-81 (Zr^{4+} & Htz [14] + H_2bdc [3]) DMA [15] (15)/120/72	MOF powder/1	10 mg	PCPs [16] (9)/waters (10 mL)	0.5–1.5 μg·L^{-1}/<13	LC [6]-UV [17]	[84]
CIM-80 (Al^{3+} & mesaconic acid) water (15)/150/3	MOF powder/3	20 mg	PAHs [18] (15) & EDCs [11] (7)/waters (10 mL)	0.75–9.3 (PAHs) & 0.11–21 μg·L^{-1} (EDCs)/<19	LC [6]-UV [17] & LC [6]-FD [19]	[54]
H_2N-MIL-53(Al) (Al^{3+} & H_2N-H_2bdc [20]) DMF [4] (30)/130/72	MOF powder/10 s	30 mg	phenols (8)/waters (10 mL)	0.4–13.3 μg·L^{-1}/<6.30	LC [6]-PDA [21]	[85]
m-μ-dSPE						
Fe_3O_4-CO_2H@MIL-101-NH_2 (Fe^{2+} & NH_2-H_2bdc [20]) DMF [4] (15)/110/24	heterogeneous composite powder/20	20 mg	fungicides (4)/waters (30)	0.04–0.4 μg·L^{-1}/<10.2	LC [6]-UV [17]	[86]
magC [22]@ZIF-8 (Zn^{2+} & 2-MIm [23]) methanol (60)/RT [24]/24	heterogeneous composite powder/10	10 mg	PAEs [25] (9)/diluted human plasma (1:3, overall volume ~9 mL)	3–10/<6.5	GC [26]-MS [7]	[87]
Fe_3O_4/GO [27]-IRMOF-3 (Zn^{2+} & NH_2-H_2bdc [20]) DMF [4] (40)/RT [24]/3	heterogeneous composite powder/30	10 mg	fungicides (5)/lettuce (10 mL aqueous extract)	0.25–1.0 μg·L^{-1}/<7.3	LC [6]-MS/MS [7]	[88]
Fe_3O_4@PDA [28]@ MIL-101(Fe)(Fe^{3+} & H_2bdc [3]) DMF [4] (80)/110/24	heterogeneous composite powder/3	60 mg	SUHs [29] (4)/waters & vegetables (25 mL sample or aqueous extract)	0.12–0.34 μg·L^{-1}/<4.8	LC [6]-PDA [21]	[89]
Fe_3O_4-NH_2/bio-MOF-1 (Zn^{2+} & adenine) DMF [4] (67.5)/130/24	heterogeneous composite powder/40	15 mg	BZTs [30] (6)/urine & waters (40 mL)	0.71–2.49/<8.8	LC [6]-MS [7]	[90]
Fe_3O_4@TMU-10 (Co^{2+} & H_2oba [31]) DMF [4] (10)/145/48	heterogeneous composite powder/4	5 mg	TCAs [32] (2)/plasma & urine (6 mL)	2–4 μg·L^{-1}/<5.2	LC [6]-UV [17]	[91]

12

Table 2. Cont.

MOF (metallic ion & ligand)[1] Synthetic solvent (mL)/T (°C)/time (h)	Microextraction format/Extraction time (min)	MOF amount	Analytes (number)/Sample (amount)	LOD [1] (ng·L^{-1})/RSD [2] (%)	Analytical technique	Ref.
SPME						
UiO-66/MoS$_2$ (Zr^{4+} & H$_2$bdc [3] + MoS$_2$) DMF [4] (20)/120/24	stainless steel arrow (MOF attached with epoxy glue)/30	~ × 25 μm thickness	PAHs [18] (16)/fish (10 mL alkaline extract)	0.11–1.40/<8.6	HS-SPME [33]-GC [26]-MS [7]	[92]
MAF-66 (Zn^{2+} & H$_2$N-Htz [34]) isopropanol (50)/RT [24]/72	stainless steel fiber (layer-by-layer deposition)/40	3.0 cm × 15 μm thickness	PAHs [18] (7)/water, potatoes & roast pork (10 mL sample or aqueous extract)	0.1–7.5/<4.2	HS-SPME [33]-GC [26]-FID [35]	[93]
JUC-72 (Ni^{2+} & 2,2'-bipyridyl) DMF [4] (80)/80/1	functionalized fused silica fiber (in-situ sol-gel method)/40	~ × 80 μm thickness	aromatic amines (2)/urine (10 mL)	0.010–0.012/<7.7	HS-SPME [33]-GC [26]-MS/MS [7]	[94]
MIL-96 (Al^{3+} & H$_3$btc [36]) water (30)/200/24	stainless steel fiber (MOF attached with epoxy glue)/30	2.0 cm × 80 μm thickness	THMs [37] (4) and TCNM [38]/waters (10 mL)	3.0–11/<10.1	HS-SPME [33]-GC [26]-MS [7]	[95]
UiO-66 (Zr^{4+} & H$_3$bdc [3]) DMF [4] (40)/120/24	stainless steel fiber (MOF attached with silicone sealant)/40	4.0 cm × 8.5 μm thickness	PAHs [18] (9)/waters (10 mL)	10–30/<5.6	HS-SPME [33]-GC [26]-FID [35]	[96]
SBSME						
UiO-66-NH$_2$ (Zr^{4+} & NH$_2$-H$_2$bdc [20]) DMF [4] (10) & acetic acid (7)/120/24	coated glass bar (in-situ polymerization method)/30	1 cm × 4000 μm thickness	SUHs [29] (5)/water & soil (5 mL sample or aqueous extract)	40–840/<13.8	LC [6]-UV [17]	[97]
MIL-68@PEEK [39]	coated dumbbell-shaped bar (covalent immobilization)/120	3 cm × 18.4 μm thickness	parabens (3)/creams & rabbit plasma (20 mL aqueous extract)	1–2/<9.74	LC [6]-MS [7]	[98]
ZIF-67 (Co^{2+} & 2-MIm [23]) methanol (80)/120/5	anodized aluminum bar (in-situ growth)/20	1 cm × 500 μm thickness	caffeine (1)/beverage & urine (10 mL)	50–100/<6.1	LC [6]-UV [17]	[99]

[1] limit of detection
[2] inter-day relative standard deviation
[3] terephthalic acid (benzene-1,4-dicarboxylic acid)
[4] N,N-dimethylformamide
[5] polypropylene
[6] liquid chromatography
[7] mass spectrometry
[8] meso-tetra(4-carboxyphenyl)porphyrin
[9] cold vapor atomic absorption spectroscopy
[10] 2-nitroterephthalic acid (2-nitrobenzene-1,4-dicarboxylic acid)
[11] endocrine disrupting chemicals
[12] diode array detector
[13] graphite furnace atomic absorption spectroscopy
[14] 1,2,4-triazole
[15] N,N-dimethylacetamide
[16] personal care products
[17] ultraviolet detector
[18] polycyclic aromatic hydrocarbons
[19] fluorescence detector
[20] 2-aminoterephthalic acid (2-aminobenzene-1,4-dicarboxylic acid)
[21] photodiode array detector
[22] magnetic graphene
[23] 2-methylimidazole
[24] room temperature
[25] phthalate esters
[26] gas chromatography
[27] graphene oxide
[28] polydopamine
[29] sulfonylurea herbicides
[30] benzodiazepines
[31] 4,4'-oxybis(benzoic acid)
[32] tricyclic antidepressants
[33] headspace solid-phase microextraction
[34] 3-amino-1,2,4-triazole
[35] flame ionization detector
[36] trimesic acid (benzene-1,3,5-tricarboxylic acid)
[37] trihalomethanes
[38] trichloronitromethane
[39] polyether ether ketone
-: non-reported

3.2. MOFs in Chromatography

MOFs, and particularly nanoMOFs, are a priori quite interesting materials as chromatographic stationary phases because their pore dimensions make it possible to significantly reduce the amount of required eluents [13]. Their applicability is still lower than that experienced in analytical sample preparation, but increasing studies are currently undertaken. As a trend, the initial studies were mostly with MOFs in gas chromatography (GC), but many MOFs are not stable at the high temperatures required in GC, and nowadays, their inclusion in liquid chromatography (LC) is much higher [19,20]. Neat MOFs' permanent porosity is clearly favorable when used as stationary phase in LC, because it permits high flow rates through it with low back-pressure and, ultimately, favors miniaturization. Easy linkage of MOFs to silica and other composites, including their incorporation in monoliths, also favor the applicability of MOFs in LC stationary phases. Furthermore, MOFs designed to present a specific and unique chiral center type are quite attractive to perform chiral separations [100], of the utmost importance in pharmaceutical, industrial and biomedicine applications. In any case, the incorporation of MOFs as stationary phases in GC and LC, or as pseudo-stationary phases in capillary electro-chromatography (CEC), benefits from the low numbers of MOFs required in such phases, together with the clear reutilization of such phases.

Table 3 includes a summary of representative analytical applications of MOFs when included as stationary of pseudo-stationary phases in different chromatographic techniques in the last three years [101–115]. Mostly, MOFs form part of composites with silica or polymers in such phases [101,102], hardly being utilized in its neat appearance. It is important to highlight the number of applications with chiral MOFs (prepared incorporating chiral organic ligands) [108–111].

Green aspects in these applications are from the need of low amounts of MOFs to prepare the phases (from 2 mg of H_2N-UiO-66 [103] in a capillary column to 1.5 g of γ-CD-MOF [106] in a packed column), reutilization of the as-prepared MOF-based stationary phases, MOF design, and in some cases, even their synthesis.

3.3. MOFs as Sensors in Spectroscopic and Alectroanalytical Methos

Many interesting features of MOFs justify their use as sensors in a number of analytical applications. For instance, some MOFs have semiconductor-like properties, and can serve as photoelectric materials [116]. These properties depend in some cases on the metal nature of the MOF, in other cases on the properties imparted by the organic ligands, and in others on the resulting crystal [117]. When incorporated to certain analytical spectroscopic applications, particularly luminescence-based, the MOF design is also important to ensure proper characteristics in the resulting material. In some cases, the luminescence phenomenon is due to the organic ligand of the MOF [118], and in other cases, to the MOF itself (carefully designing the ligand with proper coordination of the ligands to the metal) [119,120].

Table 3. Representative examples of chromatographic stationary or pseudo-stationary phases incorporating MOFs as novel materials in the period 2017–2019.

MOF (metallic ion & ligand) Synthetic solvent (mL)/T (°C)/time (h)	Stationary phase (type)/Type of column	Size & N [1] (plates·m^{-1})/k [2]/α [3]	Detector	Analytes (number)	Ref.
	LC				
UiO-67 (Zr^{4+} & H$_2$bdpc [4]) DMF [5] (40) & acetic acid (6)/120/24	UiO-67@SiO$_2$ (core-shell composite)/packed	15 cm × 4.6 mm × - & -/0–0.34/-	UV [6]	anilines (4), alkylbenzenes (5), PAHs [7] (5) & thioureas (3)	[101]
NH$_2$-MIL-101(Al) (Al^{3+} & H$_2$N-H$_2$bdc [8]) DMF [5] (40)/130/72	p(GMA-co-EDMA) [9]/NH$_2$-MIL-101(Al) (monolithic composite)/monolithic	15 cm × 100 μm i.d. [10] × 375 μm o.d. [11] & 16,000/-/-	UV [6]	PAHs [7] (4) & NSAIDs [12] (3)	[102]
UiO-66-NH$_2$ (Zr^{4+} & H$_2$N-H$_2$bdc [8]) DMF [5] (20)/120/24	pGMA [13]/UiO-66-NH$_2$ (composite)/packed	112 cm × 25 μm i.d. [10] × 365 μm o.d. [11] & 121,477/-/-	UV [6]	flavonoids (2)	[103]
UiO-66 (Zr^{4+} & H$_2$bdc [14]) DMF [5] (40) & acetic acid (4)/120/24	UiO-66@SiO$_2$ (core-shell composite)/packed	15 cm × 4.6 mm × 5–6 μm & -/-/0.2–5.6	UV [6]	xylene isomers (3)	[104]
MIL-101(Fe)-NH$_2$ (Fe^{3+} & H$_2$N-H$_2$bdc [8]) DMF [5] (13)/120/72	MIL-101(Fe)-NH$_2$@SiO$_2$ (core-shell composite)/packed	25 cm × 3 mm × 3–5 μm & 37,570/-/-	UV [6]	C$_8$ compounds (5)	[105]
γ-CD-MOF (K$^+$ & γ-CD [15]) water (150) & methanol (15)/50/12	γ-CD-MOF (composite)/packed	10 cm × 4.6 mm × 2–5 μm & 75,000/2.59–13.96/-	DAD [16]	drugs (3)	[106]
ZIF-8 (Zn^{2+} & 2-MIm [17]) DMF [5] (5.6) & methanol (14.4)/RT [18]/24	ZIF-8@SiO$_2$ (core-shell composite)/packed	5 cm × 4.6 mm × 2.2 μm & 216,202/-/-	UV [6]	xylene isomers (3)	[107]
[Nd$_3$(D-cam)$_8$(H$_2$O)$_4$Cl]$_n$ (Nd^{3+} & D-H$_2$cam [19]) methanol (60) & ACN (30) & water (30)/140/72	[Nd$_3$(D-cam)$_8$(H$_2$O)$_4$Cl]$_n$ (composite)/packed	25 cm × 2 mm × - & 9086/-/0.80–3.40	UV [6]	racemates (27)	[108]
	GC				
[Co-L-GG(H$_2$O)] (Co^{2+} & L-GG [20]) water (12) & methanol (12)/80/2	[Co-L-GG(H$_2$O)] (chiral)/coated	10 m × 0.25 mm × 2 μm & 2790/0.54–4.29/1.04–1.34	FID [21]	racemates (30)	[109]
[Zn$_2$(bdc)(L-lac)(DMF)]·DMF (Zn^{2+} & L-lactic acid & H$_2$bdc [14]) DMF [5] (10)/120/48	[Zn$_2$(bdc)(L-lac)] (chiral)/coated	10 m × 0.25 mm × 1–2 μm & -/-/-	FID [21]	biochemical compounds (12)	[110]
[Mn$_3$(HCOO)$_2$(D-cam)$_2$(DMF)$_2$]$_n$ (Mn^{2+} & D-H$_2$cam [19]) DMF [5] (5.5) & ethanol (3)/100/48	[Mn$_3$(HCOO)$_2$(D-cam)$_2$(DMF)$_2$]$_n$ (chiral)/coated	10 m × 0.25 mm × 1–2 μm & -/0.73–4.12/1.31–3.13	FID [21]	biochemical compounds (9)	[111]
	CEC				
UiO-66-NH$_2$ (Zr^{4+} & H$_2$N-H$_2$bdc [8]) DMF [5] (10) & formic acid (1)/120/24	UiO-66-NH$_2$ (functionalized)/coated	25 cm × 50 μm i.d. [10] × 365 μm o.d. [11] & -/-/-	DAD [16]	benzene derivatives (3)	[112]
Bio-MOF-1 (Zn^{2+} & H$_2$bdpc [4] & adenine) DMF [5] (13.5) & water (1)/130/24	bio-MOF-1 (functionalized)/coated	26.5 cm × 50 μm i.d. [10] × 360 μm o.d. [11] & 193000	DAD [16]	chlorobenzenes (3) and alkylbenzenes (3)	[113]
[Mn(cam)(bpy)] (Mn^{2+} & bpy & D-H$_2$cam [19]) DMF [5] (6) & ethanol (12)/100/72	[Mn(cam)(bpy)] (functionalized)/coated	21 cm × 75 μm i.d. [10] × - & 118185	DAD [16]	sulfonamides (10)	[114]
JLU-Liu23 (Cu$^+$ & TEDA [23] &1,3-bis(2-benzimidazol)benzene) DMF [5] (0.75)/105/12	JLU-Liu23 (functionalized)/coated	40 cm × 75 μm i.d. [10] × - & 194061	DAD [16]	chiral neurotransmitters (4)	[115]

[1] highest theoretical plate number reported for the column
[2] retention factor
[3] selectivity factor
[4] 4,4'-biphenyldicarboxylic acid
[5] N,N-dimethylformamide
[6] ultraviolet detector
[7] polycyclic aromatic hydrocarbons
[8] 2-aminoterephthalic acid (2-aminobenzene-1,4-dicarboxylic acid)
[9] poly(glycidil methacrylate-co-ethylene dimethacrylate)
[10] inner diameter
[11] outer diameter
[12] non-steroidal anti-inflammatory drugs
[13] poly(glycidil methacrylate)
[14] terephthalic acid (benzene-1,4-dicarboxylic acid)
[15] γ-cyclodextrin
[16] diode array detector
[17] 2-methylimidazole
[18] room temperature
[19] D-(+)-camphoric acid
[20] dipeptide H-Gly-L-Glu
[21] flame ionization detector
[22] 4,4'-bipyridine
[23] triethylenediamine
-: non-reported

Table 4 lists some representative examples of MOFs used in spectroscopic or in electroanalytical methods in the last three years, together with information of their analytical performance [42,121–128].

Table 4. Representative examples of electroanalytical and spectroscopic applications incorporating MOFs in the period 2017–2019.

MOF (metallic ion & ligand) Synthetic Solvent (mL)/T (°C)/time (h)	MOF-Based Material/amount/comment	Method/Comments	Sample Matrix (amount)	LOD [1]/RSD [2] (%)	Analytes (number)	Ref.
Electroanalytical						
ZIF-67 (Co^{2+} & 2-MIm [3]) methanol (100)/RT [4]/24	neat ZIF-67/43 mg/CPE [5] electrode	amperometry/−50 mV applied	waters (-)	1.45 μM/<3.26	hydrazine (1)	[121]
PCN-224 (Zr^{4+} & TCPP [6]) DMF [7] (50)/120/24	PCN with rGO [8]/5 mg·mL^{-1} suspension/n-type semiconductor	EIS [9]/−100 mV applied	water, swine manure & lixivium (-)	5.47 ng·L^{-1}/<3.8	p-ASA (1)	[122]
Cu-MOF (Cu^{2+} & H_3btc [10]) methanol (12) & water (12)/RT [4]/2	Cu-MOF with AuNPs [11]/60 μL 10 μM solution/aptasensor	DPV [12]/−300–500 mV applied	aCSF [13] (-)	0.45 nM/1.8	β-amyloid oligomers	[123]
Ni-BTC (Ni^{2+} & H_3btc [10]) ethanol (100) & water (100)/RT [4]/0.5	neat Ni-BTC/100 mg/CPE [5] electrode	CV [14]/400–900 mV applied	soft drinks (-)	0.08 nM/<5	Ponceau 4R (1)	[117]
H_2N-MIL-53(Al) (Al^{3+} & H_2N-H_2bdc [15]) water (150)/reflux/72	H_2N-MIL-53(Al) in polymeric matrix/40 wt % MOF/	EIS [9]/1000 mV applied	-	-/-	methanol (1) & water (1), (in gas phase)	[124]
Ce-MOF (Ce^{3+} & H_3btc [10]) ethanol (25) & water (25)/90/2	Ce-MOF@MCA [16]/1 mg/aptasensor	EIS [9]/-	milk, urine & water (-)	17.4 fg·mL^{-1}/<2.65	OTC [17] (1)	[125]
Cu-MOF (Cu^{2+} & H_3btc [10]) ethanol (7.1) & water (7.1)/150/24	Cu-MOF-GN [18]/2 mg/GCE [19] electrode	CV [14] & DPV [12]/−0.4–0.8 V & −0.2–0.6 V	water (-)	0.33–0.59 μM/<2.8	HQ [20] (1) & CT [21] (1)	[126]
Spectroscopic						
Mg-MOF-1 (Mg^{2+} & H_2ATDC [22]) DMF (4) & methanol (4) & water (2)/90/96	neat Mg-MOF-1/-/pH and thermo stability	fluorescence/K_{sv} [23] = 0.58 × 10^4	-	0.01–0.12 μM/-	Ce^{3+} (1) & NAEs [24] (8)	[118]
Pr-MOF-NFs (Pr^{3+} & AIP [25] & Phen [26]) DMF (2) & ethanol (18) & water (10)/reflux/48	neat Pr-MOF-NFs [27]/1 × 10^{-4} M/in distilled water	fluorescence/-	serum (-)	0.276 ng·mL^{-1}/<1.42	prolactin (1)	[42]
Mg-APDA (Mg^{2+} & H_2APDA [28]) DMA [29] (2) & ethanol (0.1) & water (0.1)/140/48	neat Mg-APDA/2.5 mg/in DMF [7]	fluorescence/K_{sv} [23] = 2.06 × 10^4	-	126–152 μg·L^{-1}/-	Fe^{3+} (1), pesticides (5) & antibiotics (9)	[119]
$Zr_6O_4(OH)_4(2,7-CDC_8)_6$·19$H_2$O·2DMF DMA [29] (2) & 2,7-H_2CDC [30] & acetic acid (0.27)/80/24	neat Zr-based MOF/2 mg/photostable and reusable	fluorescence/K_{sv} [23] = 5.5 × 10^4	-	0.02–0.91 μM	Fe^{3+} (1), CN^- (1) & PNP [31] (1)	[127]
Zn-MOF-1 (Zn^{2+} & H_2bdc [32] & L [33]) water (3) & ethanol (3)/170/72	neat Zn-MOF-1/-/in methanol or water	fluorescence/K_{sv} [23] = 2.36 × 10^4	-	1.90–3.84 μM/-	Fe^{3+} (1), 2,6-Dich-4-NA [34] (1) & Cr^{6+} ions (2)	[128]
Tb-MOF(Tb^{3+} & AIP [25]) ethanol (40) & water (40)/RT/1	Tb-MOF-PMMA [35]/50 mg/MOF loaded in the polymeric membrane	fluorescence/K_{sv} [23] = 4.0 × 10^4	serum & water (-)	0.30–0.35 μM/<4.7	NFAs [36] (2)	[120]

Table 4. Cont.

MOF (metallic ion & ligand) Synthetic Solvent (mL)/T (°C)/time (h)	MOF-Based Material/amount/comment	Method/Comments	Sample Matrix (amount)	LOD [1]/RSD [2] (%)	Analytes (number)	Ref.

[1] limit of detection
[2] relative standard deviation
[3] 2-methylimidazole
[4] room temperature
[5] carbon paste electrode
[6] meso-tetra(4-carboxyphenyl)porphyrin
[7] N,N-dimethylformamide
[8] reduced graphene oxide
[9] electrochemical impedance spectroscopy
[10] trimesic acid (benzene-1,3,5-tricarboxylic acid)
[11] gold nanoparticles
[12] differential pulse voltammetry
[13] artificial cerebrospinal fluid
[14] cyclic voltammetry
[15] 2-aminoterephthalic acid (2-aminobenzene-1,4-dicarboxylic acid)
[16] melamine and cyanuric acidmonomers
[17] oxytetracycline
[18] graphene
[19] glassy carbon electrode
[20] hydroquinone
[21] catechol
[22] 2′-amino-1,1′:4′,1″-terphenyl-4,4″-dicarboxylic acid
[23] highest quenching constant in M^{-1}
[24] nitro aromatic explosives
[25] 5-aminoisophthalate
[26] phenanthroline
[27] nanofibers
[28] 4,4′-(pyridine-3,5-diyl)dibenzoic acid
[29] N,N-dimethylacetamide
[30] 9H-carbazole-2,7-dicarboxylic acid
[31] p-nitrophenol
[32] terephthalic acid (benzene-1,4-dicarboxylic acid)
[33] 4-(tetrazol-5-yl)phenyl-4,2′:6′,4″-terpyridine
[34] 2,6-dichloro-4-nitroaniline
[35] poly(methyl methacrylate)
[36] nitrofuran antibiotics

The reported electroanalytical applications normally perform in electrochemical cells constituted by the electrolyte and three electrodes: a reference electrode, an auxiliary electrode (i.e., Au or Pt) and the working electrode. MOFs incorporate synthetically in the working electrode by proper modification of the support material, with the purpose of attaining ultra-functional sensors. Thus, Asadi et al. [121] and Ji et al. [117] described carbon paste electrodes (CPE) modified with neat MOFs, such as ZIF-67 and Ni-BTC. MOFs can be combined with nanoparticles with the purpose of improving the overall electrochemical behavior of the resulting composite [121]. Zhou et al. described the use of gold nanoparticles modified with aptamers (to ensure β-amyloid oligomers detection) and linked to a Cu-MOF [123]. Devices that are more complex include not only the neat MOF linked to the working electrode, but also a polymeric matrix as a part of the electrochemical system [124].

Electroanalytical applications with MOFs cover techniques like amperometry [122], differential pulse voltammetry (DPV) [123] and electrochemical impedance spectroscopy (EIS) [125].

It is important to highlight the sensitivity of the MOF-based sensors linked to the electroanalytical applications. Thus, detection limits range from 0.8 nM for Ponceau 4R in soft drinks (only requiring 43 mg of MOF) [117] to 1.45 µM for hydrazine in waters (in this case requiring 100 mg of MOF) [121]. Furthermore, electroanalytical methods are quite efficient, fast and simple, thus in accordance with GAC. In the reported applications, there is an enormous variety in the nature of the analytes determined, from drugs [126] to peptides [123]; and analyzing quite different samples, like waters [121], bio-fluids [123], and food samples [125].

Spectroscopic applications take advantage of fluorescent MOFs. Luminescent MOFs show prominent optical properties, and relatively long emission wavelength, for which the results are quite advantageous. Furthermore, luminescence sensing has attracted great attention, owing to its high sensitivity, fast response, obvious selectivity, recyclability, and simplicity regarding the instrumentation required which, in this latter case, is also linked to low costs. All these performance characteristics represent very adequate results from a GAC point of view. Regarding additional advantages of minimization of reagents consumption in these luminescence applications, low amounts of MOFs are needed: from 2 mg of neat Zr-based MOF [118] to 50 mg of Tb-MOF-PMMA [120]. Rare earth metals [42,120] and transition metals [127] normally form luminescent MOFs. In new green trends, the interest is tangible in obtaining luminescent MOFs with alkaline-earth metals in their structure [118,119], together with improvements of water stability of the resulting luminescent MOF.

As with electroanalytical applications, luminescent applications cover the determination of organic compounds [42,120], heavy metals [118,127], and even the simultaneous determination of heavy metals and organic compounds [119], as well as with samples like serum [42,120], and water [120].

4. Concluding Remarks

MOFs, as materials with almost endless applications in analytical chemistry, must comply with environmental requirements in order to follow properly GAC rules. Thus, assurance of their sustainability must begin with the MOF design (with the proper choice of the MOF constituents), followed by an adequate synthetic procedure and toxicity evaluation of the resulting material, ending up in an analytical method that can be categorized as a GAC method. This, in turn, requires an important collaboration between materials science and analytical chemistry, with an emphasis on green chemistry. Finally, and even more importantly, the rationale behind selecting a MOF for a particular application must always be the first step when setting up an MOF-based analytical method.

Author Contributions: Conceptualization, J.P. and V.P.; Formal analysis, P.R.-B. and I.T.-M.; Funding acquisition, J.P. and V.P.; Investigation, P.R.-B., I.T.-M., J.P. and V.P.; Methodology, P.R.-B., I.T.-M., J.P. and V.P.; Resources, J.P. and V.P.; Supervision, J.P. and V.P.; Writing–original draft, P.R.-B. and I.T.-M.; Writing–review & editing, J.P. and V.P.

Funding: V.P. thanks the Project Ref. MAT2017-89207-R.

Acknowledgments: I.T.-M. thanks his collaboration fellowship with the Spanish Ministry of Education (MEC) during the MS studies at ULL. J.P. thanks the "Agustín de Betancourt" Canary Program for his research associate position at ULL. V.P. acknowledges the Spanish Ministry of Economy and Competitiveness (MINECO) for the Project Ref. MAT2017-89207-R.

Conflicts of Interest: The authors declare no conflict of interest.

References

1. Matsuyama, S.; Munakata, M.; Emori, T.; Kitagawa, S. Synthesis and Crystal Structures of Novel One-dimensional - Polymers, [{M(bpen)X$_\infty$},1 [M = Cu', X = PF$_6^-$; M = Ag', X = ClO$_4$-; bpen = *trans*-1,2-bis(2-pyridyl)ethylene] and [{Cu(bpen)(CO)(CH$_3$CN)(PF$_6$)}$_\infty$]. *J. Chem. Soc. Dalton Trans.* **1991**, *1*, 2869–2874.
2. Yaghi, O.M.; Hailian, L. Hydrothermal Synthesis of a Metal-Organic Framework Containing Large Rectangular Channels. *J. Am. Chem. Soc.* **1995**, *117*, 10401–10402. [CrossRef]
3. Riou, D.; Férey, G. Hybrid open frameworks (MIL-n). Part 3: Crystal structures of the HT and LT forms of MIL-7: A new vanadium propylenediphosphonate with an open-framework. Influence of the synthesis temperature on the oxidation state of vanadium within the same structural type. *J. Mater. Chem.* **1998**, *8*, 2733–2735.
4. Farha, O.K.; Eryazici, I.; Jeong, N.C.; Hauser, B.G.; Wilmer, C.E.; Sarjeant, A.A.; Snurr, R.Q.; Nguyen, S.B.T.; Yazaydın, A.Ö.; Hupp, J.T. Metal−Organic Framework Materials with Ultrahigh Surface Areas: Is the Sky the Limit? *J. Am. Chem. Soc.* **2012**, *134*, 15016–15021. [CrossRef]
5. Khan, N.A.; Hasan, Z.; Jhung, S.H. Adsorptive removal of hazardous materials using metal-organic frameworks (MOFs): A review. *J. Hazard. Mater.* **2013**, *244–245*, 444–456. [CrossRef]
6. Moghadam, P.Z.; Li, A.; Wiggin, S.B.; Tao, A.; Maloney, A.G.P.; Wood, P.A.; Ward, S.C.; Fairen-Jimenez, D. Development of a Cambridge Structural Database Subset: A Collection of Metal–Organic Frameworks for Past, Present, and Future. *Chem. Mater.* **2017**, *29*, 2618–2625. [CrossRef]
7. Stavila, V.; Talin, A.A.; Allendorf, M.D. MOF-based electronic and opto-electronic devices. *Chem. Soc. Rev.* **2014**, *43*, 5994–6010. [CrossRef]
8. Bae, Y.S.; Snurr, R.Q. Development and evaluation of porous materials for carbon dioxide separation and capture. *Angew. Chem. Int. Ed.* **2011**, *50*, 11586–11596. [CrossRef]
9. Li, H.; Wang, K.; Sun, Y.; Lollar, C.T.; Li, J.; Zhou, H.-C. Recent advances in gas storage and separation using metal–organic frameworks. *Mater. Today* **2018**, *21*, 108–121. [CrossRef]
10. Ma, S.; Zhou, H.-C. Gas storage in porous metal–organic frameworks for clean energy applications. *Chem. Commun.* **2010**, *46*, 44–53. [CrossRef]
11. Kang, Y.-S.; Lu, Y.; Chen, K.; Zhao, Y.; Wang, P.; Sun, W.Y. Metal–organic frameworks with catalytic centers: From synthesis to catalytic application. *Coord. Chem. Rev.* **2019**, *378*, 262–280. [CrossRef]
12. Abánades-Lázaro, I.; Forgan, R.S. Application of zirconium MOFs in drug delivery and biomedicine. *Coord. Chem. Rev.* **2019**, *380*, 230–259. [CrossRef]
13. Pacheco-Fernández, I.; González-Hernández, P.; Pasán, J.; Ayala, J.H.; Pino, V. The rise of metal-organic frameworks in analytical chemistry. In *Handbook of Smart Materials in Analytical Chemistry*; De la Guardia, M., Esteve-Turrillas, F.A., Eds.; Wiley: Hoboken, NJ, USA, 2016; Volume 1, pp. 463–502.
14. Rocío-Bautista, P.; Pacheco-Fernández, I.; Pasán, J.; Pino, V. Are metal-organic frameworks able to provide a new generation of solid-phase microextraction coatings?—A review. *Anal. Chim. Acta* **2016**, *939*, 26–41. [CrossRef]
15. Rocío-Bautista, P.; González-Hernández, P.; Pino, V.; Pasán, J.; Afonso, A.M. Metal-organic frameworks as novel sorbents in dispersive-based microextraction approaches. *Trends Anal. Chem.* **2017**, *90*, 114–134. [CrossRef]
16. Maya, F.; Cabello, C.P.; Frizzarin, R.M.; Estela, J.M.; Palomino, G.T.; Cerdà, V. Magnetic solid-phase extraction using metal-organic frameworks (MOFs) and their derived carbons. *Trends Anal. Chem.* **2017**, *90*, 142–152. [CrossRef]
17. Wang, X.; Ye, N. Recent advances in metal-organic frameworks and covalent organic frameworks for sample preparation and chromatographic analysis. *Electrophoresis* **2017**, *38*, 3059–3078. [CrossRef]
18. Hashemi, B.; Zohrabi, P.; Raza, N.; Kim, K.-H. Metal-organic frameworks as advanced sorbents for the extraction and determination of pollutants from environmental, biological, and food media. *Trends Anal. Chem.* **2017**, *97*, 65–82. [CrossRef]

19. Maya, F.; Cabello, C.P.; Figuerola, A.; Palomino, G.T.; Cerdà, V. Immobilization of Metal–Organic Frameworks on Supports for Sample Preparation and Chromatographic Separation. *Chromatographia* **2019**, *82*, 361–375. [CrossRef]
20. Zhang, J.; Chen, Z. Metal-organic frameworks as stationary phase for application in chromatographic separation. *J. Chromatogr. A* **2017**, *1530*, 1–18. [CrossRef]
21. Xiaoxin, L.; Lun, S.; Sha, C. Application of Metal-Organic Frameworks in Chromatographic Separation. *Acta Chim. Sin.* **2016**, *74*, 969–979.
22. Li, A.; Liu, X.; Chai, H.; Huang, Y. Recent advances in the construction and analytical applications of metal-organic frameworks-based nanozymes. *Trends Anal. Chem.* **2018**, *105*, 391–403. [CrossRef]
23. Feldblyum, J.I.; Keenan, E.A.; Matzger, A.J.; Maldonado, S. Photoresponse Characteristics of Archetypal Metal–Organic Frameworks. *J. Phys. Chem. C* **2012**, *116*, 3112–3121. [CrossRef]
24. Yadav, D.K.; Ganesan, V.; Marken, F.; Gupta, R.; Sonkar, P.K. Metal@MOF Materials in Electroanalysis: Silver-Enhanced Oxidation Reactivity Towards Nitrophenols Adsorbed into a Zinc Metal Organic Framework—Ag@MOF-5(Zn). *Electrochim. Acta* **2016**, *219*, 482–491. [CrossRef]
25. Filippou, O.; Bitas, D.; Samanidou, V. Green approaches in sample preparation of bioanalytical samples prior to chromatographic analysis. *J. Chromatogr. B* **2017**, *1043*, 44–62. [CrossRef]
26. Sheldon, R.A. Fundamentals of green chemistry: Efficiency in reaction design. *Chem. Soc. Rev.* **2012**, *41*, 1437–1451. [CrossRef]
27. Ludwig, J.K. Green Chemistry: An Introductory Text. *Green Chem. Lett.* **2017**, *10*, 30–31. [CrossRef]
28. Valcárcel, M. *Principles of Analytical Chemistry*; Springer: Berlin, Germany, 2000; ISBN 978-3-642-57157-2.
29. Koel, M.; Kaljurand, M. *Green Analytical Chemistry*; The Royal Society of Chemistry: Cambridge, UK, 2010; ISBN 978-1-84755-872-5.
30. Namieśnik, J. Trends in Environmental Analytics and Monitoring. *Crit. Rev. Anal. Chem.* **2000**, *30*, 221–269. [CrossRef]
31. Anastas, P.T.; Warner, J.C. *Green Chemistry: Theory and Practice*; Oxford University Press: New York, NY, USA, 1998; p. 30.
32. Gałuszka, A.; Migaszewski, Z.; Namieśnik, J. The 12 principles of green analytical chemistry and the SIGNIFICANCE mnemonic of green analytical practices. *Trends Anal. Chem.* **2013**, *50*, 78–84. [CrossRef]
33. Płotka-Wasylka, J.; Szczepańska, N.; de la Guardia, M.; Namieśnik, J. Miniaturized solid-phase extraction techniques. *Trends Anal. Chem.* **2015**, *73*, 19–38. [CrossRef]
34. Tobiszewski, M.; Namieśnik, J. Greener organic solvents in analytical chemistry. *Curr. Opin. Green Sust. Chem.* **2017**, *5*, 1–4. [CrossRef]
35. Pacheco-Fernández, I.; Pino, V. Green solvents in analytical chemistry. *Curr. Opin. Green Sustain. Chem.* **2019**, *18*, 42–50. [CrossRef]
36. Yaghi, O.M.; O'Keeffe, M.; Ockwig, N.W.; Chae, H.K.; Eddaoudi, M.; Kim, J. Reticular synthesis and the design of new materials. *Nature* **2003**, *423*, 705–715. [CrossRef]
37. Furukawa, H.; Cordova, K.E.; O'Keeffe, M.; Yaghi, O.M. The Chemistry and Applications of Metal-Organic Frameworks. *Science* **2013**, *341*, 12304441–123044412. [CrossRef]
38. Bai, Y.; Dou, Y.; Xie, L.-H.; Rutledge, W.; Li, J.-R.; Zhou, H.-C. Zr-based metal–organic frameworks: Design, synthesis, structure, and applications. *Chem. Soc. Rev.* **2016**, *45*, 2327–2367. [CrossRef]
39. Rajkumar, T.; Kukkar, D.; Kim, K.-H.; Sohn, J.R.; Deed, A. Cyclodextrin-metal–organic framework (CD-MOF): From synthesis to applications. *J. Ind. Eng. Chem.* **2019**, *72*, 50–66. [CrossRef]
40. Gaab, M.; Trukhan, N.; Maurer, S.; Gummaraju, R.; Müller, U. The progression of Al-based metal-organic frameworks—From academic research to industrial production and applications. *Micropor. Mesopor. Mater.* **2012**, *157*, 131–136. [CrossRef]
41. Kim, J.; Lee, S.; Kim, J.; Lee, D. Metal–organic frameworks derived from zero-valent metal substrates: Mechanisms of formation and modulation of properties. *Adv. Funct. Mater.* **2019**, *29*, 1808466. [CrossRef]
42. Sheta, S.M.; El-Sheikh, S.M.; Abd-Elzaher, M.M. A novel optical approach for determination of prolactin based on Pr-MOF nanofibers. *Anal. Bioanal. Chem.* **2019**. [CrossRef]
43. Reinsch, H. "Green" Synthesis of Metal-Organic Frameworks. *Eur. J. Inorg. Chem.* **2016**, *27*, 4290–4299. [CrossRef]
44. Rubio-Martinez, M.; Avci-Camur, C.; Thornton, A.W.; Imaz, I.; Maspoch, D.M.; Hill, M.R. New synthetic routes towards MOF production at scale. *Chem. Soc. Rev.* **2017**, *46*, 3453–3480. [CrossRef]

45. Klimakow, M.; Klobes, P.; Thunemann, A.F.; Rademann, K.; Emmerling, F. Mechanochemical synthesis of Metal-Organic Frameworks: A fast and facile approach toward quantitative yields and high specific surface areas. *Chem. Mater.* **2010**, *22*, 5216–5221. [CrossRef]
46. Joaristi, A.M.; Juan-Alcañiz, J.; Serra-Crespo, P.; Kapteijn, F.; Gascón, J. Electrochemical synthesis of some archetypical Zn^{2+}, Cu^{2+}, and Al^{3+} Metal Organic Frameworks. *Cryst. Growth Des.* **2012**, *12*, 3489–3498. [CrossRef]
47. Julien, P.A.; Mottillo, C.; Friscic, T. Metal-Organic Frameworks meet scalable and sustainable synthesis. *Green Chem.* **2017**, *19*, 2729–2747. [CrossRef]
48. Li, P.; Cheng, F.F.; Xiong, W.W.; Zhang, Q.C. New synthetic strategies to prepare metal-organic frameworks. *Inorg. Chem. Front.* **2018**, *5*, 2693–2708. [CrossRef]
49. Khan, N.A.; Jhung, S.-H. Synthesis of Metal-Organic Frameworks (MOFs) with microwave or ultrasound: Rapid reaction, phase selectivity, and size reduction. *Coord. Chem. Rev.* **2015**, *285*, 11–23. [CrossRef]
50. Sánchez-Sánchez, M.; Getachew, N.; Díaz, K.; Díaz-García, M.; Chebude, Y.; Díaz, I. Synthesis of metal–organic frameworks in water at room temperature: Salts as linker sources. *Green Chem.* **2015**, *17*, 1500–1509. [CrossRef]
51. Zhang, J.; Bu, J.T.; Chen, S.; Wu, T.; Zheng, S.; Chen, Y.; Nieto, R.A.; Feng, P.; Bu, X. Urothermal synthesis of crystalline porous materials. *Angew. Chem. Int. Ed.* **2010**, *49*, 8876–8879. [CrossRef]
52. Tranchemontagne, D.J.; Hunt, J.R.; Yaghi, O.M. Room temperature synthesis of metal-organic frameworks: MOF-5, MOF-74, MOF-177, MOF-199, and IRMOF-0. *Tetrahedron* **2008**, *64*, 8553–8557. [CrossRef]
53. Bayliss, P.A.; Ibarra, I.A.; Pérez, E.; Yang, S.; Tang, C.C.; Poliakoff, M.; Schröder, M. Synthesis of metal–organic frameworks by continuous flow. *Green Chem.* **2014**, *16*, 3796–3802. [CrossRef]
54. Rocío-Bautista, P.; Pino, V.; Ayala, J.H.; Ruiz-Pérez, C.; Vallcorba, O.; Afonso, A.M.; Pasán, J. A green metal–organic framework to monitor water contaminants. *RSC Adv.* **2018**, *8*, 31304–31310. [CrossRef]
55. Noh, H.; Kung, C.-W.; Islamoglu, T.; Peters, A.W.; Liao, Y.; Li, P.; Garibay, S.J.; Zhang, X.; DeStefano, M.R.; Hupp, J.T.; et al. Room Temperature Synthesis of an 8-Connected Zr-Based Metal–Organic Framework for Top-Down Nanoparticle Encapsulation. *Chem. Mater.* **2018**, *30*, 2193–2197. [CrossRef]
56. Lenstra, D.C.; Rutjes, F.P.J.T. Organic synthesis in flow: Toward higher levels of sustainability. In *Sustainable Flow Chemistry: Methods and Applications*; Vaccaro, L., Ed.; Wiley: Hoboken, NJ, USA, 2017; pp. 103–134.
57. Leardi, R. Experimental design in chemistry: A tutorial. *Anal. Chim. Acta* **2009**, *652*, 161–172. [CrossRef]
58. Tamames-Tabar, C.; Cunha, D.; Imbuluzqueta, E.; Ragon, F.; Serre, C.; Blanco-Prieto, C.M.; Horcajada, P. Cytotoxicity of nanoscaled metal–organic frameworks. *J. Mater. Chem. B* **2014**, *2*, 262–271. [CrossRef]
59. Chen, G.; Leng, X.; Luo, J.; You, L.; Qu, C.; Dong, X.; Huang, H.; Yin, X.; Ni, J. In Vitro Toxicity Study of a Porous Iron (III) Metal-Organic Framework. *Molecules* **2019**, *24*, 1211. [CrossRef]
60. Horcajada, P.; Chalati, T.; Serre, C.; Gille, B.; Sebrie, C.; Baati, T.; Eubank, J.F.; Heurtau, D.; Clayette, P.; Kreuz, C.; et al. Porous metal–organic-framework nanoscale carriers as a potential platform for drug delivery and imaging. *Nat. Mater.* **2010**, *9*, 172–179. [CrossRef]
61. Su, F.; Jia, Q.; Li, Z.; Wang, M.; He, L.; Peng, D.; Song, Y.; Zhang, Z.; Fang, S. Aptamer-templated silver nanoclusters embedded in zirconium metal–organic framework for targeted antitumor drug delivery. *Micropor. Mesopor. Mater.* **2019**, *275*, 152–162. [CrossRef]
62. Vinogradov, V.V.; Drozdov, A.S.; Mingabudinova, L.R.; Shabanova, E.M.; Kolchina, N.O.; Anastasova, E.I.; Markova, A.A.; Shtil, A.A.; Milichko, V.A.; Starova, G.L.; et al. Composites based on heparin and MIL-101(Fe): The drug releasing depot for anticoagulant therapy and advanced medical nanofabrication. *J. Mater. Chem. B* **2018**, *6*, 2450–2459. [CrossRef]
63. Ma, A.; Luo, Z.; Gu, C.; Li, B.; Liu, J. Cytotoxicity of a metal–organic framework: Drug delivery. *Inorg. Chem. Commun.* **2017**, *77*, 68–71. [CrossRef]
64. Cheplakova, A.M.; Solovieva, A.O.; Pozmogova, T.N.; Vorotnikov, Y.A.; Brylev, K.A.; Vorotnikova, N.A.; Vorontsova, E.V.; Mironov, Y.V.; Poveshchenko, A.F.; Kovalenko, A.K.; et al. Nanosized mesoporous metal–organic framework MIL-101 as a nanocarrier for photoactive hexamolybdenum cluster compounds. *J. Inorg. Biochem.* **2017**, *166*, 100–107. [CrossRef]
65. Zheng, H.; Zhang, Y.; Liu, L.; Wan, W.; Guo, P.; Nyström, A.M.; Zou, X. One-pot Synthesis of Metal–Organic Frameworks with Encapsulated Target Molecules and Their Applications for Controlled Drug Delivery. *J. Am. Chem. Soc.* **2016**, *138*, 962–968. [CrossRef]

66. Sifaoui, I.; López-Arencibia, A.; Martín-Navarro, C.M.; Reyes-Batlle, M.; Wagner, C.; Chiboub, O.; Mejri, M.; Valladares, B.; Abderrabba, M.; Piñero, J.E.; et al. Programmed cell death in *Acanthamoeba castellanii* Neff induced by several molecules present in olive leaf extracts. *PLoS ONE* **2017**, *12*, 1–12. [CrossRef] [PubMed]
67. Qian, L.; Lei, D.; Duan, X.; Zhang, S.; Song, W.; Hou, C.; Tang, R. Design and preparation of metal-organic framework papers with enhanced mechanical properties and good antibacterial capacity. *Carbohydr. Polym.* **2018**, *192*, 44–51. [CrossRef] [PubMed]
68. Zhu, W.; Zhang, L.; Yang, Z.; Liu, P.; Wang, J.; Cao, J.; Shen, A.; Xu, Z.; Wang, J. An efficient tumor-inducible nanotheranostics for magnetic resonance imaging and enhanced photodynamic therapy. *Chem. Eng. J.* **2019**, *358*, 969–979. [CrossRef]
69. Xin, X.-T.; Cheng, J.-Z. A mixed-ligand approach for building a N-rich porous metal-organic framework for drug release and anticancer activity against oral squamous cell carcinoma. *J. Coord. Chem.* **2018**, *71*, 3565–3574. [CrossRef]
70. Lázaro, I.A.; Haddad, S.; Rodrigo-Muñoz, J.M.; Marshall, R.J.; Sastre, B.; del Pozo, V.; Fairen-Jimenez, D.; Forgan, R.S. Surface-Functionalization of Zr-Fumarate MOF for Selective Cytotoxicity and Immune System Compatibility in Nanoscale Drug Delivery. *ACS Appl. Mater. Interf.* **2018**, *10*, 31146–31157. [CrossRef] [PubMed]
71. Shen, S.; Li, L.; Li, S.; Bai, Y.; Liu, H. Metal–organic frameworks induce autophagy in mouse embryonic fibroblast cells. *Nanoscale* **2018**, *10*, 18161–18168. [CrossRef] [PubMed]
72. Armenta, S.; Garrigues, S.; Esteve-Turrillas, F.A.; Guardia, M. Green extraction techniques in green analytical chemistry. *Trends Anal. Chem.* **2019**. [CrossRef]
73. Pacheco-Fernández, I.; González-Hernández, P.; Rocío-Bautista, P.; Trujillo-Rodríguez, M.J.; Pino, V. Main uses of Microwaves and Ultrasounds in Analytical Extraction Schemes: An Overview. In *Analytical Separation Science*; Anderson, J.L., Stalcup, A., Berthod, A., Pino, V., Eds.; Wiley: Hoboken, NJ, USA, 2016; Volume 5, pp. 1469–1501.
74. Yamini, Y.; Rezazadeh, M.; Seidi, S. Liquid-phase microextraction—The different principles and configurations. *Trends Anal. Chem.* **2019**, *112*, 264–272. [CrossRef]
75. Mogaddam, M.R.A.; Mohebbi, A.; Pazhohan, A.; Khodadadeian, F.; Farajzadeh, M.A. Headspace mode of liquid phase microextraction: A review. *Trends Anal. Chem.* **2019**, *110*, 8–14. [CrossRef]
76. Chisvert, A.; Cárdenas, S.; Lucena, R. Dispersive micro-solid phase extraction. *Trends Anal. Chem.* **2019**, *112*, 224–247. [CrossRef]
77. Olcer, Y.A.; Tascon, M.; Eroglu, A.E.; Boyacı, E. Thin film microextraction: Towards faster and more sensitive microextraction. *Trends Anal. Chem.* **2019**, *113*, 97–101. [CrossRef]
78. Llompart, M.; Celeiro, M.; García-Jares, C.; Dagnac, T. Environmental applications of solid-phase microextraction. *Trends Anal. Chem.* **2019**, *112*, 224–247. [CrossRef]
79. Gao, G.; Xing, X.; Liu, T.; Wang, J.; Hou, X. UiO-66(Zr) as sorbent for porous membrane protected micro-solid-phase extraction androgens and progestogens in environmental water samples coupled with LC-MS/MS analysis: The application of experimental and molecular simulation method. *Microchem. J.* **2019**, *146*, 126–133. [CrossRef]
80. Kahkha, M.R.R.; Daliran, S.; Oveisi, A.R.; Kaykhaii, M.; Sepehri, Z. The Mesoporous Porphyrinic Zirconium Metal-Organic Framework for Pipette-Tip Solid-Phase Extraction of Mercury from Fish Samples Followed by Cold Vapor Atomic Absorption Spectrometric Determination. *Food Anal. Methods.* **2017**, *10*, 2175–2184. [CrossRef]
81. Gao, G.; Li, S.; Li, S.; Wang, Y.; Zhao, P.; Zhang, X.; Hou, X. A combination of computational–experimental study on metal-organic frameworks MIL-53(Al) as sorbent for simultaneous determination of estrogens and glucocorticoids in water and urine samples by dispersive micro-solid-phase extraction coupled to UPLC-MS/MS. *Talanta* **2018**, *180*, 358–367. [CrossRef] [PubMed]
82. Taima-Mancera, I.; Rocío-Bautista, P.; Pasán, J.; Ayala, J.H.; Ruiz-Pérez, C.; Afonso, A.M.; Lago, A.B.; Pino, V. Influence of ligand functionalization of UiO-66-based Metal-Organic Frameworks when used as sorbents in dispersive solid-phase analytical microextraction for different aqueous organic pollutants. *Molecules* **2018**, *23*, 2869. [CrossRef]
83. Kashanaki, R.; Ebrahimzadeh, H.; Moradi, M. Metal–organic framework based micro solid phase extraction coupled with supramolecular solvent microextraction to determine copper in water and food samples. *New J. Chem.* **2018**, *42*, 5806–5813. [CrossRef]

84. González-Hernández, P.; Lago, A.B.; Pasán, J.; Ruiz-Pérez, C.; Ayala, J.H.; Afonso, A.M.; Pino, V. Application of a pillared-layer Zn-triazolate metal-organic framework in the dispersive miniaturized solid-phase extraction of personal care products from wastewater samples. *Molecules* **2019**, *24*, 690. [CrossRef]
85. Boontongto, T.; Siriwong, K.; Burakham, R. Amine-Functionalized Metal–Organic Framework as a New Sorbent for Vortex-Assisted Dispersive Micro-Solid Phase Extraction of Phenol Residues in Water Samples Prior to HPLC Analysis: Experimental and Computational Studies. *Chromatographia* **2018**, *81*, 735–747. [CrossRef]
86. Huang, Y.-F.; Liu, Q.-H.; Li, K.; Li, Y.; Chang, N. Magnetic iron(III)-based framework composites for the magnetic solid-phase extraction of fungicides from environmental water samples. *J. Sep. Sci.* **2018**, *41*, 1129–1137. [CrossRef]
87. Lu, Y.; Wang, B.; Yan, Y.; Liang, H.; Wu, D. Silica Protection–Sacrifice Functionalization of Magnetic Graphene with a Metal–Organic Framework (ZIF-8) to Provide a Solid-Phase Extraction Composite for Recognization of Phthalate Easers from Human Plasma Samples. *Chromatographia* **2019**, *82*, 625–634. [CrossRef]
88. Liu, G.; Huang, X.; Lu, M.; Li, L.; Li, T.; Xu, D. Facile synthesis of magnetic zinc metal-organic framework for extraction of nitrogen-containing heterocyclic fungicides from lettuce vegetable samples. *J. Sep. Sci.* **2019**, *1*, 1–8. [CrossRef] [PubMed]
89. Deng, Y.; Zhang, R.; Li, D.; Sun, P.; Su, P.; Yang, Y. Preparation of iron-based MIL-101 functionalized polydopamine@Fe$_3$O$_4$ magnetic composites for extracting sulfonylurea herbicides from environmental water and vegetable samples. *J. Sep. Sci.* **2018**, *41*, 2046–2055. [CrossRef] [PubMed]
90. Zhang, S.; Yao, W.; Fu, D.; Zhang, C.; Zhao, H. Fabrication of magnetic zinc adeninate metal–organic frameworks for the extraction of benzodiazepines from urine and wastewater. *J. Sep. Sci.* **2018**, *41*, 1864–1870. [CrossRef] [PubMed]
91. Safari, M.; Shahlaei, M.; Yamini, Y.; Shakorian, M.; Arkan, E. Magnetic framework composite as sorbent for magnetic solid phase extraction coupled with high performance liquid chromatography for simultaneous extraction and determination of tricyclic antidepressants. *Anal. Chim. Acta* **2018**, *1034*, 204–213. [CrossRef] [PubMed]
92. Yuan, Y.; Lin, X.; Li, T.; Pang, T.; Dong, Y.; Zhuo, R.; Wang, Q.; Cao, Y.; Gan, N. A solid phase microextraction Arrow with zirconium metal–organic framework/molybdenum disulfide coating coupled with gas chromatography–mass spectrometer for the determination of polycyclic aromatic hydrocarbons in fish samples. *J. Chromatogr. A* **2019**, *1592*, 9–18. [CrossRef] [PubMed]
93. Liu, M.; Liu, J.; Guo, C.; Li, Y. Metal azolate framework-66-coated fiber for headspace solid-phase microextraction of polycyclic aromatic hydrocarbons. *J. Chromatogr. A* **2019**, *1592*, 57–63. [CrossRef] [PubMed]
94. Niu, J.; Zhao, X.; Jin, Y.; Yang, G.; Li, Z.; Wang, J.; Zhao, R.; Li, Z. Determination of aromatic amines in the urine of smokers using a porous organic framework (JUC-Z2)-coated solid-phase microextraction fiber. *J. Chromatogr. A* **2018**, *1555*, 37–44. [CrossRef]
95. Zhang, B.; Xu, G.; Li, L.; Wang, X.; Li, N.; Zhao, R.-S.; Lin, J. Facile fabrication of MIL-96 as coating fiber for solid-phase microextraction of trihalomethanes and halonitromethanes in water samples. *Chem. Eng. J.* **2018**, *350*, 240–247. [CrossRef]
96. Tian, Y.; Sun, M.; Wang, X.; Luo, C.; Feng, J. A Nanospherical Metal–Organic Framework UiO-66 for Solid-Phase Microextraction of Polycyclic Aromatic Hydrocarbons. *Chromatographia* **2018**, *81*, 1053–1061. [CrossRef]
97. Yang, J.-H.; Cui, C.-X.; Qu, L.-B.; Chen, J.; Zhou, X.-M.; Zhang, Y.-P. Preparation of a monolithic magnetic stir bar for the determination of sulfonylurea herbicides coupled with HPLC. *Microchem. J.* **2018**, *141*, 369–376. [CrossRef]
98. Wang, C.; Zhou, W.; Liao, X.; Wang, X.; Chen, Z. Covalent immobilization of metal organic frameworks onto chemical resistant poly(ether ether ketone) jacket for stir bar extraction. *Anal. Chim. Acta* **2018**, *1025*, 124–133. [CrossRef] [PubMed]
99. Ghani, M.; Ghoreishi, S.M.; Azamati, M. In-situ growth of zeolitic imidazole framework-67 on nanoporousanodized aluminum bar as stir-bar sorptive extraction sorbent fordetermining caffeine. *J. Chromatogr. A* **2018**, *1577*, 15–23. [CrossRef] [PubMed]
100. Bhattacharjee, S.; Khan, M.I.; Li, X.; Zhu, Q.-L.; Wu, X.-T. Recent Progress in Asymmetric Catalysis and Chromatographic Separation by Chiral Metal–Organic Frameworks. *Catalysts* **2018**, *8*, 120–148. [CrossRef]

101. Li, X.; Li, B.; Liu, M.; Zhou, Y.; Zhang, L.; Qiao, X. Core–Shell Metal–Organic Frameworks as the Mixed-Mode Stationary Phase for Hydrophilic Interaction/Reversed-Phase Chromatography. *ACS Appl. Mater. Interf.* **2019**, *11*, 10320–10327. [CrossRef] [PubMed]
102. Pérez-Cejuela, H.M.; Carrasco-Correa, E.J.; Shahat, A.; Simó-Alfonso, E.F.; Herrero-Martínez, J.M. Incorporation of metal-organic framework amino-modified MIL-101 into glycidyl methacrylate monoliths for nano LC Separation. *J. Sep. Sci.* **2019**, *42*, 834–842. [CrossRef] [PubMed]
103. Chen, K.; Zhang, L.; Zhang, W. Preparation and evaluation of open-tubular capillary column combining a metal–organic framework and a brush-shaped polymer for liquid chromatography. *J. Sep. Sci.* **2018**, *41*, 2347–2353. [CrossRef] [PubMed]
104. Gao, B.; Huang, M.; Zhang, Z.; Yang, Q.; Su, B.; Yang, Y.; Ren, Q.; Bao, Z. Hybridization of metal–organic framework and monodisperse spherical silica for chromatographic separation of xylene isomers. *Chin. J. Chem. Eng.* **2019**, *27*, 818–826. [CrossRef]
105. Ehrling, S.; Kutzscher, C.; Freund, P.; Müller, P.; Senkovska, I.; Kaskel, S. MOF@SiO$_2$ core-shell composites as stationary phase in high performance liquid chromatography. *Micropor. Mesopor. Mat.* **2018**, *263*, 268–274. [CrossRef]
106. Xu, X.; Wang, C.; Li, H.; Li, X.; Liu, B.; Singh, V.; Wang, S.; Sun, L.; Gref, R.; Zhan, J. Evaluation of drug loading capabilities of γ-cyclodextrin-metal organic frameworks by high performance liquid chromatography. *J. Chromatogr. A* **2017**, *1488*, 37–44. [CrossRef]
107. Qu, Q.; Xuan, H.; Zhang, K.; Chen, X.; Ding, Y.; Feng, S.; Xu, Q. Core-shell silica particles with dendritic pore channels impregnated with zeolite imidazolate framework-8 for high performance liquid chromatography separation. *J. Chromatogr. A* **2017**, *1505*, 63–68. [CrossRef] [PubMed]
108. Zhang, P.; Wang, L.; Zhang, J.-H.; He, Y.-J.; Li, Q.; Luo, L.; Zhang, M.; Yuan, L.-M. Homochiral metal-organic framework immobilized on silica gel by the interfacial polymerization for HPLC enantioseparations. *J. Liq. Chromatogr. Relat. Technol.* **2018**, *1*, 1–7. [CrossRef]
109. Lang, L.; Shengming, X.; Junhui, Z.; Ling, C.; Pengjing, Z.; Liming, Y. A Gas Chromatographic Stationary of Homochiral Metal-peptide Framework Material and Its Applications. *Chem. Res. Chin. Univ.* **2017**, *33*, 24–30.
110. Zheng, D.-D.; Wang, L.; Yang, T.; Zhang, Y.; Wang, Q.; Kurmoo, M.; Zeng, M.-H. A Porous Metal–Organic Framework [Zn$_2$(bdc)(L-lac)] as a Coating Material for Capillary Columns of Gas Chromatography. *Inorg. Chem.* **2017**, *56*, 11043–11049. [CrossRef] [PubMed]
111. Zheng, D.-D.; Zhang, Y.; Wang, L.; Kurmoo, M.; Zeng, M.-H. A rod-spacer mixed ligands MOF [Mn$_3$(HCOO)$_2$(Dcam)$_2$(DMF)$_2$]$_n$ as coating material for gas chromatography capillary column. *Inorg. Chem. Com.* **2017**, *82*, 34–38. [CrossRef]
112. Tang, P.; Wang, R.; Chen, Z. In situ growth of Zr-based metal-organic framework UiO-66-NH$_2$ for open-tubular capillary electrochromatography. *Electrophoresis* **2018**, *39*, 2619–2625. [CrossRef] [PubMed]
113. Li, Z.; Mao, Z.; Chen, Z. In-situ growth of a metal organic framework composed of zinc(II), adeninate and biphenyldicarboxylate as a stationary phase for open-tubular capillary electrochromatography. *Microchim. Acta* **2019**, *186*, 53–61. [CrossRef]
114. Wang, X.; Ye, N.; Hu, X.; Liu, Q.; Li, J.; Peng, L.; Ma, X. Open-tubular capillary electrochromatographic determination of ten sulfonamides in tap water and milk by a metal-organic framework-coated capillary column. *Electrophoresis* **2018**, *39*, 2236–2245. [CrossRef]
115. Pan, C.; Lv, W.; Niu, X.; Wang, G.; Chen, H.; Chen, X. Homochiral zeolite-like metal-organic framework with DNA like double-helicity structure as stationary phase for capillary electrochromatography enantioseparation. *J. Chromatogr. A* **2018**, *1541*, 31–38. [CrossRef]
116. Tachikawa, T.; Choi, J.R.; Fujitsuka, M.; Majima, T. Photoinduced charge-transfer processes on MOF-5 nanoparticles: Elucidating differences between metal-organic frameworks and semiconductor metal oxides. *J. Phys. Chem. C* **2008**, *112*, 14090–14101. [CrossRef]
117. Ji, L.; Jin, Y.; Wu, K.; Wan, C.; Yang, N.; Tang, Y. Morphology-dependent electrochemical sensing performance of metal (Ni, Co, Zn)-organic frameworks. *Anal. Chim. Acta* **2018**, *1031*, 60–66. [CrossRef] [PubMed]
118. Hu, J.-S.; Dong, S.-J.; Wu, K.; Zhang, X.-L.; Jiang, J.; Yuan, J.; Zheng, M.-D. An ultrastable magnesium-organic framework as multi-responsive luminescent sensor for detecting trinitrotoluene and metal ions with high selectivity and sensitivity. *Sens. Actuat. B* **2019**, *283*, 255–261. [CrossRef]

119. Xu, N.; Zhang, Q.; Hou, B.; Cheng, Q.; Zhang, G. A Novel Magnesium Metal–Organic Framework as a Multiresponsive Luminescent Sensor for Fe(III) Ions, Pesticides, and Antibiotics with High Selectivity and Sensitivity. *Inorg. Chem.* **2018**, *57*, 13330–13340. [CrossRef] [PubMed]
120. Zhang, F.; Yao, H.; Zhao, Y.; Li, X.; Zhang, G.; Yang, Y. Mixed matrix membranes incorporated with Ln-MOF for selective and sensitive detection of nitrofuran antibiotics based on inner filter effect. *Talanta* **2017**, *174*, 660–666. [CrossRef] [PubMed]
121. Asadi, F.; Azizi, S.N.; Ghasemi, S. Preparation of Ag nanoparticles on nano cobalt-based metal organic framework (ZIF-67) as catalyst support for electrochemical determination of hydrazine. *J. Mater. Sci. Mater. Electron.* **2019**, *30*, 5410–5420. [CrossRef]
122. Peng, M.; Guan, G.; Deng, H.; Han, B.; Tian, C.; Zhuang, J.; Xu, Y.; Liu, W.; Lin, Z. PCN-224/rGO nanocomposite based photoelectrochemical sensor with intrinsic recognition ability for efficient p-arsanilic acid detection. *Environ. Nano* **2019**, *6*, 207–215. [CrossRef]
123. Zhou, Y.; Li, C.; Li, X.; Zhu, X.; Ye, B.; Xu, M. A sensitive aptasensor for the detection of b-amyloid oligomers based on metal–organic frameworks as electrochemical signal probes. *Anal. Methods* **2018**, *10*, 4430–4437. [CrossRef]
124. Sachdeva, S.; Koper, S.J.H.; Sabetghadam, A.; Soccol, D.; Gravesteijn, D.J.; Kapteijn, F.; Sudhölter, E.J.R.; Gascon, J.; de Smet, L.C.P.M. Gas Phase Sensing of Alcohols by Metal Organic Framework–Polymer Composite Materials. *ACS Appl. Mater. Interf.* **2017**, *9*, 24926–24935. [CrossRef]
125. Zhou, N.; Ma, Y.; Hu, B.; He, L.; Wang, S.; Zhang, Z.; Luc, S. Construction of Ce-MOF@COF hybrid nanostructure: Label-free aptasensor for the ultrasensitive detection of oxytetracycline residues in aqueous solution environments. *Biosens. Bioelectron.* **2019**, *127*, 92–100. [CrossRef]
126. Li, J.; Xia, J.; Zhang, F.; Wang, Z.; Liu, Q. An electrochemical sensor based on copper-based metal-organic frameworks-graphene composites for determination of dihydroxybenzene isomers in water. *Talanta* **2018**, *181*, 80–86. [CrossRef]
127. Das, A.; Biswas, S. A multi-responsive carbazole-functionalized Zr(IV)-based metal-organic framework for selective sensing of Fe(III), cyanide and p-nitrophenol. *Sens. Actuat. B* **2017**, *250*, 121–131. [CrossRef]
128. Guo, X.-Y.; Dong, Z.-P.; Zhao, F.; Liu, Z.-L.; Wang, Y.-Q. Zinc(II)–organic framework as a multi-responsive photoluminescence sensor for efficient and recyclable detection of pesticide 2,6-dichloro- 4-nitroaniline, Fe(III) and Cr(VI). *New J. Chem.* **2019**, *43*, 2353–2361. [CrossRef]

© 2019 by the authors. Licensee MDPI, Basel, Switzerland. This article is an open access article distributed under the terms and conditions of the Creative Commons Attribution (CC BY) license (http://creativecommons.org/licenses/by/4.0/).

Review

Alternative Green Extraction Phases Applied to Microextraction Techniques for Organic Compound Determination

Eduardo Carasek *, Gabrieli Bernardi, Sângela N. do Carmo and Camila M.S. Vieira

Department of Chemistry, Federal University of Santa Catarina, Santa Catarina, SC 88040-900, Brazil
* Correspondence: eduardo.carasek@ufsc.br

Received: 29 May 2019; Accepted: 9 July 2019; Published: 16 July 2019

Abstract: The use of green extraction phases has gained much attention in different fields of study, including in sample preparation for the determination of organic compounds by chromatography techniques. Green extraction phases are considered as an alternative to conventional phases due to several advantages such as non-toxicity, biodegradability, low cost and ease of preparation. In addition, the use of greener extraction phases reinforces the environmentally-friendly features of microextraction techniques. Thus, this work presents a review about new materials that have been used in extraction phases applied to liquid and sorbent-based microextractions of organic compounds in different matrices.

Keywords: biosorbents; microextraction; organic compounds; green extraction phases

1. Introduction

Sample preparation is a crucial step in analytical methods for determining organic compounds. The isolation of the analytes from the matrix is a major task to ensure the quantification and unambiguous identification of such compounds [1]. Classical sample preparation techniques, such as liquid–liquid extraction (LLE) and solid-phase extraction (SPE), are usually time-consuming and labor-intensive. These techniques usually use large volumes of organic solvents, which are expensive and generate a considerable amount of waste that is harmful for human health and the environment [2].

Microextraction techniques such as those based on sorbent microextraction and liquid-phase microextraction are considered of great importance, since they represent an environmentally friendly alternative to classical extraction methods [3]. There are different microextraction configurations and modes of use. Sorbent microextraction may be considered as an advanced and miniaturized solid phase extraction (SPE) technique. Solid phase microextraction (SPME) [4] and thin film microextraction (TFME) [5] belong to this category. Similarly, liquid phase microextraction (LPME) can be considered as miniaturized liquid–liquid extraction procedures [6]. Most LPME techniques used include dispersive liquid–liquid microextraction (DLLME) [7] and single drop microextraction (SDME) [8,9].

In general, microextractions are carried out using an appropriate extraction phase, which can be a liquid [6] or a solid material [10], depending on the technique chosen. There is a large variety of extraction phases commercially available. However, in the last decade, efforts have been devoted to the development of new materials to be used as "greener" extraction phases. The green aspects of these alternative materials contribute to a less harmful and lower-cost analysis [3]. Furthermore, their usage reinforces the environmentally friendly character of microextraction techniques. In some specific cases, it increases selectivity and hence applicability for treating complex samples. For example, ionic liquids (ILs) and their tunable properties meet the criteria for extracting some compounds.

Based on that, the aim of this work is to review the new materials used as green extraction phases for the determination of organic compounds by microextraction and chromatographic techniques. Furthermore, some recent applications of these materials in various matrices are presented.

2. Biosorbent-Based Extraction Phases

Natural, renewable and biodegradable sorbents are denominated biosorbents and have attracted a great deal of attention in the sample preparation area, due to their low cost, non-toxicity and high availability [11]. There are several materials from different sources that can be used as biosorbents, such as agricultural waste products, industrial by-products and biomass derived from usually discarded materials [12]. Some materials such as cork, bamboo charcoal, bract, and recycled diatomaceous earth have already been applied in extraction phases to a large number of microextraction techniques based on solid phase extraction. Thus, in the following topics, the characteristics and some relevant recent applications of these materials in the biosorbent-based extraction phase will be discussed. More information about the applications and validation parameters of the reported methods are summarized in Table 1.

2.1. Cork as a Biosorbent

Cork is the bark of the cork oak tree (*Quercus suber* L.) and as a lignocellulosic material, it is composed of 40% suberin, 24% lignin, 20% cellulose and hemicellulose and 15% of other extractives [13]. In 2013, Dias et al. [14] proposed, for the first time, the use of a cork-based biosorbent as a coating for the solid phase microextraction technique (SPME) introduced by Pawliszyn et al. in 1990 as a miniaturized technique [4]. The procedure to obtain SPME cork fiber involves immobilization of the cork powder (approx 200 mech) on a nitinol wire of 0.2 mm thickness and approximately 2 cm length. After this, the wires with biosorbent are heated at a temperature of 180 °C for 90 min. Before use, the cork fibers produced are conditioned at 260 °C for 60 min in a gas chromatograph (GC) injection port [14].

The fiber characterization conducted using Fourier transform infrared spectroscopy (FTIR) and scanning electron microscopy (SEM) showed a heterogeneous chemical composition. Lignin presents several aromatic rings that may allow π-π interactions between sorbent phase and analytes, mainly the non-polar compounds. On the other hand, cellulose and hemicellulose exhibit a number of O–H groups in their structure, allowing for hydrogen-bonding and dipole–dipole interactions with the compounds presenting intermediate polarity. Furthermore, a homogeneously distributed coating and a porous structure are reported for the surface of the fiber. The coating thickness obtained for the proposed fiber was about 55 μm [14].

The SPME biosorbent-based fiber has already been successfully applied for the determination of polycyclic aromatic hydrocarbons (PAH) [14], organochloride pesticides (OCPs) [15] and UV filters such as 3-(4-methylbenzylidene) camphor (4-MBC) and 2-ethylhexyl 4-(dimethylamino) benzoate (OD-PABA) [16] (Table 1). In the work proposed by Dias et al. [14], the cork fiber extracted the PAHs by adsorption through π-π interactions and suberin was reported to play a more important role than lignin, in this case. The cork fiber was compared to commercially available fibers such as polydimethrylsiloxane/divinylbenzene (PMDS/DVB), divinylbenzene/carboxen/polydimethylsiloxane (DVB/CAR/PDMS) and polydimethylsiloxane (PDMS), presenting similar or better extraction efficiency for most compounds. An advantage reported by the authors was the lifetime of the coating layer, which was higher than the commercial ones, 50–100 times against 40 times, respectively.

Table 1. Applications and validation parameters of different microextraction techniques using biosorbents as the extraction phase for organic compound determination. SPME—solid phase microextraction; TFME—thin film microextraction; BAµE—bar-adsorptive microextraction; SPE—solid-phase extraction.

Biosorbent	Technique	Analyte	Matrix	LOQ	LOD	Linear Range	Recovery (%)	Precision (RSD%)	Method	Ref.
Cork	SPME	polycyclic aromatic hydrocarbons (PAHs)	River water	0.1 µg L^{-1}	0.03 µg L^{-1}	0.1–10 µg L^{-1}	70–103	1.9–15.7	GC-MS	[14]
		Organochlorine pesticides	River water	1–10 ng L^{-1}	0.3–3 ng L^{-1}	1–50 ng L^{-1}	60–112	0.5–25.5	GC-ECD	[15]
		UV filters	River water	0.01–0.1 µg L^{-1}	0.004–0.03 µg L^{-1}	0.01–0.5 µg L^{-1}	67–107	3–18	GC-MS	[16]
	TFME	Emerging contaminants	River water	0.8–15 µg L^{-1}	0.3–5.5 µg L^{-1}	5–400 µg L^{-1}	72–125	4–18	HPLC-DAD	[17]
	BAµE	Parabens, benzophenone and triclocarban	Lake water, effluent, wastewater	1.6–20 µg L^{-1} (15 mm) 0.64–8 µg L^{-1} (7.5 mm)	0.5–6.5 µg L^{-1} (15 mm) 0.2–2.5 µg L^{-1} (7.5 mm)	1.6–500 µg L^{-1} (15 mm) 0.64–400 µg L^{-1} (7.5 mm)	65–123 (7.5 mm)	3–22 (7.5 mm)	HPLC-DAD	[18]
		Hexanal and heptanal	Human urine	2.19–3 µmol L^{-1}	0.73–1 µmol L^{-1}	2.19–8 µmol L^{-1}	88–111	3–7	HPLC-DAD	[19]
Bract	SPME	Organochlorine pesticides	River and lake water	0.65–2.38 ng L^{-1}	0.19–0.71 ng L^{-1}	5–100 ng L^{-}	60–110	5–19	GC-ECD	[20]
		PAHs	Lake water	0.01–0.1 µg L^{-1}	0.003–0.03 µg L^{-1}	0.01–4 µg L^{-1}	68–117	0.6–17	GC-MS	[21]
	TFME	Steroid estrogens	Human urine	0.1–10 µg L^{-1}	0.3–3 µg L^{-1}	0.1–400 µg L^{-1}	71–105	1–17	HPLC-FLD	[22]
	SPME	PAHs	River water	0.1–0.5 µg L^{-1}	0.03–0.16 µg L^{-1}	0.1–25 µg L^{-1}	83–100	2–15	GC-MS	[23]
Diatomaceous earth	TFME	Endocrine disruptors	River water	3–23 µg L^{-1}	1–8 µg L^{-1}	5–285 µg L^{-1}	70–117	1–21	HPLC-DAD	[24]
	BAµE	Methyl and ethyl paraben, benzophenone, triclocarban	Lake water	0.63–6.9 µg L^{-1}	0.19–2 µg L^{-1}	0.63–100 µg L^{-1}	63–124	1–20	HPLC-DAD	[25]
Bamboo charcoal	SPME	Phthalate esters	Tap and river water		0.004–0.023 µg L^{-1}	0.1–100 µg L^{-1}	61–87	1.89–9.85	GC-MS	[26]
Moringa oleifera seeds	µ-SPE	Phthalate esters	Milk	0.1–3.7 µg L^{-1}	0.01–1.2 µg L^{-1}	1–100 µg L^{-1}	77–103	3.6–9.4	GC-MS	[27]
MMT clay	RDSE	polychlorinated biphenyl (PCB)	Wastewater	6.5–103.8 ng L^{-1}	3 ng L^{-1} to 43 ng L^{-1}		80–86	2–24	GC-ECD	[28]
Cork and MMT clay	RDSE	Parabens	River and tap water	0.8 µg L^{-1} (cork) 3 µg L^{-1} (MMT clay)	0.24 µg L^{-1} (cork) 0.90 µg L^{-1} (MMT clay)	0.8–75 µg L^{-1} (cork) 3–100 µg L^{-1} (MMT clay)	80–118 (cork) 80–119 (MMT clay)	1.15–14.29 (cork) 3.24–18.14 (MMT clay)	LC-MS/MS	[29]

When used for OCP determination [15] the extraction efficiency of the cork fiber was attributed mostly to dipole–dipole interactions with the analytes. The authors also reported the occurrence of hydrogen bonds with the compounds containing oxygen atoms. In the work of Silva et al. [16], the cork fiber extraction efficiency for 4-MCB and OD-PABA was compared with PDMS/DVB and PDMS fibers, and the results showed a better extraction efficiency when the cork fiber was used, for both analytes.

Cork has also been used with thin film microextraction (TFME) [17]. TFME comprises a new geometry for SPME, aiming to provide more sensitivity for this technique. The device used in TFME consists of a support coated with a thin layer of a sorbent phase that can be used in headspace or immersion mode. Moreover, this technique has been designed to fit a commercially available 96-well plate system providing high-throughput analyses. To date, there has been only one study published using cork with TFME, by Morés et al. in 2017 [17]. The TFME cork coating coupled with 96-well plate system was used as a high-throughput method for the extraction of emerging contaminants in a water sample by high-performance liquid chromatography-diode-array detector HPLC-DAD (Table 1). In this work, analytes with the log K_{ow} ranging from 2.49 to 5.92 were successfully extracted by the cork sorbent phase.

Another microextraction technique that used cork as the sorbent phase is called bar-adsorptive microextraction (BAμE) [18]. The BAμE device consists of a finely divided powder (up to 5 mg) fixed with suitable adhesives in polypropylene supports with cylindrical bar format. In the experimental procedure, the adsorbing bars are placed in direct contact with the sample, under constant stirring. Due to the low density of the polypropylene support, this floats just below the vortex formed by agitation, preventing direct contact of the bar with the flask's walls containing the sample, thus increasing the useful life of the device. After extraction, the desorption step consists of completely inserting the bar into vials containing a few microliters of a suitable extraction solvent.

Cork biosorbent has been used twice with BAμE. It was first used in 2015 for determination of polar and intermediate polarity compounds (parabens, benzophenone and triclocarban) in water samples by HPLC-DAD [18]. In this study, bars of 7.5 and 15 mm in length were used. Hollow cylindrical polypropylene tubes (15 mm length and 3 mm diameter) were coated with adhesive films followed by a layer of the cork powder (200 mesh). Before use, the bars were conditioned under ultrasound agitation with 250 μL of acetonitrile (ACN) for 15 min. The half bars (length of 7.5 mm) were obtained by cutting the 15 mm length bar in half. As shown in Table 1, the quantification limits ranged from 1.6 to 20 μg L^{-1} using a bar of 15 mm and 0.64 to 8 μg L^{-1} using a bar of 7.5 mm.

The second study was published in 2017. At this time, the use of cork BAμE bars was extended to biological samples for determination of two potential lung cancer biomarkers (hexanal and heptanal) in human urine by HPLC-DAD [19]. In this study, the adsorptive bar surface was impregnated with 2,4-dinitrophenylhydrazine (DNPH) so that derivatization and extraction were accomplished simultaneously on the surface of the bar under acidic conditions. Relative recoveries in urine samples varied from 88 to 111% (Table 1). Figure 1 illustrates a scheme of the biobased BAμE procedure used. According to the authors, one of the main advantages was the low cost of the method, since polypropylene tubes, adhesive and cork obtained from cork stoppers were used to produce the devices.

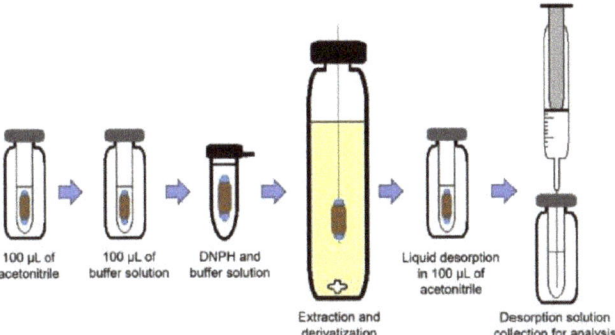

Figure 1. Scheme of the biobased bar-adsorptive microextraction (BAµE) procedure to determine hexanal and heptanal in human urine by HPLC-DAD. Reproduced with permission from [19], Copyright Elsevier, 2017.

2.2. Bract as a Biosorbent

Another lignocellulosic material was reported in 2017 as a green extraction phase for SPME [20]. The material, called bract, is the non-developed seeds obtained from the tree *Araucaria angustifolia* (Bert) O. Kuntze, a conifer found in the south and southeast of Brazil and in eastern Argentina. This material is composed of 45% lignin, 46% holocellulose (cellulose and hemicellulose) and 15% total extractives. The process for obtaining bract-based fibers is similar to those already described for cork fibers. Both materials are similar; however, the cork powder presents a better attachment to the nitinol wire, so it is easier to handle. Bract has been used as an environmentally friendly and low-cost biosorbent coating for SPME for the determination of OCPs in river and lake water [20] and PAH's in river water [21] (Table 1). The characterization of bract fiber carried out by thermogravimetric analysis (TGA), SEM and FTIR showed that the fiber offers satisfactory thermal stability with no decomposition observed up to 260 °C. SEM micrographs presented a highly porous and rough morphology and a film thickness of approximately 60 µm [20]. Like cork, bract is also a lignocellulosic material. The FTIR spectrum revealed peaks related to O–H bond and C–H stretching assigned to polysaccharides and lignin. C=C stretching from the aromatic rings and a peak related to C–O–C bond were also identified. Figure 2 shows a scheme of the preparation (2A) and SEM micrographs obtained for bract SPME fiber (2B).

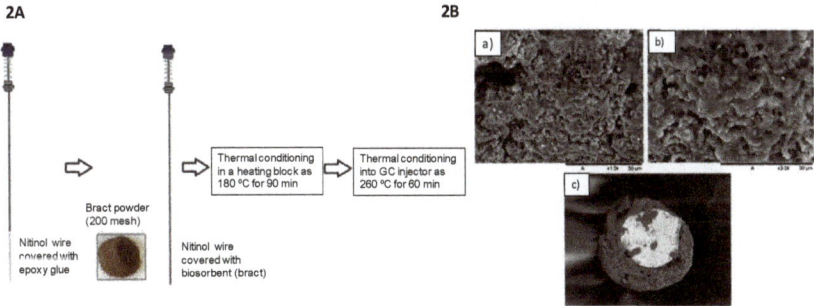

Figure 2. (**2A**) Scheme of the preparation of SPME fibers and (**2B**) SEM micrographs obtained for bract fiber (a) magnification of 1500× (b) magnification of 3000× and (c) a cross-section of the proposed fiber at magnification of 300×. Reproduced with permission from [20], Copyright Elsevier, 2017.

When applied to the determination of OCPs in water samples by gas chromatography–electron capture detection (GC-ECD), a satisfactory analytical performance was reported with limits of detection

(LODs) ranging from 0.19 to 0.71 ng L^{-1}. In addition, the biosorbent-based fiber provided efficient extractions when compared with the commercial mixed coating fiber DVB/Car/PDMS. In 2018, bract fiber was used for the determination of PAH's in water samples by gas chromatography–mass spectrometry (GC-MS) [21]. In this study, the LODs varied from 0.003 to 0.03 µg L^{-1} (Table 1).

Bract has also been used with TFME combined with a 96-well plate for the determination of steroid estrogens in human urine by liquid chromatography fluorescence detector (HPLC-FLD) [22]. At this time, the target compounds presented intermediate polarity with log P ≤ 4.12. The extraction efficiency of the bract layer was explained by the π-π interactions between lignin and the analytes. The LODs of the method varied between 0.3 µg L^{-1} for 17-β-estradiol and 3 µg L^{-1} for estrone (Table 1). As an advantage, in this study, the authors reported the use of the 96-well plate system, allowing for 1.7 min/sample turnaround times for the proposed method.

2.3. Recycled Diatomaceous Earth as a Biosorbent Material

Diatomaceous earth is an amorphous silicate sediment originating from fossilized unicellular microorganisms on algae of the class Bacillariophyceae centricae. This material is composed mainly of silica dioxide and small amounts of aluminum, iron, calcium, magnesium, sodium and potassium. After being used for the filtration and clarification of beer in a brewery, the diatomaceous earth was subject to thermal treatment and then used as the extraction phase for SPME [23]. The FTIR characterization of the material revealed O–H bonds from silanol groups. Moreover, asymmetric stretching was reported assigned to Si–O–Si bonds, frequently found in silicate materials. This biosorbent has been used along with SPME for the determination of PAH's in river water samples by GC–MS [23]. In the comparison with the commercial fibers PDMS/DVB and PDMS, the biosorbent showed better extraction efficiency for most compounds, except for acenaphthylene ($C_{12}H_8$), fluorene ($C_{13}H_{10}$) phenanthrene ($C_{14}H_{10}$) and pyrene ($C_{16}H_{10}$), for which PDMS/DVB was better. In this work, the authors did not provide a possible explanation for the interactions between analytes and the biosorbent. However, a fiber limitation was reported regarding the use of salt in the optimization step. According to the authors, salt particles added to the samples can remain adsorbed in the surface of the extraction phase, causing fiber damage. In this case, if the salt addition is necessary to improve extraction efficiency, a cleaning step with water may be done before the fiber insertion into the GC injection port.

Other applications of this sorbent include TFME with 96-well plate system for the determination of endocrine disruptors in water samples by HPLC-DAD [24] and with BAµE in the determination of methyl and ethyl paraben, benzophenone and triclocarban in water by HPLC-DAD [25] (Table 1). In the work of Kirschner et al. [24], bisphenol A (BPA), benzophenone (BzP), triclocarban (TCC), 4-methylbenzylidene camphor (4-MBC) and 2-ethylhexyl-p-methoxycinnamate (EHMC) were successfully determined from environmental water samples. Considering the analyte structure, the authors attribute the extraction efficiency of diatomaceous earth to the O–H moieties presented in the sorbent and the O–H and N–H groups in the target compounds. In this work, information about the extraction phase stability in the presence of organic solvents was provided. After successive extractions, the biosorbent blades were able to be used without expressive loss in the extraction efficiency for at least 20 extraction/desorption cycles. The proposed method exhibited satisfactory analytical performance, with LODs varying between 1 and 8 µg L^{-1} and determination coefficient ranging from 0.9926 to 0.9988.

2.4. Other Materials Used as Biosorbents

A range of other materials characterized as biosorbents have also been used in combination with microextraction techniques for organic compound determination. Although there are still few applications involving these materials, a brief description is provided, along with the existent applications.

Bamboo plants are characterized by rapid growth and are widely distributed in China. Bamboo charcoal is obtained by submitting the bamboo to high temperatures (over 800 °C), producing a material with high density, porous structure and a large surface area. Bamboo charcoal was proposed as a novel and inexpensive SPME coating material for determination of 11 phthalate esters (PAE) in water samples by GC-MS [26].

Another material used as a biosorbent was obtained from powdered seeds of the *Moringa oleifera* tree. This material is considered to possess a highly fibrous and naturally functionalized surface. The characterization of the moringa-based biosorbent using SEM and FITR showed a porous framework of interconnected fibers, and various functional moieties were identified such as O–H, N–H and C–H and CH_2 groups. The first application for organic compound determination was in 2016, as a sorbent for the determination of 13 phthalate esters (PE) in a milk sample by micro-solid phase extraction (μ-SPE) and GC-MS [27]. The relative polar PEs interacted with the sorbent through the functional moieties identified. The more the alkyl chain of PEs increased, the lower the extraction efficiency became.

An eco-material denominated montmorillonite (MMT) clay, modified through the intercalation of ionic liquids (IL), has also been applied in the extraction phase [28,29]. MMT is a clay mineral composed of structural layers consisting of an octahedral alumina sheet sandwiched between two tetrahedral silica sheets. MMT is found in sediments, soils or rock and has been modified to adsorb organic compounds of low polarity from aqueous solutions. In 2016, Fiscal-Ladino et al. [28] used rotating-disk sorptive extraction (RDSE) and MMT in the extraction phase for the determination of polychlorinated biphenyl (PCB) compounds in water samples with GC-ECD (Table 1). The RDSE device consists of a rotating Teflon disk containing an embedded miniature magnetic stirring bar. In this study, SEM was employed to characterize the novel sorbent, and the results showed clusters of particles with a narrow size distribution of approximately 25 mm. The extraction efficiency achieved for the MMT modified with 1-hexadecyl-3-methylimidazolium bromide (HDMIM-Br) phase was compared with commercial phases and showed the highest response for all the studied analytes.

Very recently, the viability of MMT-HDMIM-Br as a green sorbent for RDSE was again investigated [29]. In this study, cork and montmorillonite clay modified with ionic liquid were explored for the determination of parabens in water samples by high-performance liquid chromatography—tandem mass spectrometry (LC-MS/MS). The proposed method presented limits of detection of 0.24 μg L^{-1} for the cork and 0.90 μg L^{-1} for the MMT-HDMIM-Br with correlation coefficients higher than 0.9939 for both biosorbents.

2.5. Concluding Remarks about Biosorbents

In general, biosorbents demonstrated great versatility for the extraction of the different classes of compounds. Lignocellulosic biosorbents, such as cork, bract and *Moringa oleifera* seeds, are mainly composed of lignin, cellulose and hemicellulose. These macromolecules have a number of chemical groups that are capable of interacting with a wide range of analytes with different polarities. The works reported in this review showed studies in which cork was able to satisfactorily extract non-polar compounds, such as PAH's, and compounds with intermediate polarity, such as parabens, benzophenone and triclocarban. Bract biosorbent demonstrated similar behavior, presenting satisfactory extraction for compounds with low polarity, such as OCPs, and for those with intermediate polarity, such as steroid estrogens.

The extraction efficiency of these lignocellulosic materials is mostly explained by the π-π interactions between lignin and the analytes, or through hydrogen-bonding and dipole-dipole interactions between O–H groups presented in the cellulose and hemicellulose with the O–H, N–H bonds and Cl present in the analytes. When biosorbents were used with SPME for PAH extraction in water samples, bract fiber showed lower LODs than cork and diatomaceous earth fiber. The same was observed for OCP determination in water samples. Bract has a higher percentage of lignin in its structure than cork, which could explain the higher extraction efficiency for the non-polar compounds.

By using BAμE as the microextraction technique, the diatomaceous earth bar provided lower LODs than the cork bar for the extraction of parabens, benzophenone and triclocarban in water samples. Although diatomaceous earth has been used for PAHs determination, it has shown good extraction efficiency for compounds with intermediate polarity, which was mainly due to interactions through O–H groups. It is also worth mentioning that the porous structure of these biosorbents plays an important role in the extraction through physical interaction with the analytes.

As a final remark, cork has been the material most used with different microextraction techniques and for a large variety of compounds. This fact can be related to the ease with which it is obtained through the reuse of wine bottle corks. The other biosorbents are more limited, such as bract, which is obtained from trees in southern Brazil and in eastern Argentina. Similarly, diatomaceous earth is a sub-product from the beer filtration and clarification process. Most of the works report the comparison with commercial extraction phases. In general, the results are similar or even better, in some cases. However, the procedures employed in the preparation of the devices, in particular for SPME, may be a limitation for the widespread use of these bio-based extraction phases.

3. Ionic Liquids (ILs) as Green Extraction Phase

Ionic liquids (ILs) are non-molecular solvents with melting points below 100 °C, negligible vapor pressure at room temperature, high thermal stability and variable viscosity. The ILs' miscibility in water and organic solvents can be controlled by selecting the cation or anion combination or by the addition of certain functional groups in the IL molecule. Most often, ILs are composed of large asymmetric organic cations and inorganic or organic anions. The most usually employed IL anions are polyatomic inorganic species, such as PF_6^- and BF_4^-, and the most relevant cations are a pyridinium and imidazolium ring with one or more alkyl groups attached to the nitrogen or carbon atoms [30]. ILs have been successfully applied to the liquid phase microextraction technique (LPME) as a less toxic alternative to conventional organic solvents. Considering these most notable properties, the potential usage of ILs as the extraction phase for LPME and applications has already been extensively reviewed by different authors [31–35]. The successful use of ILs in extraction phases is related to their structure. In addition to the common interactions existing in conventional solvents, ILs also have ionic interactions which confer miscibility when dissolved in polar substances. At the same time, the presence of alkyl chains in the cation determines the solubility in less polar substances. A review by Han and coworkers in 2012 presents the physical properties of some of the most commonly used ILs [32].

In 2003, Liu et al. [36] reported the first application of ILs in the extraction phase in single drop microextraction (SDME). SDME is a simple, easy-to-operate and reliable LPME-based method developed in the 1990s [8]. In this report, IL-based SDME coupled with HPLC was applied for the preconcentration and analysis of polycyclic aromatic hydrocarbons (PAHs) using the IL 1-octyl-3-methylimidazolium PF_6 [C_8C_1IM-PF_6] as the extraction solvent. Compared with 1-octanol, ILs provided higher enrichment factors (EFs), enabling the use of extended extraction times and larger drop volumes. In 2015, Marcinkowski et al. reviewed the analytical potential of ILs in SDME [37].

One of the most relevant applications of ionic liquids concerns their use as the extraction phase in dispersive liquid-liquid microextraction (DLLME). DLLME is a powerful extraction technique in which microliter volumes of an extraction solvent are dispersed in the sample to extract and preconcentrate the analytes [7]. The tunable properties of ILs have made these solvents particularly attractive for DLLME applications. Trujillo-Rodríguez et al. [38] reviewed in 2013 the use of ILs in the different types of DLLME and Rykowska et al. [39] recently reviewed modern approaches for IL-DLLME. In a recent application, IL-DLLME was used for the first time in the extraction phase for cortisone and cortisol determination from human saliva samples by HPLC-UV. The method provided high selectivity and EFs to achieve biological levels [40].

3.1. Magnetic Ionic Liquids (MILs) as Green Extraction Phase

A subclass of the ILs, denominated magnetic ionic liquids (MILs), has also been used as a green alternative to conventional organic solvents in LPME applications. Their physicochemical properties are similar to conventional ILs; nonetheless, MILs exhibit a strong response to external magnetic fields. MILs are obtained by the introduction of a paramagnetic component into the cation or anion of the IL structure. Often the paramagnetic component is comprised of a transition or lanthanide metal ions [41]. Synthesis, properties and analytical applications of MILs, including micro extractions, have been already reviewed [42].

MILs have been applied to many LPME techniques. Table 2 shows the most recent applications (since 2017). However, most of the MIL-based extraction approaches are performed using DLLME. In this case, a mixture containing the MIL, dissolved in a small amount of an organic solvent, is dispersed in the sample and then recovered with a magnetic rod. The first application of MIL-DLLME was described in 2014 for the extraction of triazine herbicides in vegetable oils using 1-hexyl-3-methylimidazolium tetrachloroferrate ([C6mim] [FeCl$_4$]) as the extraction phase [43]. Very recently, Sajid et al. [44] reviewed significant milestones of employing MILs for analytical extraction application and the main drawbacks of using MILs with DLLME.

Among the most recent applications, one in particular has attracted significant attention, since a new generation of MILs suitable for in situ DLLME were presented. MILs comprising paramagnetic cations containing Ni(II) metal centers coordinated with four N-alkylimidazole ligands and chloride anions were used for in situ DLLME and extraction of both polar and non-polar pollutants in aqueous samples. In this work, a metathesis reaction was originated by mixing a water-soluble MIL into the aqueous sample followed by the addition of bis [(trifluoromethyl) sulfonyl] imide ([NTf2-]) anion. This reaction produced a water-immiscible extraction solvent containing the preconcentrated analytes. The MIL was then isolated by magnetic separation and subjected to analysis using reversed-phase HPLC-DAD. The proposed methodology achieved higher extraction efficiency when compared to the conventional MIL-dispersive liquid-liquid microextraction. Extraction efficiencies ranging from 46.8 to 88.6% and 65.4 to 97.0% for the [Ni(C$_4$IM)$_4$$^{2+}$]2[Cl$^-$] and the [Ni(BeIM)$_4$$^{2+}$]2[Cl$^-$] MILs were obtained [45].

MILs have also been successfully applied to the SDME technique. In a recent study, a high-throughput parallel-single-drop microextraction (Pa-SDME) was developed [46]. According to the authors, Pa-SDME combines some advantageous features of trihexyl (tetradecyl) phosphonium tetrachloro manganite (II) ([P6, 6, 6, 14$^+$]$_2$[MnCl$_4$$^{2-}$]) MIL such as drop stability and extraction capacity with the 96-well plate advantages for obtaining high-throughput analysis. In this study, the determination of parabens, bisphenol A, benzophenone and triclocarban was conducted from environmental aqueous samples by HPLC-DAD. The method validation was carried out after the optimization step, and LODs ranging from 1.5 to 3 µg L^{-1} were achieved. Coefficients of determination were higher than 0.994, and intraday and interday precision ranged from 0.6 to 21.3% (n = 3) and 10.4–20.2% (n = 9), respectively. Relative recovery ranged between 63% and 126%. Figure 3 shows the Pa-SDME lab-made extraction apparatus used for the extractions.

Table 2. Recent applications of magnetic ionic liquids (MILs) for extraction of different analytes from various matrices. DLLME—liquid–liquid dispersive microextraction.

Method	MIL	Analyte	Matrix	LOD	Instrumentation	Ref.
In situ DLLME	[P$_{6,6,6,14}$$^+$]$_2$[CoCl$_4$$^{2-}$]	Biogenic amines	Wine fish	1.3–3.9 µg L^{-1} 1.2–3.8 µg kg^{-1}	HPLC-UV	[47]
DLLME	[P$_{6,6,6,14}$$^+$] [Cl$^-$]	Estriol Estrone Parabens Carbamazepine Diazepam Ketoprofen Ibuprofen 17α-Ethynylestradiol Triclocarban Aldicarb Methyl parathion Metolachlor Diuron Bisphenol A	River water	1.5–15 µg L^{-1}	HPLC-DAD	[48]

Table 2. Cont.

Method	MIL	Analyte	Matrix	LOD	Instrumentation	Ref.
In situ SB-DLLME	$[Ni(C_4IM)_4^{2+}]2[Cl^-]$ $[Ni(C_8IM)_4^{2+}]2[Cl^-]$ $[Co(C8IM)_4^{2+}]2[Cl^-]$	Naphthalene Acenaphthene Fluorene 1-chloro-4-nitrobenzene Biphenyl 5-Bromoacenaphthene 3-Tert-butylphenol	Tap and mineral water	4.8–15 µg L^{-1} 1–10 µg L^{-1} 5.9–30 µg L^{-1}	HS-GC-MS	[49]
DLLME	P_{66614}^+] [Dy(III)(hfacac)$_4^-$]	Triazines and sulfonamides	Lake water, effluent wastewater	0.011–0.03 µg L^{-1}	HPLC-DAD	[50]
DLLME	$[P_{6,6,6,14}^+]_2[MnCl_4^{2-}]$	Estrogens	Human urine	2 ng mL^{-1}	HPLC-DAD	[51]
SB-DLME	$[P_{6,6,6,14}^+]$ [Ni(II)(hfaca)$_3^-$]	PAHs	River water and rain water	1.7–28.7 ng L^{-1}	GC-MS	[52]
HS-SDME and DLLME	$[P_{6,6,6,14}^+]_2[MnCl_4^{2-}]$	Aromatic compounds	Lake water	0.005–1 µg L^{-1} and 0.04–1 µg L^{-1}	HPLC-DAD	[53]
SB-DLME	$[P_{6,6,6,14}^+]$ [Ni(hfacac)$_3^-$]	UV filters	River and sea water	9.9–26.7 ng L^{-1}	GC-MS	[54]
Vacum-HS-SDME	$[P_{6,6,6,14}^+]$ [Mn(hfacac)$_3^-$]	Free fatty acids	Milk	14.5–216 µg L^{-1}	GC-MS	[55]

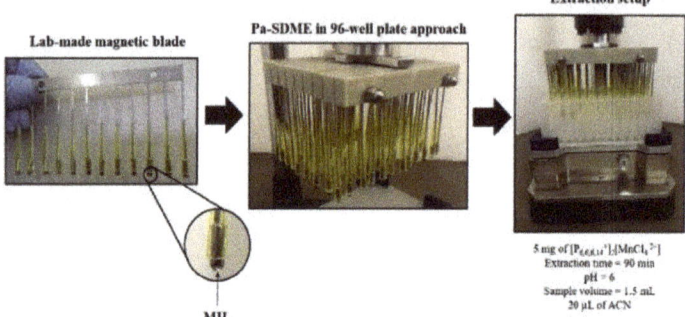

Figure 3. An overview of the extraction procedure using the novel parallel-single-drop microextraction (Pa-SDME)/MIL-based approach. Reproduced with permission from [46], Copyright Elsevier, 2017.

3.2. Deep Eutectic Solvent (DES) and Natural Deep Eutectic Solvents (NADES) as a Green Extraction Solvent

The concept of deep eutectic solvents (DESs) was first introduced by Abbot et al. in 2003 [56]. DESs consist of two solid compounds interacting via hydrogen bonds to form a liquid phase with a lower melting point compared to each individual component [57]. The most popular DES involves the combination of choline chloride (ChCl) with urea, carboxylic acids (e.g., citric, succinic, and oxalic acids) and glycerol acting as hydrogen bond donors (HBDs). The use of ChCl has been related to some advantages for DES production, including ease of preparation, biocompatibility, non-toxicity and biodegradability. Although DESs are considered a subclass of IL, they are cheaper and easier to prepare due to the lower cost of the raw materials. Also, they present less toxicity and are often biodegradable, which makes them valuable alternative solvents. One of the most attractive features of these solvents is that, like ILs, their chemical properties can be tuned through the manipulation of their chemical structures (HBA and HBD) to interact more effectively with the target analytes. Florindo et al. (2018) [58] provide a closer look into DES intermolecular interactions; however, there is still a lack of knowledge regarding this topic.

In 2017, Shishov et al. published a review of the applications of DES in analytical chemistry, including their use in the extraction phase in microextraction techniques [59]. Nowadays, these solvents represent a very promising alternative in the sample preparation area, mainly due to their easy acquisition and versatility for extract different classes of compounds. In the work of Farajzadeh et al., a gas-assisted DLLME method using a mixture of ChCl and 4-chlorophenol (1:2 molar ratio) as the extraction solvent was developed for pesticide residue determination in vegetable and fruit by GC-FID [60]. The proposed method was optimized, and enrichment factors and extraction recoveries were achieved in the range of 247–355 and 49–71%, respectively.

Two hydrophobic deep eutectic solvents were synthetized and used as the extraction solvent with air-assisted DLLME (AA-DLLME) for pre-concentration and extraction of benzophenone-type UV filters from aqueous samples and determination by HPLC-DAD [61]. DESs were obtained by mixing $_{DL}$-menthol and quaternary ammonium salts with a straight-chain monobasic acid. After optimization, a DES consisting of $_{DL}$-menthol and decanoic acid mixture (1:1) was chosen for UV-filter extraction. Analytical parameters of merit were evaluated, and the developed method exhibited low limits of detection (0.5 to 0.02 ng mL^{-1}) and repeatability in the range of 1.5–4.9 and 0.6–5.6% for intraday ($n = 6$) and interday ($n = 6$) determinations, respectively. The method was applied to determine the benzophenone-type filters in environmental water samples, and relative recoveries ranged from 88.8 to 105.9%.

Recently, a novel approach for effective liquid-liquid microextraction based on DES decomposition was reported [62]. In this work, DESs were synthesized from tetrabutylammonium bromide and long-chain alcohols. Afterwards, they were decomposed in the aqueous phase, resulting in an in-situ dispersion of organic phase and extraction of hydrophobic analytes. The method was applied to 17b-estradiol microextraction from transdermal gel samples. Efficient extraction of 95 ± 5% and reproducibility of 6% were obtained. Figure 4 shows a scheme of liquid-liquid microextraction based on in-situ decomposition of deep eutectic solvent.

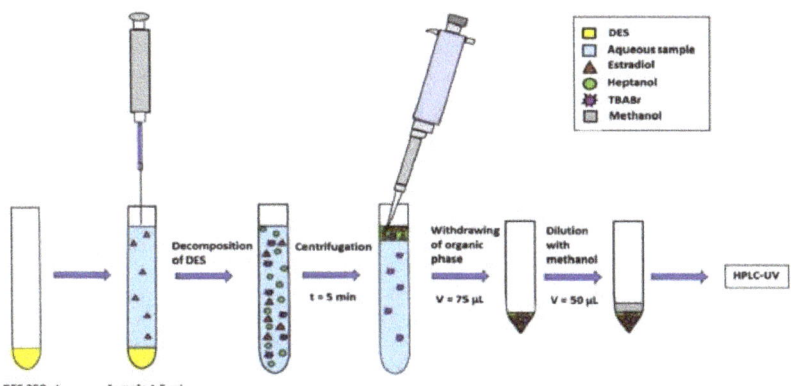

Figure 4. Schema of liquid-liquid microextraction based on in situ decomposition of deep eutectic solvent. Reproduced with permission from [62], Copyright Elsevier, 2019.

Natural deep eutectic solvents (NADESs) are considered a sub-class of DESs, and they consist of a mixture of cheap and natural compounds such as sugars, alcohols, organic acids, and amino acids [63]. The most significant features of NADESs include adjustable viscosity, since they are liquid at temperature below 0 °C, sustainability, and the capability of dissolving a diverse range of analytes with different polarities. In a review by Hashemi et al., (2018) [63] the authors show the most common compounds for preparation of DES and the chemical structure of the most used NADES components. Cunha et al., 2018 [64] presented the main LPME techniques using DES or NADES as extraction solvents to determine several polar volatile and non-volatile compounds from food and water matrices. Table 3 shows some recent applications of DES/NADES (since 2017) with microextraction techniques for the determination of organic compounds in various matrices by chromatographic methods.

Table 3. Recent applications of deep eutectic solvents (DES)/natural deep eutectic solvents (NADES) for extraction of different analytes from various matrices.

Technique	DES/NADES Composition	Analytes	Matrix	LOD	Instrumentation	Ref.
DES-ALLME	Ch-Cl: TNO [1]	Methadone	Water and biologic	0.7 µg L^{-1}	GC-FID	[65]
UA-DLLME	trioctylmethylammonium chloride: decanoic acid	UV filters	Water	0.15–0.30 ng mL^{-1}	HPLC-UV	[66]
VA-LLME	Decanoic acid: Methyltrioctylammonium bromide	Malondialdehyde (MDA) and Formaldehyde (FA)	Human urine, apple juice and rain water	2.0 and 10.0 ng mL^{-1}	HPLC-UV	[67]
AA-EME	ChCl: Ph-EtOH	Amphetamine-type stimulants (Ats)	Human plasma and pharmaceutical wastewater	2.0–5.0 ng mL^{-1}	HPLC-UV	[68]
SFO–AALLME	Ch-Cl: n-butyric acid	Aromatic amines	Aqueous samples	1.8–6.0 ng L^{-1}	GC-MS	[69]
UA-DLLME	thymol, ±camphor, decanoic: 10-undecylenic acids	PAHs	Industrial effluents	0.0039–0.0098 µg L^{-1}	GC-MS	[70]
DSPE-DES-AALLME	ChCl: 4-chlorophenol	Tricyclic antidepressant drugs	Human urine and plasma	8–15 and 32–60 ng L^{-1}	GC-MS	[71]
DES-GALLME	Mixture of two or three different carboxylic acids (C8, C9, C10, C11 and C12)	Phenolic compounds	Water	0.22–0.53 µg L^{-1}	HPLC-UV	[72]
VA-RP-LLME	[N4444]Cl, TBA [2]: ethylene glycol (EG)	Triazine herbicides	Vegetable oil samples	0.60–1.50 µg L^{-1}	HPLC-UV	[73]
MA-in syringe DLLME	ChCl: phenol and ChCl: butyric acid	Herbicides	Wheat	1.6–12 ng kg^{-1}	GC-MS	[74]
DLLME	Hexafluoro isopropanol: l-carnitine/betaine	Pyrethroids	Tea beverages and fruit juices	0.06–0.17 ng mL^{-1}	HPLC	[75]
UA-DLLME-DES	Quaternary phosphonium salts: straight-chain monobasic acids	Pyrethroids	Water	0.30–0.60 µg/L	HPLC-UV	[76]

1—5, 6, 7, 8-Tetrahydro-5, 5, 8, 8-tetramethylnaphthalen-2-ol; 2—tetrabutylammonium chloride ([N4444] Cl, TBA).

4. Supramolecular Solvent (SUPRAS) as Extraction Phase

Supramolecular solvents (SUPRASs) are nano-structured liquids in which the spontaneous association of different molecules self-organizes in a biphasic system formed by a continuous and a dispersed phase. SUPRASs' amphiphilic nature highlights one of their main advantages, providing excellent solvation for a wide range of organic and inorganic compounds. This characteristic is due to the presence of supramolecular aggregates, promoting a solvent with different degrees of polarity [77–79]. Supramolecular solvents have already been used in the determination of several classes of compounds such as parabens, pesticides, polycyclic aromatic hydrocarbons and bisphenols [80–83].

A very recent SUPRAS paper showed SUPRAS applicability with LPME as the extraction solvent. In this work, cationic surfactants didodecyldimethylammonium bromide (DDAB) and dodecyltrimethylammonium bromide (DTAB) were used in the extraction solvent mixture. SUPRAS-based liquid phase microextraction (SUPRAS-LPME) was also used in the preconcentration of five TCs (tetracycline, oxytetracycline, chlortetracycline, methacycline and doxycycline) in milk, egg and honey samples. An alkaline solution of the analytes was preconcentrated via electrostatic and hydrophobic interactions in the presence of the SUPRAS extraction solvent. The extraction mechanism was confirmed by the exploration of SUPRAS' Zeta potential and particle size. According to the authors, the results showed an excellent quantification method using SUPRAS and LPME for the determination of TCs in various matrices [84].

Recently, an innovative study proposed the application of a novel hexafluoroisopropanol (HFIP)/Brij-35 based SUPRAS in the determination of some organic compounds in water samples also using LPME. Brij-35 is a budget-friendly and non-toxic anionic surfactant that has a high cloud point (>100 °C). Presenting characteristics of a strong hydrogen-bond donor, elevated density and high hydrophobicity, HFIP was used in Brij-35's density regulation and cloud-point reduction to below room temperature. The HFIP/Brij-35 SUPRAS-based LPME procedure allowed its preparation at room temperature with centrifugation only, making it very simple. Quantification of parabens with HFIP/Brij-35 showed good linearity and correlation coefficients higher than 0.9990. Spiked samples provided recoveries from 90.2% to 112.4% and relative standard deviation of lower than 9% [85].

5. Bio-Based Solvents

Bio-based solvents are a group of green solvents that have several advantageous characteristics such as low toxicity and non-flammability, besides being biodegradable and renewable, as they are produced from biomass and agricultural materials [86]. One example is ethanol, which has been used for decades in classical liquid-liquid extraction. Among others, glycerol, 2-methyl tetrahydrofuran (meTHF), ethyl lactate, p-cymene, and terpenes are part of the bio-based solvent groups that have attracted interest for applications in separation methodologies due to their characteristics [87]. A very interesting bio-based solvent is D-Limonene. It is derived from citrus peel and, like other bio-solvents, is low-cost and biodegradable and exhibits low toxicity. Its main attraction has been in the substitution of traditional solvents such as acetone, toluene, and chlorinated and fluorinated solvents in several applications. Due to its characteristics as a degreaser, this solvent has been applied to the removal of oils and fats [88]. In addition, D-Limonene may be a substitute for toxic organic solvents in Soxhlet extraction procedures [89].

Recently, a liquid–liquid dispersive microextraction (DLLME) method applying D-Limonene and B-carotene for the determination of b-cyclodextrin (b-CD) was developed [90]. These two mixed bio-solvents show a strong adsorption characteristic. When placed in the presence of b-CD, B-carotene forms a complex that increases the absorbance of the extracted phase, generating an excellent analytical signal for the determination of the target compound. The validation of the method presented an excellent limit of detection (0.00004 mol L^{-1}) with a linear range from 0.0004 to 0.006 mol L^{-1}. The method was applied for the determination of b-CD in water and pharmaceutical samples, obtaining recovery values between 94.2 and 108.0%, confirming the efficiency of the method.

Despite the advantages, these solvents are still poorly explored as the extraction phase for microextraction techniques, mainly with chromatography techniques. Some drawbacks may have to be overcome, such as the high viscosity that causes poor analyte mass transfer and also the incompatibility with analytical instrumentation [91]. Nevertheless, an alternative would be their combination with other green solvents or their use as modifiers of solid sorbents, as already proposed by Hashemi et al. (2018) [63].

6. Conclusions

The development of alternative green extraction phases represents an important research field in chemical analysis for the determination of different analytes from various matrices. This approach has been exploited in several recent publications and highlighted in this article. The use of biosorbents in analytical chemistry, mainly applied to microextraction techniques, is a very promising eco-friendly and cheap alternative. Cork, bract and diatomaceous earth have shown tremendous potential as alternative sorbents to commercial phases (PDMS/DVB, DVB/Car/PDMS and PDMS). However, preparation of the fibers may limit their use. Additional efforts need to be made in order to expand the applicability of the existing biosorbents to different groups of analytes. The use of green extraction phases as an alternative to conventional organic solvents has led to remarkable improvements with regard to environmentally friendly aspects. DES/NADES and SUPRAS are a very promising alternative in sample preparation. Additional research is needed to exploit their interactions with the analytes and also to expand their applicability. The application of bio-based solvents in the extraction phase should be further investigated, since there are only a few reports regarding these subjects.

Funding: This research received no external funding.

Acknowledgments: The authors are grateful to the Brazilian Government Agency Conselho Nacional de Desenvolvimento Científico e Tecnológico (CNPq) and Coordenação de Aperfeiçoamento de Pessoal de Nível Superior (CAPES), for the financial support which made this research possible.

Conflicts of Interest: The authors declare no conflicts of interest.

References

1. Filippou, O.; Dimitrios, B.; Samanidou, V. Green approaches in sample preparation of bioanalytical samples prior to chromatographic analysis. *J. Chromatogr. B* **2017**, *1043*, 4–62. [CrossRef] [PubMed]
2. Stocka, J.; Tankiewicz, M.; Biziuk, M.; Namieśnik, J. Green Aspects of Techniques for the Determination of Currently Used Pesticides in Environmental Samples. *Int. J. Mol. Sci.* **2011**, *12*, 7785–7805. [CrossRef] [PubMed]
3. Armenta, S.; Garrigues, S.; Esteve-Turrillas, F.A.; de la Guardia, M. Green extraction techniques in green analytical chemistry. *Trac-Trend Anal. Chem.* **2019**, *116*, 248–253. [CrossRef]
4. Arthur, C.L.; Pawliszyn, J. Solid phase microextraction with thermal desorption using fused silica optical fibers. *Anal. Chem.* **1990**, *62*, 2145–2148. [CrossRef]
5. Olcer, Y.A.; Tascon, M.; Eroglu, A.E.; Boyaci, E. Thin film microextraction: Towards faster and more sensitive microextraction. *Trac-Trend Anal. Chem.* **2019**, *113*, 93–101. [CrossRef]
6. Carasek, E.; Merib, J.; Mafra, G.; Spudeit, D. A recent overview of the application of liquid-phase microextraction to the determination of organic micro-pollutants. *Trac-Trend Anal. Chem.* **2018**, *108*, 203–209. [CrossRef]
7. Rezaee, M.; Assadi, Y.; Hosseinia, M.R.M.; Aghaee, E.; Ahmadi, F.; Berijani, S. Determination of organic compounds in water using dispersive liquid-liquid microextraction. *J. Chromatogr. A.* **2006**, *1116*, 1–9. [CrossRef] [PubMed]
8. Jeannot, M.A.; Cantwell, F.F. Solvent microextraction into a single drop. *Anal. Chem.* **1996**, *68*, 2236–2240. [CrossRef]
9. Moreda-Piñeiro, J.; Moreda-Piñeiro, A. Green Extraction Techniques Principles, Advances and Applications. In *Comprehensive Analytical Chemistry*, 1st ed.; Ibanez, E., Cifuentes, A., Eds.; Elsevier: New York, NY, USA, 2017; Volume 76, pp. 519–573.
10. Armenta, S.; de la Guardia, M.; Mamiesnik, J. Green Microextractions. In *Analytical Microextraction Techniques*; Valcárcel, M., Cárdenas, S., Lucena, R., Eds.; Bentham Science Publishers: Sharjah, UAE, 2017; pp. 4–22. [CrossRef]
11. Demirbas, A. Heavy metal adsorption onto agro-based waste materials: A review. *J. Hazard. Mater.* **2008**, *157*, 220–229. [CrossRef]
12. Araujo, L.A.; Bezerra, C.O.; Cusioli, L.F.; Silva, M.F.; Nishi, L.; Gomes, R.G.; Bergamasco, R. Moringa oleifera biomass residue for the removal of pharmaceuticals from water. *J. Environ. Chem. Eng.* **2018**, *6*, 7192–7199. [CrossRef]
13. Neto, C.P.; Rocha, J.; Gil, A.; Cordeiro, N.; Esculcas, A.P.; Rocha, S.; Delgadillo, I.; De Jesus, J.D.P.; Correia, A.J.F. 13C solid-state nuclear magnetic resonance and Fourier transform infrared studies of the thermal decomposition of cork. *Solid State Nucl. Mag. Reson.* **1995**, *4*, 143–151. [CrossRef]
14. Dias, A.N.; Simão, V.; Merib, J.; Carasek, E. Cork as a new (green) coating for solid-phase microextraction: Determination of polycyclic aromatic hydrocarbons in water samples by gas chromatography–mass spectrometry. *Anal. Chim. Acta* **2013**, *772*, 33–39. [CrossRef] [PubMed]
15. Dias, A.N.; Simão, V.; Merib, J.; Carasek, E. Use of green coating (cork) in solid-phase microextraction for the determination of organochlorine pesticides in water by gas chromatography-electron capture detection. *Talanta* **2015**, *134*, 409–414. [CrossRef] [PubMed]
16. Silva, A.C.; Dias, A.N.; Carasek, E. Exploiting Cork as Biosorbent Extraction Phase for Solid-Phase Microextraction to Determine 3-(4-Methylbenzylidene)camphor and 2-Ethylhexyl 4-(Dimethylamino)benzoate in River Water by Gas Chromatography-Mass Spectrometry. *J. Braz. Chem. Soc.* **2017**, *28*, 2341–2347. [CrossRef]
17. Morés, L.; Dias, A.N.; Carasek, E. Development of a high-throughput method based on thin-film microextraction using a 96-well plate system with a cork coating for the extraction of emerging contaminants in river water samples. *J. Sep. Sci.* **2018**, *41*, 697–703. [CrossRef] [PubMed]

18. Dias, A.N.; da Silva, A.C.; Simão, V.; Merib, J.; Carasek, E. A novel approach to bar adsorptive microextraction: Cork as extractor phase for determination of benzophenone, triclocarban and parabens in aqueous samples. *Anal. Chim. Acta* **2015**, *888*, 59–66. [CrossRef] [PubMed]
19. Oenning, A.L.; Morés, L.; Dias, A.N.; Carasek, E. A new configuration for bar adsorptive microextraction (BAμE) for the quantification of biomarkers (hexanal and heptanal) in human urine by HPLC providing an alternative for early lung cancer diagnosis. *Anal. Chim. Acta* **2017**, *965*, 54–62. [CrossRef] [PubMed]
20. Do Carmo, S.N.; Merib, J.; Dias, A.N.; Stolberg, J.; Budziak, D.; Carasek, E. A low-cost biosorbent-based coating for the highly sensitive determination of organochlorine pesticides by solid-phase microextraction and gas chromatography-electron capture detection. *J. Chromatogr. A* **2017**, *1525*, 23–31. [CrossRef]
21. Suterio, N.; do Carmo, S.; Budziak, D.; Merib, J.; Carasek, E. Use of a Natural Sorbent as Alternative Solid-Phase Microextraction Coating for the Determination of Polycyclic Aromatic Hydrocarbons in Water Samples by Gas Chromatography-Mass Spectrometry. *J. Chromatogr. Sci.* **2000**, *38*, 55–60. [CrossRef]
22. Do Carmo, S.N.; Merib, J.; Carasek, E. Bract as a novel extraction phase in thin-film SPME combined with 96-well plate system for the high-throughput determination of estrogens in human urine by liquid chromatography coupled to fluorescence detection. *J. Chromatogr. B* **2019**, *1118–1119*, 17–24. [CrossRef]
23. Reinert, N.P.; Vieira, C.M.S.; Da Silveira, C.B.; Budziak, D.; Carasek, E. A Low-Cost Approach Using Diatomaceous Earth Biosorbent as Alternative SPME Coating for the Determination of PAHs in Water Samples by GC-MS. *Separations* **2018**, *5*, 55. [CrossRef]
24. Kirschner, N.; Dias, A.N.; Budziak, D.; da Silveira, C.B..; Merib, J.; Carasek, E. Novel approach to high-throughput determination of endocrine disruptors using recycled diatomaceous earth as a green sorbent phase for thin-film solid-phase microextraction combined with 96-well plate system. *Anal. Chim. Acta* **2017**, *996*, 29–37. [CrossRef] [PubMed]
25. Mafra, G.; Oenning, A.L.; Dias, A.N.; Merib, J.; Budziak, D.; da Silveira, C.B.; Carasek, E. Low-cost approach to increase the analysis throughput of bar adsorptive microextraction (BAμE) combined with environmentally-friendly renewable sorbent phase of recycled diatomaceous earth. *Talanta* **2018**, *178*, 886–893. [CrossRef] [PubMed]
26. Zhao, R.S.; Liu, Y.L.; Zhou, J.B.; Chen, X.F.; Wang, X. Bamboo charcoal as a novel solid-phase microextraction coating material for enrichment and determination of eleven phthalate esters in environmental water samples. *Anal. Bioanal. Chem.* **2013**, *405*, 4993–4996. [CrossRef] [PubMed]
27. Sajid, M.; Basheer, C.; Alshara, A.; Narasimhan, K.; Buhmeida, A.; Al Qahtani, M.; Al-Ahwal, M.S. Development of natural sorbent based micro-solid-phase extraction for determination of phthalate esters in milk samples. *Anal. Chim. Acta* **2016**, *924*, 35–44. [CrossRef] [PubMed]
28. Fiscal-Ladino, J.A.; Obando-Ceballos, M.; Rosero-Moreano, M.; Montaño, D.F.; Cardona, W.; Giraldo, L.F.; Richter, P. Ionic liquids intercalated in montmorillonite as the sorptive phase for the extraction of low-polarity organic compounds from water by rotating-disk sorptive extraction. *Anal. Chim. Acta* **2017**, *953*, 23–31. [CrossRef] [PubMed]
29. Vieira, C.M.S.; Mazurkievicz, M.; Lopez, A.; Debatin, V.; Micke, G.; Richter, P.; Rosero Moreano, M.; Carasek, E. Exploiting green sorbents in rotating-disk sorptive extraction for the determination of parabens by high performance liquid chromatography tandem electrospray ionization triple quadrupole mass spectrometry. *J. Sep. Sci.* **2018**, *41*, 4047–4054. [CrossRef] [PubMed]
30. Ho, T.D.; Zhang, C.; Hantao, L.W.; Anderson, J.L. Ionic Liquids in Analytical Chemistry: Fundamentals, Advances, and Perspectives. *Anal. Chem.* **2014**, *86*, 262–285. [CrossRef]
31. Sun, P.; Armstrong, D.W. Ionic liquids in analytical chemistry. *Anal. Chim. Acta* **2010**, *661*, 1–16. [CrossRef]
32. Han, D.; Tang, B.; Lee, Y.R.; Row, K.H. Application of ionic liquid in liquid phase microextraction technology. *J. Sep. Sci.* **2012**, *35*, 2949–2961. [CrossRef]
33. Zhang, P.; Hu, L.; Lu, R.; Zhou, W.; Gao, H. Application of ionic liquids for liquid–liquid microextraction. *Anal. Methods* **2013**, *5*, 5376–5385. [CrossRef]
34. Escudero, L.B.; Grijalba, A.C.; Martinis, E.M.; Wuilloud, R.G. Bioanalytical separation and preconcentration using ionic liquids. *Anal. Bioanal. Chem.* **2013**, *405*, 7597–7613. [CrossRef] [PubMed]
35. Clark, K.D.; Emaus, M.N.; Varona, M.; Bowers, A.N.; Anderson, J.L. Ionic Liquids: Solvents and Sorbents in Sample Preparation. *J. Sep. Sci.* **2018**, *41*, 209–235. [CrossRef] [PubMed]
36. Liu, J.; Jiang, G.; Chi, Y.; Cai, Y.; Zhou, Q.; Hu, J.-T. Use of Ionic Liquids for Liquid-Phase Microextraction of Polycyclic Aromatic Hydrocarbons. *Anal. Chem.* **2003**, *75*, 5870–5876. [CrossRef] [PubMed]

37. Marcinkowski, L.; Pena-Pereira, F.; Kloskowski, A.; Namiesnik, J. Opportunities and shortcomings of ionic liquids in single-drop microextraction. *Trac-Trend Anal. Chem.* **2015**, *72*, 153–168. [CrossRef]
38. Trujillo-Rodríguez, M.J.; Rocío-Bautista, P.; Pino, V.; Afonso, A.M. Ionic liquids in dispersive liquid-liquid microextraction. *Trac-Trend Anal. Chem.* **2013**, *51*, 87–106. [CrossRef]
39. Rykowska, I.; Ziemblińska, J.; Nowak, I. Modern approaches in dispersive liquid-liquid microextraction (DLLME) based on ionic liquids: A review. *J. Mol. Liq.* **2018**, *259*, 319–339. [CrossRef]
40. Abujaber, F.; Ricardo, A.I.C.; Ríos, A.; Bernardo, F.J.G.; Martín-Doimeadios, R.C.R. Ionic liquid dispersive liquid-liquid microextraction combined with LC-UV-Vis for the fast and simultaneous determination of cortisone and cortisol in human saliva samples. *J. Pharm. Biomed. Anal.* **2019**, *165*, 141–146. [CrossRef]
41. Clark, K.D.; Nacham, O.; Purslow, J.A.; Pierson, S.A.; Anderson, J.L. Magnetic ionic liquids in analytical chemistry: A review. *Anal. Chim. Acta* **2016**, *934*, 9–21. [CrossRef]
42. Santos, E.; Albo, J.; Irabien, A. Magnetic ionic liquids: Synthesis, properties and applications. *RSC Adv.* **2014**, *4*, 40008–40018. [CrossRef]
43. Wang, Y.; Sun, Y.; Xu, B.; Li, X.; Jin, R.; Zhang, H.; Song, D. Magnetic ionic liquid-based dispersive liquid-liquid microextraction for the determination of triazine herbicides in vegetable oils by liquid chromatography. *J. Chromatogr. A* **2014**, *1373*, 9–16. [CrossRef] [PubMed]
44. Sajid, M. Magnetic ionic liquids in analytical sample preparation: A literature review. *Trends Anal. Chem.* **2019**, *113*, 210–223. [CrossRef]
45. Trujillo-Rodríguez, M.J.; Anderson, J.L. In situ formation of hydrophobic magnetic ionic liquids for dispersive liquid-liquid microextraction. *J. Chromatogr. A* **2019**, *1588*, 8–16. [CrossRef] [PubMed]
46. Mafra, G.; Vieira, A.A.; Merib, J.; Anderson, J.L.; Carasek, E. Single drop microextraction in a 96-well plate format: A step toward automated and high-throughput analysis. *Anal. Chim. Acta* **2019**, *1063*, 159–166. [CrossRef]
47. Cao, D.; Xua, X.; Xuea, S.; Fenga, X.; Zhanga, L. An in situ derivatization combined with magnetic ionic liquid-based fast dispersive liquid-liquid microextraction for determination of biogenic amines in food samples. *Talanta* **2019**, *199*, 212–219. [CrossRef]
48. Silva, A.C.; Mafra, G.; Spudeit, D.; Merib, J.; Carasek, E. Magnetic ionic liquids as an efficient tool for the multiresidue screening of organic contaminants in river water samples. *Sep. Sci. Plus* **2019**, *2*, 51–58. [CrossRef]
49. Trujillo-Rodríguez, M.J.; Anderson, J.L. In situ generation of hydrophobic magnetic ionic liquids in stir bar dispersive liquid-liquid microextraction coupled with headspace gas chromatography. *Talanta* **2019**, *196*, 420–428. [CrossRef]
50. Chatzimitakos, T.G.; Pierson, S.A.; Anderson, J.L.; Stalikas, C.D. Enhanced magnetic ionic liquid-based dispersive liquid-liquid microextraction of triazines and sulfonamides through a one-pot, pH-modulated approach. *J. Chromatogr. A* **2018**, *1571*, 47–54. [CrossRef]
51. Merib, J.; Spudeit, D.A.; Corazza, G.; Carasek, E.; Anderson, J.L. Magnetic ionic liquids as versatile extraction phases for the rapid determination of estrogens in human urine by dispersive liquid-liquid microextraction coupled with high-performance liquid chromatography-diode array detection. *Anal. Bioanal. Chem.* **2018**, *410*, 4689–4699. [CrossRef]
52. Benedé, J.L.; Anderson, J.L.; Chisvert, A. Trace determination of volatile polycyclic aromatic hydrocarbons in natural waters by magnetic ionic liquid-based stir bar dispersive liquid microextraction. *Talanta* **2018**, *176*, 253–261. [CrossRef]
53. An, J.; Rahn, K.L.; Anderson, J.L. Headspace single drop microextraction versus dispersive liquid-liquid microextraction using magnetic ionic liquid extraction solvents. *Talanta* **2017**, *167*, 268–278. [CrossRef] [PubMed]
54. Chisvert, A.; Benedé, J.L.; Anderson, J.L.; Pierson, S.A.; Salvador, A. Introducing a new and rapid microextraction approach based on magnetic ionic liquids: Stir bar dispersive liquid microextraction. *Anal. Chim. Acta* **2017**, *983*, 130–140. [CrossRef] [PubMed]
55. Trujillo-Rodríguez, M.J.; Pino, V.; Anderson, J.L. Magnetic ionic liquids as extraction solvents in vacuum headspace single drop microextraction. *Talanta* **2017**, *172*, 86–94. [CrossRef] [PubMed]
56. Abbott, A.P.; Capper, A.P.G.; Davies, D.L.; Rasheed, R.K.; Tambyrajah, V. Novel solvent properties of choline chloride/urea mixtures. *Chem. Commun.* **2003**, *1*, 70–71. [CrossRef]

57. Zhang, Q.; Vigier, K.D.O.; Royer, S.; Jerome, F. Deep eutectic solvents: Syntheses, properties and applications. *Chem. Soc. Rev.* **2012**, *41*, 7108–7146. [CrossRef] [PubMed]
58. Florindo, C.; McIntosh, A.J.S.; Welton, T.; Branco, L.C.; Marrucho, I.M. A closer look into deep eutectic solvents: Exploring intermolecular interactions using solvatochromic probes. *Phys. Chem. Chem. Phys.* **2018**, *20*, 206–213. [CrossRef] [PubMed]
59. Shishov, A.; Bulatov, A.; Locatelli, M.; Carradori, S.; Andruch, V. Application of deep eutectic solvents in analytical chemistry. A review. *Microchem. J.* **2017**, *135*, 33–38. [CrossRef]
60. Farajzadeh, M.A.; Sattari, D.M.; Yadeghari, A. Deep eutectic solvent based gas-assisted dispersive liquid-phase microextraction combined with gas chromatography and flame ionization detection for the determination of some pesticide residues in fruit and vegetable samples. *J. Sep. Sci.* **2017**, *40*, 2253–2260. [CrossRef]
61. Ge, D.; Zhang, Y.; Dai, Y.; Yang, S. Air-assisted dispersive liquid-liquid microextraction based on a new hydrophobic deep eutectic solvent for the preconcentration of benzophenone-type UV filters from aqueous samples. *J. Sep. Sci.* **2018**, *41*, 1635–1643. [CrossRef]
62. Shishov, A.; Chromá, R.; Vakha, C.; Kuchár, J.; Simon, A.; Andruch, A.; Bulatov, A. In situ decomposition of deep eutectic solvent as a novel approach in liquid-liquid microextraction. *Anal. Chim. Acta* **2019**, *1065*, 49–55. [CrossRef]
63. Hashemi, B.; Zohrabi, P.; Dehdashtian, S. Application of green solvents as sorbent modifiers in sorptive-based extraction techniques for extraction of environmental pollutants. *Trac-Trend Anal. Chem.* **2018**, *109*, 50–61. [CrossRef]
64. Cunha, S.C.; Fernandes, J.O. Extraction techniques with deep eutectic solvents. *Trac-Trend Anal. Chem.* **2018**, *105*, 225–239. [CrossRef]
65. Lamei, N.; Ezoddin, M.; Abdi, K. Air assisted emulsification liquid-liquid microextraction based on deep eutectic solvent for preconcentration of methadone in water and biological samples. *Talanta* **2017**, *165*, 176–181. [CrossRef] [PubMed]
66. Wang, H.; Hu, L.; Liu, X.; Yin, S.; Lu, R.; Zhang, S.; Zhou, W.; Gao, H. Deep eutectic solvent-based ultrasound-assisted dispersive liquid-liquid microextraction coupled with high-performance liquid chromatography for the determination of ultraviolet filters in water samples. *J. Chromatogr. A* **2017**, *1516*, 1–8. [CrossRef] [PubMed]
67. Safavi, A.; Ahmadi, R.; Ramezani, A.M. Vortex-assisted liquid-liquid microextraction based on hydrophobic deep eutectic solvent for determination of malondialdehyde and formaldehyde by HPLC-UV approach. *Microchem. J.* **2018**, *143*, 166–174. [CrossRef]
68. Rajabi, M.; Ghassa, N.; Hemmati, M.; Asghari, A. Emulsification microextraction of amphetamine and methamphetamine in complex matrices using an up-to-date generation of eco-friendly and relatively hydrophobic deep eutectic solvent. *J. Chromatogr. A* **2018**, *1576*, 11–19. [CrossRef] [PubMed]
69. Torbatia, M.; Mohebbi, A.; Farajzadeh, M.A.; Mogaddam, M.R.A. Simultaneous derivatization and air–assisted liquid–liquid microextraction based on solidification of lighter than water deep eutectic solvent followed by gas chromatography–mass spectrometry: An efficient and rapid method for trace analysis of aromatic amines in aqueous samples. *Anal. Chim. Acta* **2018**, *1032*, 48–55. [CrossRef]
70. Makoś, P.; Przyjazny, A.; Boczkj, G. Hydrophobic deep eutectic solvents as "green" extraction media for polycyclic aromatic hydrocarbons in aqueous samples. *J. Chromatogr. A* **2018**, *1570*, 28–37. [CrossRef]
71. Yaripour, A.M.S.; Farajzadeh, M.A.; Mogaddam, M.R.A. Combination of dispersive solid phase extraction and deep eutectic solvent–based air-assisted liquid–liquid microextraction followed by gas chromatography–mass spectrometry as an efficient analytical method for the quantification of some tricyclic antidepressant drugs in biological fluids. *J. Chromatogr. A* **2018**, *1571*, 84–93. [CrossRef]
72. Yanga, D.; Wang, Y.; Peng, J.; Xun, C.; Yang, Y. A green deep eutectic solvents microextraction coupled with acid-base induction for extraction of trace phenolic compounds in large volume water samples. *Ecotoxicol. Environ. Safe* **2019**, *178*, 130–136. [CrossRef]
73. Wang, H.; Huang, X.; Qian, H.; Lu, R.; Zhang, S.; Zhou, W.; Gao, H.; Xu, D. Vortex-assisted deep eutectic solvent reversed-phase liquid–liquid microextraction of triazine herbicides in edible vegetable oils. *J. Chromatogr. A* **2019**, *1589*, 10–17. [CrossRef] [PubMed]

74. Torbati, M.; Farajzadeh, M.A.; Mogaddam, M.R.A.; Torbati, M. Development of microwave-assisted liquid-liquid extraction combined with lighter than water in syringe dispersive liquid-liquid microextraction using deep eutectic solvents: Application in extraction of some herbicides from wheat. *Microchem. J.* **2019**, *147*, 1103–1108. [CrossRef]
75. Deng, W.; Yu, L.; Li Ji, X.; Xuanxuan, C.; Zixin, W.; Xiao, D.Y. Hexafluoroisopropanol-based hydrophobic deep eutectic solvents for dispersive liquid-liquid microextraction of pyrethroids in tea beverages and fruit juices. *Food Chem.* **2019**, *274*, 891–899. [CrossRef] [PubMed]
76. Liu, X.; Liu, C.; Qian, H.; Qu, Y.; Zhang, S.; Lu, R.; Gao, H.; Zhou, W. Ultrasound-assisted dispersive liquid-liquid microextraction based on a hydrophobic deep eutectic solvent for the preconcentration of pyrethroid insecticides prior to determination by high-performance liquid chromatography. *Microchem. J.* **2019**, *146*, 614–621. [CrossRef]
77. Soylak, M.; Khan, M.; Yilmaz, E. Switchable solvent based liquid phase micro- extraction of uranium in environmental samples: A green approach. *Anal. Methods* **2016**, *8*, 979–986. [CrossRef]
78. Costi, E.M.; Sicilia, M.D.; Rubio, S. Supramolecular solvents in solid sample microextractions: Application to the determination of residues of oxolinic acid and flumequine in fish and shellfish. *J. Chromatogr. A* **2010**, *1217*, 1447–1454. [CrossRef] [PubMed]
79. Moral, A.; Sicilia, M.D.; Rubio, S. Determination of benzimidazolic fungicides in fruits and vegetables by supramolecular solvent-based microextraction/liquid chromatography/fluorescence detection. *Anal. Chim. Acta* **2009**, *650*, 207–213. [CrossRef]
80. Feizi, N.Y.; Moradi, Y.M.; Karimi, M.; Salamata, Q.; Amanzadeh, H. A new generation of nano–structured supramolecular solvents based on propanol/gemini surfactant for liquid phase microextraction. *Anal. Chim. Acta* **2017**, *953*, 1–9. [CrossRef]
81. Caballero-Casero, N.; Çabuk, H.; Martínez-Sagarra, G.; Devesa, J.A.; Rubio, S. Nanostructured alkyl carboxylic acid–based restricted access solvents: Application to the combined microextraction and cleanup of polycyclic aromatic hydrocarbons in mosses. *Anal. Chim. Acta* **2015**, *890*, 124–133. [CrossRef]
82. Zohrabi, P.; Shamsipur, M.; Hashemi, M.; Hashemi, B. Liquid–phase microextraction of organophosphorus pesticides using supramolecular solvent as a carrier for ferrofluid. *Talanta* **2016**, *160*, 340–346. [CrossRef]
83. Salatti-Dorado, J.A.; Caballero-Casero, N.; Sicilia, M.D.; Lunar, M.L.; Rubio, S. The use of a restricted access volatile supramolecular solvent for the LC/MS–MS assay of bisphenol A in urine with a significant reduction of phospholipid–based matrix effects. *Anal. Chim. Acta* **2017**, *950*, 71–79. [CrossRef] [PubMed]
84. Gissawong, N.; Boonchiangma, S.; Mukdasai, S.; Srijaranai, S. Vesicular supramolecular solvent-based microextraction followed by high performance liquid chromatographic analysis of tetracyclines. *Talanta* **2019**, *200*, 203–211. [CrossRef] [PubMed]
85. Chen, J.; Deng, W.; Li, X.; Wang, X.; Xiao, Y. Hexafluoroisopropanol/Brij-35 based supramolecular solvent for liquid-phase microextraction of parabens in different matrix samples. *J. Chromatogr. A* **2019**, *1591*, 33–43. [CrossRef] [PubMed]
86. Chemat, F.; Vian, M.A.; Cravotto, G. Green extraction of natural products: Concept and principles. *Int. J. Mol. Sci.* **2012**, *13*, 8615–8627. [CrossRef] [PubMed]
87. Li, Z.; Smith, K.H.; Stevens, G.W. The use of environmentally sustainable bio- derived solvents in solvent extraction applications: A review. *Chin. J. Chem. Eng.* **2016**, *24*, 215–220. [CrossRef]
88. Chemat, S.; Tomao, V.; Chemat, F. Limonene as Green Solvent for Extraction of Natural Products. In *Green Solvents I*; Mohammad, A., Inamuddin, Eds.; Springer: Dordrecht, The Netherlands, 2012; pp. 175–186. [CrossRef]
89. Virot, M.; Tomao, V.; Ginies, C.; Chemat, F. Total lipid extraction of food using d- limonene as an alternative to n-hexane. *Chromatographia* **2008**, *68*, 311–313. [CrossRef]
90. Pourreza, N.; Naghdi, T. D-Limonene as a green bio-solvent for dispersive liquid–liquid microextraction of b-cyclodextrin followed by spectrophotometric determination. *J. Ind. Eng. Chem.* **2017**, *51*, 71–76. [CrossRef]
91. Tobiszewski, M. Analytical chemistry with biosolvents. *Anal. Bioanal. Chem.* **2019**, 1–6. [CrossRef]

© 2019 by the authors. Licensee MDPI, Basel, Switzerland. This article is an open access article distributed under the terms and conditions of the Creative Commons Attribution (CC BY) license (http://creativecommons.org/licenses/by/4.0/).

Review

Hunting Molecules in Complex Matrices with SPME Arrows: A Review

Jason S. Herrington *,†, German A. Gómez-Ríos *,†, Colton Myers, Gary Stidsen and David S. Bell

Restek Corporation, Bellefonte, PA 16823, USA; colton.myers@restek.com (C.M.); gary.stidsen@restek.com (G.S.); david.bell@restek.com (D.S.B.)
* Correspondence: jason.herrington@restek.com (J.S.H.); german.gomez@restek.com (G.A.G.-R.)
† Those authors contributed equally to this work.

Received: 1 August 2019; Accepted: 28 November 2019; Published: 15 February 2020

Abstract: Thirty years since the invention and public disclosure of solid phase microextraction (SPME), the technology continues evolving and inspiring several other green extraction technologies amenable for the collection of small molecules present in complex matrices. In this manuscript, we review the fundamental and operational aspects of a novel SPME geometry that can be used to "hunt" target molecules in complex matrices: the SPME Arrow. In addition, a series of applications in environmental, food, cannabis and forensic analysis are succinctly covered. Finally, special emphasis is placed on novel interfaces to analytical instrumentation, as well as recent developments in coating materials for the SPME Arrow.

Keywords: SPME; green chemistry; air sampling; complex matrices; mass spectrometry

---

1. Introduction

Solid phase microextraction (SPME) is a concept that embraces an array of technologies, or devices, with several common features:

- First, all SPME technologies comprise a minute amount of extraction phase or sorbent material. This sorbent is typically adhered to a solid substrate, and said substrate can take multiple geometries [1]. The purpose of the sorptive material or coating, is to collect/enrich analytes of interest present in a complex matrix while preventing other matrix components from adhering to said surface [2].
- Second, analyte collection is based on partitioning between the extraction phase and the matrix. Thus, controlling the extraction conditions (e.g., temperature, ionic strength, and humidity) and the extraction times is critical to assure reproducible results and use this tool for quantitative applications [3].
- Third, a SPME device can carry out multiple steps of the analytical workflow such as analyte collection (e.g., sampling), sample preparation (e.g., the clean-up of analytes of interest from other matrix components), analyte transportation (e.g., when the sampling is performed outside of a laboratory), and analyte transfer into an analytical instrument (e.g., when thermal desorption is used on gas chromatography (GC)) [4].
- Fourth, most SPME technologies reduce/eliminate the use of solvents/additives during the sample preparation step and can consequently be considered green analytical chemistry technologies [5].
- Fifth, analyte elution can be performed via thermal, liquid, or laser desorption, depending on the characteristics of the extraction phase, and it can be introduced into an analytical instrument such a mass spectrometer via a chromatographic separation technique [6]. In the case of gas chromatography, analyte introduction onto the instrument is typically performed via the direct thermal desorption of the SPME device.

SPME, originally conceived and patented by researchers at the University of Waterloo (UW) in the late 1980s, was licensed and commercialized at the beginning of the 1990s by Supelco Inc. (now Millipore-Sigma, Bellefonte, PA, USA). The first peer-reviewed manuscript, published in 1989 [7], brought to light the most know configuration of SPME: "the fiber" [8]. As shown in Figure 1, a traditional SPME device is composed of the following parts: a color coded screw hub (A), a sealing septum (B), a septum piercing needle (C), a fiber attachment needle (D), and a coated fused silica fiber (E). Though multifarious SPME devices have been developed since the mid-1990s for thermal, liquid, and inclusive laser desorption [1,4,9], the thin cylindrical geometry described by Bellardi et al. [7] is the most well-known and the leader in sales worldwide. Indeed, GC coupled to several detection systems (e.g., ECD, FID, and MS) is the most commonly used instrument to interface SPME devices.

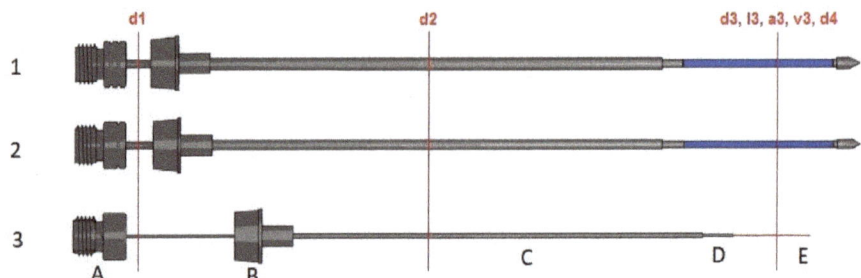

Figure 1. Solid phase microextraction (SPME) Arrows and a traditional SPME fiber; (**1**) 1.5 mm SPME Arrow; (**2**) 1.1 mm SPME Arrow; and (**3**) 23 gauge traditional SPME fiber. d1: Support tubing; d2: septum piercing needle; d3: phase diameter; d4: phase support tubing diameter; l3: phase length; a3: phase area; and v3: phase volume.

Even though the SPME patent did not thwart academia/industry from conducting research on SPME devices and extraction phases [10–12], it categorically prevented corporations from commercializing improved versions of the "traditional" fiber (e.g., enhancements to the substrate to make a more robust technology). The expiration of intellectual property a few years ago enabled commercial vendors to not only offer the "traditional" SPME fibers, but to also mechanically and chemically enhance versions of this technology. An example of these enhancements includes the first large volume SPME fiber developed by CTC Analytics AG for GC applications, known as the SPME Arrow (see Figures 1 and 2) [13].

In parallel with the development of the SPME Arrow, other green chemistry technologies aiming to overcome the drawbacks of the traditional SPME fiber have also appeared [14]. Among them, one can highlight the thin film microextraction (TFME), the stir bar sorptive extraction (SBSE), and the in-tube extraction (ITEX). As recently reviewed by Dugheri and Olcer [9,14], most of these technologies offer better analytical features over the traditional SPME fiber. However, some of them are harder to automate [15] or are not compatible for direct immersion experiments [16]. Thus, the focus of the current review article is to summarize the fundamental and operational aspects of SPME Arrows, as well as recent developments and future directions [17,18].

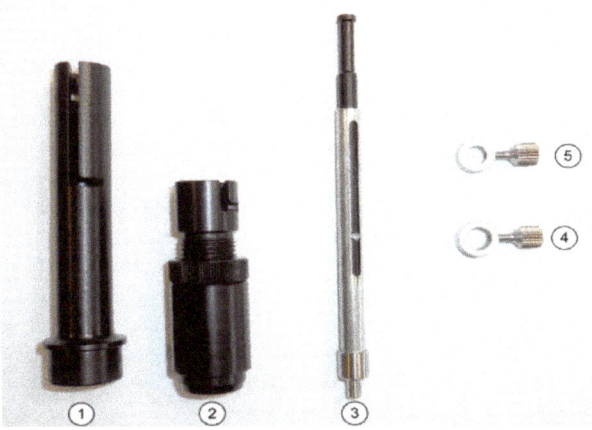

Figure 2. Restek PAL traditional SPME fiber and SPME Arrow manual extraction and injection kit. (**1**) Extraction guide; (**2**) injection guide; (**3**) Arrow/fiber syringe; (**4**) large inner diameter (ID) locking screw; and (**5**) small ID locking screw.

2. SPME Arrow Design

As portrayed in Figure 3, the SPME device workflow comprises several steps including the following: 1. piercing the septum of the vial with the outer needle; 2. exposing the extraction phase to the sample and collecting the analytes of interest for a fixed period of time; 3. withdrawing the extraction phase into the needle of the SPME device; 4. transporting the SPME device to the instrument station, and 5. transferring the SPME device into the injection port of the instrument, so the analytes are eluted from the extraction phase via thermal desorption.

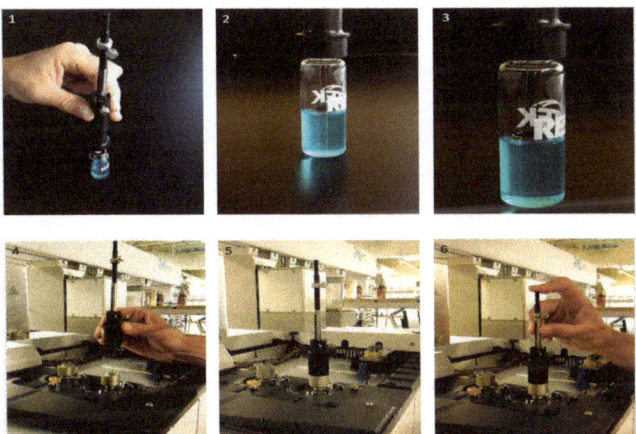

Figure 3. SPME (Arrow and traditional fiber) workflow for gas chromatography (GC) analysis. Manual SPME extraction (**1–3**) and injection (**4–6**). (**1**) Load SPME in manual syringe and extraction holder and position over sample vial. (**2**) Penetrate sample vial septum. (**3**) Press down on manual holder plunger to expose SPME extraction phase and begin sampling. (**4**) Load SPME in manual injection holder and position over GC inlet. (**5**) Penetrate GC inlet septum. (**6**) Press down on manual holder plunger to expose SPME extraction phase and begin desorption.

SPME Arrows were designed to overcome short-comings associated with the workflow of traditional SPME fibers, such as limited mechanical robustness ([19]), poor inter-device reproducibility, and small extraction phase volumes ([9]). The body of an SPME fiber is made of stainless steel, and the extraction phase is typically coated on fused silica. Consequently, one of the most common complaints of traditional SPME fiber end-users is a lack of physical durability of the fused silica ([20]). Though this problem can be partially alleviated by coating the extraction phase on a metal core (e.g., nitinol) [20,21], the body of a traditional SPME fiber is also commonly reported as fragile (see Figure 4). Consequently, commercial vendors have begun offering pre-drilled GC inlet septa and thin-walled vial septa to help mitigate issues with the damage of the core during the extraction and desorption steps. It is not uncommon for a SPME device to become injured, even within the first use. Failure rates appear to be highest amongst new end users; however, experienced end users are not immune to these issues either. Due to the delicate nature of traditional SPME fiber devices, most of them cannot be repaired. Manual extractions and desorptions appear to increase traditional SPME fiber failure rates, so the adoption of robotic autosamplers (e.g., CTC PAL or more commonly referred to as "rail" systems) has helped alleviate this problem to some degree. However, regardless of manual or automated injection, traditional SPME fiber devices typically fail to reach their true potential lifetime due to some sort of physical damage. "Potential" lifetime is stated because the majority of these mechanical failures appear to take place well before the fiber phase has been exhausted (i.e., it has not reached the life time of the coating).

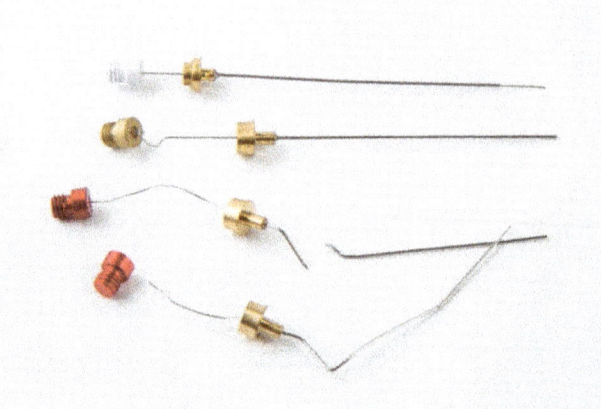

Figure 4. Mechanical failures common to traditional SPME fibers.

2.1. Physical Attributes

SPME Arrows only share the following physical attributes with traditional SPME fibers: 1. the dimensions and thread types on the color-coded hubs; 2. the dimensions of the needle ferrule; and 3. a stainless steel composition of the support tubing and septum piercing needle. Beyond the aforementioned, SPME Arrows diverge from traditional SPME fibers in an attempt to increase mechanical durability. Traditional SPME fibers have 23 or 24 gauge (i.e., 0.573 or 0.511 mm, respectively) external diameters on their septum piercing needle. The SPME Arrow was initially developed with 1.1 and 1.5 mm (i.e., ~17 and 15 gauge) external diameters on their septum piercing needle, which is approximately 2 and 3 times the diameter of traditional SPME fibers, respectively. Figure 1 provides a scale image for visual comparison of the physical attributes of SPME Arrows and a 23 gauge traditional SPME fiber. Furthermore, Table 1 breaks down the divergent physical attributes of the SPME Arrows and a 23 gauge traditional SPME fiber. Most notably, the support tubing (i.e., plunger), septum piercing needle, and phase support tubing of the 1.1 and 1.5 mm SPME Arrows have increased external diameters by 264–332%, 174–237%, and 583–449%, respectively over a 23 gauge traditional SPME fiber. These increased external diameters are largely responsible for the increased mechanical robustness of SPME Arrows. In particular, the increase in the support tubing appears to contribute the most improvement in durability, as this seems to be the most common failure point associated with traditional SPME fibers [13].

Table 1. Dimension breakdown of 23 gauge traditional SPME fiber (100 µm polydimethylsiloxane (PDMS), 1.1 mm SPME Arrow (100 µm PDMS), and 1.5 mm SPME Arrow (250 µm PDMS).

Label	Description	Units	Traditional SPME	1.1 mm SPME Arrow	1.5 mm SPME Arrow	1.1 mm% Increase	1.5 mm% Increase
d1	Support tubing	mm	0.304	0.804	1.01	264	332
d2	Septum piercing needle	mm	0.634	1.10	1.50	174	237
d3	Phase diameter	mm	0.285	0.721	0.912	253	320
d4	Phase support tubing diameter	mm	0.111	0.647	0.498	583	449
l3	Phase length	mm	10.0	20.0	20.0	200	200
a3	Phase area	mm^2	9.40	44.0	62.8	468	668
v3	Phase volume	µL	0.600	3.80	11.8	633	1967

It is important to note that the SPME Arrows have "arrow" shaped tips from which they garner their name. These arrow tips have the same external diameter as the SPME Arrows' septum piercing needle (i.e., 1.1 or 1.5 mm). The arrow tip helps increase the mechanical robustness of the SPME Arrow, as the force required to penetrate the vial and/or GC inlet septa is less on an SPME Arrow when compared to a traditional SPME fiber, despite the increase in diameter. It has been demonstrated that a 1.1 mm SPME Arrow only requires 799 g of force to penetrate and headspace vial septum, whereas, a 0.63 mm traditional SPME fiber requires 1188 g (~50% more than the SPME Arrow) of force to penetrate the same configuration [22]. Furthermore, the SPME Arrow tip has been demonstrated to cut slits in vial and GC inlet septa, as opposed to coring septa. Therefore, septa lifetime appears to be as good, if not better, than traditional SPME fibers [23]. Furthermore, when retracted (i.e., not extracting/desorbing), the Arrow tip serves the purpose of a protective cap, thereby minimizing the diffusion of compounds into the septum piercing needle and ultimately reaching the phase. This helps minimize background contamination in between analyses or losses of analytes while the fibers are store for analysis on the tray of the GC system [21].

More recently, the SPME Arrow has been advanced with a 0.804 mm support tubing housed inside a 1.5 mm septum piercing needle (denoted as 1.5* mm in Table 2). This design was released to overcome problems associated with using SPME Arrows for direct immersion (DI) extractions, such as phase swelling and the subsequently sloughing off and/or damaged when the SPME Arrow support tubing is retracted inside a 1.1 mm septum piercing needle. The wider 1.5 mm septum piercing needle provides enough clearance for the swollen phase on the 0.804 mm support tubing, thereby mitigating the sloughing issues.

Table 2. Current SPME Arrow phase configurations.

Analytes	Molecular Weight *	Stationary Phase	Thickness (µm)	Needle Diameter (mm)
Volatile	60–275	Polydimethylsiloxane (PDMS)	100	1.1 and 1.5 *
Volatile (high capacity)	60–275	PDMS	250	1.5
Polar, semi-volatile	80–300	Polyacrylate	100	1.1
Very volatile	30–225	Carbon Wide Range (WR)/PDMS	120	1.1 and 1.5 *
Aromatic, semi-volatile	60–300	Divinylbenzene (DVB)/PDMS	120	1.1 and 1.5 *
Volatile and semi-volatile	40–275	DVB/Carbon WR/PDMS	120	1.1 and 1.5 *

* 0.804 mm support tubing housed inside of a 1.5 mm septum piercing needle to allow for phase swelling during direction immersion (DI) extractions.

2.2. Phase

SPME Arrows were also designed to overcome the small phase areas and volumes associated with traditional SPME fibers. As shown in the Table 1, a 23 gauge traditional SPME fiber with 100 µm of polydimethylsiloxane (PDMS) has a 9.40 mm^2 and 0.600 µL phase. The 1.1 mm Arrow with 100 µm of PDMS, which represents the most direct comparison to the aforementioned, has a 44.0 mm^2 and 3.80 µL phase; which is a 468% and 633% increase in area and volume, respectively. This increase in phase area and volume has resulted in an increase in sensitivity and/or capacity, which is addressed later on in the performance and applications sections of this manuscript.

As shown in Table 2, SPME Arrows have been developed with most of the traditional SPME fiber phase offerings. The carbowax-polyethylene glycol (PEG) and carbowax/templated resin (CW/TPR) phase are the only phases not currently available on the SPME Arrow platform. Though the SPME Arrow has most of the phases, it is important to note that not all phase thickness/configurations have been replicated in the SPME Arrow. For example, although the SPME arrow is available in PDMS, it is only available in 100 and 250 µm PDMS configurations. The 7 and 30 µm PDMS configurations found in the traditional SPME fiber offerings are not available in the SPME Arrow. Additional phase thickness deviations may be observed when looking at the other phases (e.g., 75/85 µm of carbon on the traditional SPME fiber compared to 120 µm of carbon on the SPME Arrow).

3. SPME Arrow Accommodations

End users may not purchase a SPME Arrow and begin using it as direct replacement for the traditional SPME fiber. There are several things an analyst will need to modify/replace in order to accommodate the SPME Arrows' larger external diameters, including: 1. the injection tool (both manual and robotic); 2. the GC inlet; and 3. the GC inlet liner.

First, the manual tool may need modification or replacement (see Figure 2). For example, the traditional SPME fiber holder [19] does not accommodate SPME Arrows. Despite the SPME Arrows sharing a similar length and the same dimensions on the color-coded hubs, the Supelco manual holder tip diameter is too small. Some users have worked around this problem by drilling out the manual holder with a 3/64" metal compatible drill bit. Alternatively, other users have purchased the Restek PAL manual injection kit (Restek Corporation, Bellefonte, PA, USA), which accommodates both traditional SPME fibers and SPME Arrows without modification.

In terms of robotic platforms, CTC Analytics AG does not support the use of SPME Arrows on their second generation PAL systems (e.g., PAL/PAL-xt). Therefore, users with these systems will need to modify their existing SPME holder or acquire an appropriate SPME Arrow holder from Chromtech (Bad Camberg, Germany). Newer generations of robotic arms, such as PAL3 systems (e.g., RTC and RSI), fully support the SPME Arrow as long as users acquire the appropriate tool from

CTC Analytics AG. In the case of rail systems commercialized by other vendors such as Agilent (Santa Clara, CA, USA), Gerstel (Linthicum Heights, MD, USA), Shimadzu (Kyoto, Kyoto Prefecture, Japan), and Thermo (Waltham, MA, USA), users need to contact the manufacturer to acquire the appropriate tool for their robotic system. Furthermore, the fiber conditioner found on second and third generation PAL systems does not accommodate the SPME Arrow diameters. Therefore, users need to directly condition the SPME devices in the GC inlet or acquire an SPME Arrow-specific conditioner module from the appropriate manufacturer.

In terms of the GC system, the inlets that are currently installed on the instruments made by major manufacturers (e.g., Agilent, Shimadzu, and Thermo) do not accommodate SPME Arrows. However, factory-modified GC inlets are commercially available for those manufacturers. Alternatively, users can modify their existing GC inlets by drilling out the excess of metal with a 3/64" metal drill bit. Likewise, traditional SPME fibers require the use of 0.75 mm inlet liners or greater. With the smallest outer needle diameter of 1.1 mm, SPME Arrows require the use of lager inlet liners. Several commercial vendors provide straight-walled inlet liners capable of accommodating SPME Arrows. For example, a 1.8 mm ID straight/inlet liner can accommodate both the 1.1 and 1.5 mm SPME Arrows. It is important to note, that in lieu of SPME Arrow-specific liners, end users may be able to use a standard 2.0 mm straight-walled liner and not suffer any significant performance losses [24].

4. SPME Arrow Performance

4.1. Benchmarking

In the previous sections, several advantages and disadvantages of SPME Arrows when compared to traditional SPME fibers were reviewed. However, those were mostly focused around the physical attributes associated with SPME Arrows. The current section focuses on how the aforementioned physical differences in phase area and volume translate into analytical performance. Several benchmarking applications have been chosen to compare SPME Arrows with traditional SPME fibers. It is important to note that the 1.1 mm SPME Arrow is the focus for the remainder of this section and for most of the manuscript, as this configuration is available in the most popular SPME phases (e.g., polydimethylsiloxane (PDMS) and divinylbenzene (DVB)), as shown in Table 2, whereas the 1.5 mm SPME Arrow is only available in a 250 μm polydimethylsiloxane (PDMS) configuration, which is anticipated to have less applications. It is important to point out that the 1.5* mm SPME Arrows commercially launched a few weeks prior to the writing of this manuscript, so there are little data to warrant a discussion beyond what was mentioned previously.

When comparing a 100 μm PDMS 1.1 mm SPME Arrow against a 100 μm PDMS 23 gauge SPME fiber after the headspace (HS) extraction of volatile organic compounds (VOCs) in drinking water [spiked at 2.5 ppb as per International Organization for Standardization (ISO) method 17943] via GC coupled to mass spectrometry, it was found that equilibration times, extraction times, and desorption temperatures were equivalent among the two devices. Figure 5 presents a chromatogram overlaying instrumental response of the SPME Arrow and SPME fiber for the 92 VOCs. The SPME Arrow's response is higher than the traditional SPME fiber. On average, the SPME Arrow demonstrated a ~4× increase in response over traditional SPME fibers. This observation is consistent with the SPME Arrow's larger phase volume, which correlates to a greater volume/mass of target analyte collected and thereby an increased analytical response. It is important to note that "on average" was stated in the previous point, because the increase in response for very volatile compounds like vinyl chloride was ~10× on the SPME Arrow vs. the traditional SPME fiber. However, a semi-volatile compound like naphthalene only saw ~2× increase in sensitivity on the SPME Arrow compared to the traditional SPME fiber. Such differences in extraction recoveries are correlated to different vapor pressures (i.e., Henry's law constants) of said analytes and, expectedly, the availability in the headspace being the rate-limiting step. Likewise, it was observed that the SPME Arrow generated linear results

(0.998 median R^2) over a wide calibration range (0.0025–166 µg/L); with excellent precision (3.24% median RSD); and low sensitivity (30.0 ng/L median MDL) for all 92 ISO 17943 VOCs [25].

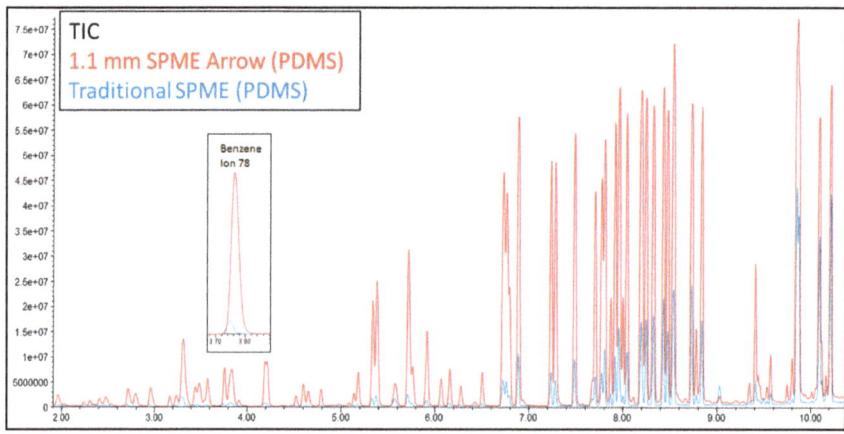

Figure 5. SPME Arrow overlay (TIC) on traditional SPME fiber for ISO 71943 headspace- volatile organic compounds (HS-VOCs) in drinking water. Acquired on Agilent 7890B/5977B GC-MS.

4.2. Method Development

Initial studies with SPME Arrows indicate that method development should follow the same logic and approach already demonstrated to be optimum for traditional SPME fibers [26]. For example, Herrington et al. evaluated SPME Arrows to see if there was a deviation in the extraction times required for SPME Arrows given the increase in phase volume. For this study, 100 µm PDMS 1.1 mm SPME Arrows and 100 µm PDMS 23 gauge traditional SPME fibers were evaluated for 92 HS VOCs, which had been spiked in drinking water at 2.5 ppb per method ISO 17943 [25]. Everything was equivalent (e.g., equilibration times and desorption temperatures), except for the extraction times. Extraction times of 15, 30, 60, 120, 240, 480, 960, and 1920 s were evaluated for each SPME (n = 3 for each SPME and each extraction time). Figure S1 shows the results from which the following two observations were made: 1. the SPME Arrow continued to demonstrate an increase in response (i.e., amount collected per unit of time) [9], which was again attributed to the increase in phase area over traditional SPME fibers; 2. both SPME types equilibrated at the same time (~120 s) for most of the 92 VOCs evaluated. This observation was consistent with the fact that both SPME types were 100 µm PDMS. Since the phase thickness was the same, it was deemed reasonable that the gas phase kinetics was the same; therefore, the equilibrium times were equivalent.

4.3. Troubleshooting

An extensive literature search only produced one reported issue associated with the SPME Arrow. Hartonen et al. reported chromatographic issues of double amine peaks when using SPME Arrows for extracting HS amines from water [27]. The root cause was determined to be an increased amount of water vapor extracted from the samples, due to the SPME Arrows' increased surface area and volume. However, it is important to note that other work on HS volatiles from drinking water [27] did not report any chromatographic issues and/or water issues. However, this work was conducted on thermally conditioned fibers with the use of sodium chloride in samples and split mode during desorption: the latter two were important conclusions Hartonen et al. arrived at [27]. In addition, Helin et al. did not report any issues when using SPME Arrow for HS amines in water with an SPME Arrow [17]. Finally, Gionfriddo et al. demonstrated short-chain aliphatic amines are best analyzed with the use of derivatization [28].

Beyond the aforementioned works, it is important to note that other work on HS volatiles from drinking water investigated the use of the 1.5 mm SPME Arrow with 250 μm of PDMS over a 1.1 mm SPME Arrow with 100 μm of PDMS [25]. As shown in Figure S2, light gases like chloroethane had ~2.5 times the response on the 1.5 mm SPME Arrow compared to the 1.1 mm SPME Arrow. However, the chromatography began to tail and split. A split injection could overcome this poor chromatography; however, then any sensitivity gains would be lost. It is believed that this tailing was a product of the fact that the phase was so thick on the 1.5 mm SPME Arrow that these lighter gases deeply penetrated into the thick phase, which then caused tailing up desorption. This work, Hartonen et al.'s work, and work not shown here have tended to suggest that SPME Arrows perform best with the use of a small (e.g., 2:1 or 5:1) split during thermal desorption. However, more extensive work will have to investigate this in the future. For instance, forthcoming work should consider the use of programmed temperature vaporization (PTV)-type inlets for SPME Arrows [14].

5. SPME Arrow Applications

5.1. Environmental Analysis

SPME technology has applicability for environmental pollutants in air, water, soil and sediment, and it can be used in the field or in the laboratory for sample preparation [17,18]. An advantage of the SPME Arrow is the larger phase volume that allows for the collection of a higher amount of the target compounds, so reporting requirements for environmental data users can be met.

One drawback of the traditional SPME fibers for headspace analysis is that the amount of analyte enriched on the coating is not sufficient to meet the reporting limits established by environmental agencies. The SPME Arrow design allows for four-to-five times higher analytical responses of compounds than a traditional SPME fiber [29]. For instance, the work performed by Kremser et al. presented a comparison of several techniques for determining volatile organic compounds including: purge and trap (P and T), ITEX, sample loop, traditional SPME fiber, gas-tight syringe, and SPME Arrow. As can be seen in Table 3, the detection limits for the SPME Arrow are in the same range as P and T and ITEX techniques, which are representative of exhaustive extraction techniques. However, the automation of workflow for the SPME Arrow is not only simpler than for P and T but also compatible with direct immersion experiments, which are not doable by ITEX [14].

In another study, Kaziur and collaborators developed a method capable of detecting picogram-per-liter (0.05 and 0.6 ng L^{-1}) levels of water taste and odor compounds (i.e., isopropyl-3-methoxypyrazine, 2-isobutyl-3-methoxypyrazine, geosmin, 2-methylisoborneol, 2,4,6-trichloroanisole, 2,4,6-bromoanisole, and beta-ionone) in water samples [18]. As a matter of fact, this fully automated workflow was more sensitive than existing methodologies (see Table 4).

Table 3. Comparison of different headspace technologies towards the analysis of VOCs. Table reprinted with permission of Springer from Reference [29], 2019.

Method Name	Sources with Reported Detection Limits in ng L^{-1}										
	This Work	[30]	[31]	[32]	[33]	[34]	[35]	[36]	[37]	[38]	[39]	[40]
Syringe	25–143	/	/	1–2 × 10^3	/	/	25–53	/	66–570 × 10^3	/	/	/
Loop	25–168	/	/	/	/	/	/	/	/	/	8–20 × 10^3	5–20 × 10^3
SPME	0.5–79.6	5.0–50	2.0–550	80–600	/	/	/	8.0–12	/	/	2–26 × 10^3	/
PAL SPME Arrow	0.7–4.9	/	/	/	/	/	/	/	/	/	/	/
Trap	0.6–9.6	/	/	/	/	1.0–10	/	/	7–149 × 10^3	/	/	/
ITEX	0.9–9.1	/	/	/	28–799	1.0–70	/	/	/	0.5–91	/	/
Sample matrix	Lab Water ^{a}	Spirit ^{b}	Wastewater NS	Water ^{b}	Lab water ^{a}	Lab water ^{a}	Urine NS	Meconium NS	Spirit ^{c}	Water ^{d}	Beer ^{e}	Wet rice ^{b}
MDL method												

Sample matrices and MDL determination methods are indicated as well (NS = not specified): ^{a} MDL according to EPA [41], ^{b} baseline standard deviation × 3, ^{c} according to German standard procedure (DIN 32645) [38], ^{d} according to IUPAC, ^{e} extrapolated from standard curve.

Table 4. Comparison of the developed PAL SPME Arrow with similar methods for determination of taste and odor compounds in water samples. Table reprinted with permission of Springer from Reference [18], 2019.

Method Name	Analytes	Sample Volume (mL)	Calibration Range (ng L^{-1})	LOD (ng L^{-1})	RSD (%)	Ref.
Purge and trap	IPMP, IBMP, GSM, MIB, TCA	20	10–200	0.2–2	<8	[42]
Solvent microextraction	GSM	5	5–900	0.8	<5	[43]
SBSE	GSM, MIB, TCA	60	0.1–100	0.02–0.16	<3.7	[44]
DLLME	GSM, MIB	12	10–1000	2 and 9	<11	[45]
SPME	IPMP, GSM, MIB, TCA, BIN	40	5–100	0.2–0.5	<7	[46]
PAL SPME Arrow	IPMP, IBMP, GSM, MIB, TCA, BIN, TBA	10	1 (2.6)–1000 (2600)	0.05–0.6	<11	-

Headspace work with SPME Arrow has been performed for the analysis of volatile amines in waste water. For example, after careful optimization of the extraction conditions, Helin et al. observed that the SPME Arrow produced lower detection limits, and higher recoveries of dimethyl amine (DMA) over the traditional SPME fiber (88% vs. 57%, respectively) [17].

The SPME Arrow has also been used for the analysis of semi-volatile compounds using immersion extraction. Typically, semi-volatile sample preparation for water samples has used liquid-liquid extraction (LLE) or solid phase extraction (SPE). Boyaci et al. did a comparison of SPME technology versus LLE and SPE, and it was found that the reduction of unwanted matrix interferences, the solventless extraction, the reusability, and the feasibility for high throughput sample analysis made SPME a more attractive technique [47]. Kremser et al. did extensive work with freely dissolved polycyclic aromatic hydrocarbons (PAHs) in water [48]. Extraction times and stirring rates were optimized to determine the method detection limits, and these were compared to previously published data for traditional SPME and SBSE (see Table 5). Direct immersion experiments using the SPME Arrow for freely dissolved semi-organic compounds showed detection limits five times lower than the traditional SPME fiber and similar results to stir bar sorptive extraction (SBSE). When comparing the SPME Arrow to SBSE, the authors highlighted the easiness of automation and utilization of shorter extraction times.

The SPME Arrow has also been used for environmental air sampling with the in-field analysis of biogenic volatile organic compounds (BVOCs). Monoterpenes (gamma-pinene and d3-carene) and aliphatic aldehydes (octanal and decanal) were selected by Barreira et al. to represent expected compounds to be found in field testing [49]. In-laboratory extraction efficiencies showed that the SPME Arrow had two-to-three times more area count than the traditional SPME fiber. Sampling was then performed in a boreal forest in Hyytiälä, Finland. Barreira et al. determined that the extraction efficiency of the SPME Arrow was two-times higher or greater, depending on compound, than the traditional SPME fiber, allowing for more sensitive testing (see Figure 6) [49].

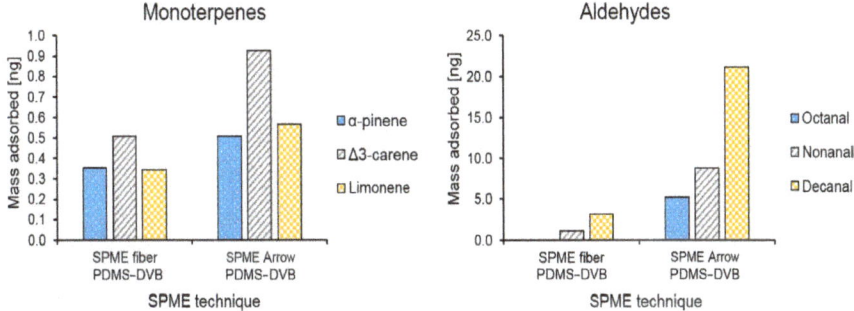

Figure 6. Comparison between the average mass of identified monoterpenes (α-pinene, 13-carene, and limonene) and aldehydes (octanal, nonanal, and decanal) collected with different polydimethylsiloxane-divinylbenzene (PDMS-DVB) SPME devices (fiber and Arrow) from ambient air and measured by GC-MS. Figure reprinted with permission of Elsevier from Reference [48], 2019.

Table 5. Comparison of SPME, SPME Arrow and stir bar sorbtive extraction (SBSE) for determination of polycyclic aromatic hydrocarbons (PAHs) in water via direct immersion. MDL and RSD results were obtained with PAL SPME Arrow (250 µm × 20 mm, 10.2 µL) for PAHs in water in comparison with literature data for classical SPME fibers and SBSE bars. Table reprinted from Reference [48] (open access).

Compound	PAL SPME Arrow		SPME (Cheng et al.) [50]		SBSE (Carrera et al.) [51]	
	MDL (ng L^{-1})	RSD (%) (at 10 ng L^{-1})	LOD (SD × 3)	RSD (conc. at S/N = 3 × 3)	LOD (conc. at S/N = 3 × 3)	RSD (%) (at 50 ng L^{-1})
Naphthalene	0.3	5.7	2.7	9	/	/
Acenaphthylene	0.2	6	1.8	6	0.1	/
Acenaphthene	0.1	7.1	0.9	3	/	/
Fluorene	0.2	5.6	3	10	0.1	8.3
Phenanthrene	0.2	5.5	2.1	7	0.1	1.1
Anthracene	0.3	7.6	2.1	7	0.2	2.1
Pyrene	0.2	6.4	3.6	12	0.2	/
Fluoranthene	0.2	6.2	2.1	7	0.2	/
1,2-Benzanthracene	0.1	6.2	2.1	7	0.2	6
Chrysene	0.1	11	1.5	5	0.2	10.6
Benzo(b)fluoranthene	0.2	10.5	2.7	9	0.1	/
Benzo(k)fluoranthene	0.2	8.6	1.8	6	0.1	/
Benzo[a]pyrene	0.3	7.2	3.6	12	0.1	/
Indeno(1,2,3 cd)pyrene	0.8	9.2	3.6	12	0.3	/
Dibenz(ah)anthracene	0.6	11.3	/	/	0.3	/
Benzo(ghi)perylene	0.8	11.9	1.8	6	0.3	/

MDL values calculated with a 99% confidence interval/not determined

5.2. Food Analysis

There are not as many publications on SPME Arrow in food analysis, as compared to traditional SPME fibers [8]. However, in the past couple of years, several applications have spanned the analysis of diverse matrices including fish and rice [52,53]. For instance, Song and coworkers compared carboxen/polydimethylsiloxane (CAR/PDMS) SPME sorbents in the fiber format to the arrow format for the HS extraction of volatiles present in salt-fermented sand lance fish sauce [53]. The researchers reported that alcohols, aldehydes and pyrazines, with the exception of 1-pentenol, were more effectively extracted using the SPME Arrow. Some compounds that are believed to be important to the flavor profile were only observed using the arrow device. Lan and coworkers described a modified zeolitic imidizolate framework (ZIF-8) as a solid phase microextraction support on the arrow construct [54]. The group compared the novel adsorbent to a commercially available carboxen/polydimethylsiloxane device for the sampling of volatile amines in wastewater and food samples (salmon and mushrooms). The researchers found that the commercial ZIF-8 material exhibited a small pore size (5.6 Å), which likely excluded the model amine compounds, resulting in a low extraction efficiency. The acidification of the material significantly increased the extraction of small volatile amines, presumably due to an increase in pore size. The modified ZIF-8 arrow design provided comparable extraction efficiencies for small, volatile amines from several matrices as compared to commercial carboxen/polydimethylsiloxane SPME Arrow devices. Yuan and coworkers investigated metal organic framework (MOF) sorbents applied to arrow SPME devices for the determination of PAH contaminants in fish samples [55]. The authors employed a zirconia based UiO-66-molybdenum disulfide composite and compared the recovery of PAH contaminants from seafood samples to both commercially-available arrow and fibers coated with PDMS/carboxen/DVB coatings. The combination of the MOF composite sorbent and the arrow format resulted in and increased number of PAH species detected. Lan et. al. described the analysis of low molecular weight aliphatic amines from various matrices using several different modifications of silica sorbent applied to the arrow format [56]. The scientists demonstrated that dimethylamine and trimethylamine could be effectively detected and quantitated in mushroom samples using these devices.

5.3. Terpenes in Cannabis

With the rapid growth of the cannabis market taking place, the need for analytical testing has become more critical. An area of interest in this market is being able to identify and place cannabis flowers into the correct chemical variety, otherwise known as chemovar. To properly classify cannabis chemovars, a comprehensive chemical profile examining compounds, such as terpenes and cannabinoids, is collected [57]. An analysis of terpenes is typically done via headspace—gas chromatography—mass spectrometry (HS-GC-MS). Herein, a 1.1 mm 120 μm divinylbenzene (DVB)/PDMS SPME Arrow was used to analyze the terpene content in *Humulus lupulus* (hops), and individual terpene responses were compared to that of a HS-syringe method typically used for this application and a 65 μm DVB/PDMS traditional SPME fiber. Hops were used in place of cannabis, as cannabis could not be legally obtained. Hops were ground using drying ice and then stored in the freezer until needed. Ten-to-fifteen milligrams of ground hops were added to a 20 mL crimp top HS vial.

Current methodologies recommend extraction times of 10 min for the traditional SPME fiber analysis of terpenes [58]. Improvements to this parameter alone have the potential to decrease instrument runtimes and increasing the number of samples that contract laboratories are able to test. As can be seen in Figure 7, average responses for a 10 min extraction time using a traditional SPME fiber are equivalent to that of a 2 min extraction time with the SPME Arrow. This can be done without sacrificing reproducibility, as both techniques showed % RSDs under ≤10%. Lighter terpenes (monoterpenes) gave better responses on the SPME Arrow, while the heavier terpenes (sesquiterpenes) were more comparable. However, the traditional fiber did provide slightly better responses for the sesquiterpenes.

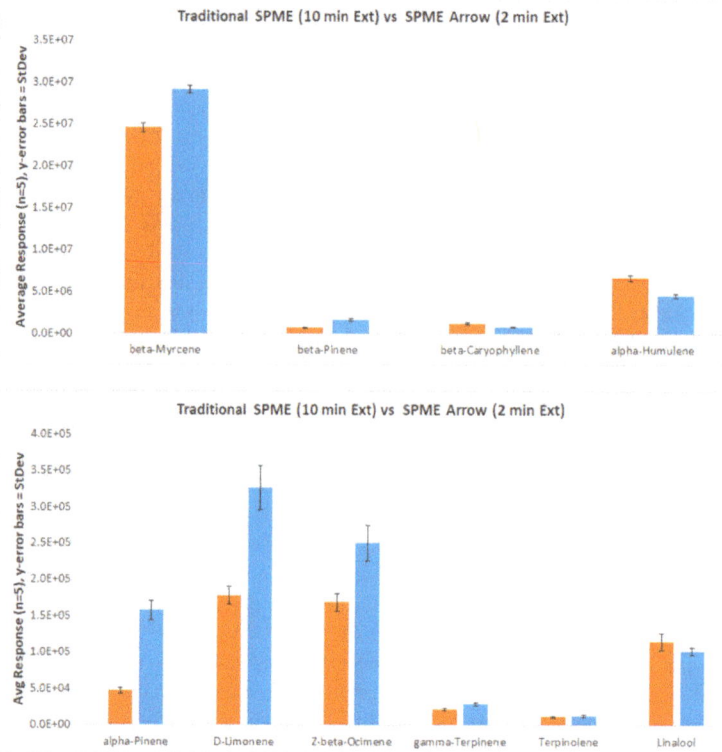

Figure 7. Extraction of VOCs in cannabis-related applications. Comparison of the traditional SPME fiber vs. SPME Arrow. Acquired on Agilent 7890B/5977B GC-MS.

6. Future Directions

The future of the SPME Arrow is promising, and among the multiple research and application avenues, this new geometry of SPME may grow rapidly. The following three areas are anticipated to expand: the development of new extraction phases [26,55,56], the Arrow's direct interface with MS instrumentation [59], and the Arrow's evolution towards smarter devices.

The first premise beyond the development of novel extraction phases is that the performance of these materials is better than those already commercially available by improving either robustness, extraction capabilities or device ruggedness [6,14,60]. For instance, the lack of inter-device reproducibility of one of the most popular extraction phases in SPME for GC analysis, the triple phase (DVB/Car/PDMS), has been its Achilles' heel [61]. Therefore, the recently launched SPME Arrow with a blended triple phase is expected to make a greater impact in the food and fragrance industry [62]. The main issue with the "original" tri-phasic SPME fibers relies on the process used for its manufacturing given that, if the two extraction phases (DVB-PDMS and Car-PDMS) are not properly aligned, this can lead to differences in the amount of extraction phase per coating length, thus leading to significant differences in the amount of analyte extracted—particularly volatile compounds [61]. In tri-phasic SPME Arrows, the particles of DVB and Car are embedded on the PDMS, resolving the issue of multiphase irregularities and consequently leading to better inter-device reproducibility.

As recently reviewed by Gomez-Rios and Mirabelli [4], the direct interface of SPME devices to MS has been increasing in the last five years, and SPME Arrows have not been an exception. For instance, research from the Zenobi's group demonstrated the applicability of an automated SPME

Arrow workflow, hyphened with a dielectric barrier desorption ionization (DBDI) coupled to an LTQ Orbitrap, to analyze common organic contaminants in treated wastewater (see Figure 8) [59]. Limits of detection as low as 3 ng/L were attained with a total analysis time per sample of less than 10 min (see Figure S3). Given that the ionization mechanism by DBDI, as well as by other direct-to-MS technologies, primarily relies on proton transfer for analyte ionization, a good response is typically attained for polar compounds with high proton affinity, whereas low ionization efficiencies are observed for nonpolar compounds. To overcome this limitation, a dopant-assisted DBDI ionization approach has been proposed to more efficiently ionize PAHs extracted from water. Though this strategy allowed for an up-to-an-order-of-magnitude signal enhancement for the analytes monitored, others go in detriment, and, consequently, the use (or not) of dopant must be driven by the target analytes. Certainly, DBDI is only one of the possible means of interfacing SPME Arrows to MS instrumentation. Other options include but are not limited to direct analysis in real time (DART) [63–65], atmospheric pressure photon ionization (APPI) [66], and thermal desorption-electrospray ionization (TD-ESI) [67]. It is anticipated that SPME Arrow-like geometries [11] compatible with liquid desorption and complex biological matrices will be developed in the near future and potentially coupled to direct-to-MS technologies such nano-electrospray ionization (nano-ESI) [68], desorption electrospray ionization [69], and the microfluidic open interface (also known as Open Probe Sampling Interface, OPSI) [70–72], as means to enhance the speed of analysis and the sensitivity [73]. Areas where such devices can make a great impact include but are not limited to, food fraud [74], volatomics [64], and rapid screening in the clinical chemistry realm [75].

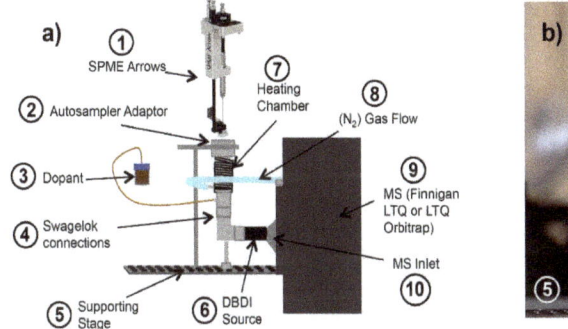
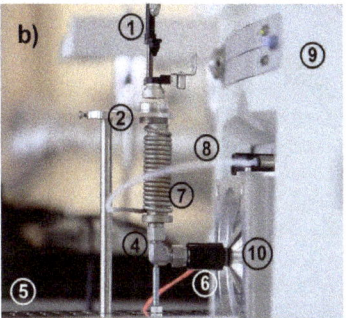

Figure 8. Schematic drawing (**a**) and image (**b**) of the sample introduction system (with and without dopant), source, and dielectric barrier desorption ionization (DBDI) interface with the mass spectrometer. Figure reprinted with permission of Elsevier from Reference [59], 2020.

Finally, as part of the advances on analytical chemistry towards the Internet of Things and other intelligent platforms [76], the development of "smart" SPME devices has begun. The so-called "smart" SPME devices, currently commercialized by PAL, have been exclusively designed for their rail systems and comprise a chip that automatically informs instruments about the phase coated on the device and its usage history. Though the smartness of the device is strictly related to physical/mechanical aspects of the device, the development of Arrows with "smarter" coatings [6] (e.g., ionic liquids [77], MOF [78], and carbon nanotubes (CNT) [79]) and "smarter" geometries and substrates components [80] is expected.

Supplementary Materials: The following are available online at http://www.mdpi.com/2297-8739/7/1/12/s1, Figure S1: Extraction times for SPME Arrow and traditional SPME fiber for ISO 17943 HS-VOCs. Acquired on Agilent 7890B/5977B GC-MS, Figure S2: 1.5 mm SPME Arrow (red trace) versus 1.1 mm SPME Arrow (black trace) for analysis of chloroethane in water samples according to method ISO 17943 HS-VOCs. Acquired on Agilent 7890B/5977B GC-MS., Figure S3: SPME-Arrow and DBDI for determination of ppt of contaminants in waste water. Quantitation of four compounds of varying polarities and contaminant classes: (a) DEET, (b) Tamoxifen, (c) Pyrene, (d) Metolachlor. Figure reprinted with permission of Elsevier from Reference [59], 2020.

Author Contributions: C.M., G.S., D.S.B., J.S.H. and G.A.G.-R. wrote the review. J.S.H. and G.A.G.-R. planned and revised the document. All authors have read and agree to the published version of the manuscript.

Funding: This research received no external funding.

Conflicts of Interest: All the authors work at Restek Corporation and Restek Corporation commercializes SPME Arrows.

References

1. Reyes-Garcés, N.; Gionfriddo, E.; Gómez-Ríos, G.A.; Alam, M.N.; Boyaci, E.; Bojko, B.; Singh, V.; Grandy, J.J.; Pawliszyn, J. Advances in solid phase microextraction and perspective on future directions. *Anal. Chem.* **2018**, *90*, 302–360. [CrossRef] [PubMed]
2. Reyes-Garcés, N.; Gionfriddo, E. Recent developments and applications of solid phase microextraction as a sample preparation approach for mass-spectrometry-based metabolomics and lipidomics. *TrAC Trends Anal. Chem.* **2019**, *113*, 172–181. [CrossRef]
3. Alam, M.N.; Nazdrajić, E.; Singh, V.; Tascon, M.; Pawliszyn, J. Effect of Transport Parameters and Device Geometry on Extraction Kinetics and Efficiency in Direct Immersion Solid-phase Microextraction. *Anal. Chem.* **2018**. [CrossRef] [PubMed]
4. Gómez-Ríos, G.A.; Mirabelli, M.F. Solid Phase Microextraction-Mass Spectrometry: Metanoia. *TrAC Trends Anal. Chem.* **2019**, *112*, 201–211. [CrossRef]
5. Armenta, S.; Garrigues, S.; de la Guardia, M. The role of green extraction techniques in Green Analytical Chemistry. *TrAC Trends Anal. Chem.* **2015**. [CrossRef]
6. Gómez-Ríos, G.A.; Garcés, N.R.; Tascon, M. Smart Materials in Solid Phase Microextraction (SPME). In *Handbook of Smart Materials in Analytical Chemistry*; John Wiley & Sons, Ltd.: Chichester, UK, 2019; pp. 581–620. [CrossRef]
7. Belardi, R.P.; Pawliszyn, J.B. The application of chemically modified fused silica fibers in the extraction of organics from water matrix samples and their rapid transfer to capillary columns. *Water Pollut. Res. J. Can.* **1989**, *24*, 179–191. [CrossRef]
8. Godage, N.H.; Gionfriddo, E. A critical outlook on recent developments and applications of matrix compatible coatings for solid phase microextraction. *TrAC Trends Anal. Chem.* **2019**, *111*, 220–228. [CrossRef]
9. Olcer, Y.A.; Tascon, M.; Eroglu, A.E.; Boyacı, E. Thin film microextraction: Towards faster and more sensitive microextraction. *TrAC Trends Anal. Chem.* **2019**, *113*, 93–101. [CrossRef]
10. Gutiérrez-Serpa, A.; Schorn-García, D.; Jiménez-Moreno, F.; Jiménez-Abizanda, A.I.; Pino, V. Braid solid-phase microextraction of polycyclic aromatic hydrocarbons by using fibers coated with silver-based nanomaterials in combination with HPLC with fluorometric detection. *Microchim. Acta* **2019**, *186*, 311. [CrossRef]
11. Poole, J.J.; Grandy, J.J.; Yu, M.; Boyaci, E.; Gómez-Ríos, G.A.; Reyes-Garcés, N.; Bojko, B.; Heide, H.V.; Pawliszyn, J. Deposition of a Sorbent into a Recession on a Solid Support to Provide a New, Mechanically Robust Solid-Phase Microextraction Device. *Anal. Chem.* **2017**, *89*, 8021–8026. [CrossRef]
12. Truong, T.V.; Lee, E.D.; Black, B.D.; Truong, T.X.; Lee, M.L. Coiled wire filament sample introduction for gas chromatography–mass spectrometry. *Int. J. Mass Spectrom.* **2018**, *427*, 123–132. [CrossRef]
13. Ziegler, M.; Schmarr, H.-G. Comparison of Solid-Phase Microextraction Using Classical Fibers Versus Mini-Arrows Applying Multiple Headspace Extraction and Various Agitation Techniques. *Chromatographia* **2019**, *82*, 635–640. [CrossRef]
14. Dugheri, S.; Mucci, N.; Bonari, A.; Marrubini, G.; Cappelli, G.; Ubiali, D.; Campagna, M.; Montalti, M.; Arcangeli, G. Solid phase microextraction techniques used for gas chromatography: A review. *Acta Chromatogr.* **2019**, 1–9. [CrossRef]
15. Grandy, J.J.; Boyacı, E.; Pawliszyn, J. Development of a Carbon Mesh Supported Thin Film Microextraction Membrane As a Means to Lower the Detection Limits of Benchtop and Portable GC/MS Instrumentation. *Anal. Chem.* **2016**, *88*, 1760–1767. [CrossRef] [PubMed]
16. Ruiz-Jimenez, J.; Zanca, N.; Lan, H.; Jussila, M.; Hartonen, K.; Riekkola, M.-L. Aerial drone as a carrier for miniaturized air sampling systems. *J. Chromatogr. A* **2019**, *1597*, 202–208. [CrossRef] [PubMed]
17. Helin, A.; Rönkkö, T.; Parshintsev, J.; Hartonen, K.; Schilling, B.; Läubli, T.; Riekkola, M.-L. Solid phase microextraction Arrow for the sampling of volatile amines in wastewater and atmosphere. *J. Chromatogr. A* **2015**, *1426*, 56–63. [CrossRef] [PubMed]

18. Kaziur, W.; Salemi, A.; Jochmann, M.A.; Schmidt, T.C. Automated determination of picogram-per-liter level of water taste and odor compounds using solid-phase microextraction arrow coupled with gas chromatography-mass spectrometry. *Anal. Bioanal. Chem.* **2019**, *411*, 2653–2662. [CrossRef]
19. Risticevic, S.; Lord, H.; Górecki, T.; Arthur, C.L.; Pawliszyn, J. Protocol for solid-phase microextraction method development. *Nat. Protoc.* **2010**, *5*, 122–139. [CrossRef]
20. Setkova, L.; Risticevic, S.; Linton, C.M.; Ouyang, G.; Bragg, L.M.; Pawliszyn, J. Solid-phase microextraction-gas chromatography-time-of-flight mass spectrometry utilized for the evaluation of the new-generation super elastic fiber assemblies. *Anal. Chim. Acta* **2007**, *581*, 221–231. [CrossRef]
21. Gómez-Ríos, G.A.; Reyes-Garcés, N.; Pawliszyn, J. Evaluation of a multi-fiber exchange solid-phase microextraction system and its application to on-site sampling. *J. Sep. Sci.* **2015**, *38*. [CrossRef]
22. Herrington, J. SPME Arrow Blog 7 ChromaBLOGraphy: Restek's Chromatography Blog. Available online: https://blog.restek.com/?p=41704 (accessed on 31 July 2019).
23. Herrington, J. SPME Arrow Blog 2 ChromaBLOGraphy: Restek's Chromatography Blog. Available online: https://blog.restek.com/?p=37889 (accessed on 31 July 2019).
24. Herrington, J.S. Solid-Phase Microextraction Liners for Headspace Volatile Organic Compounds. *Column* **2018**, *14*, 36–40. Available online: http://files.alfresco.mjh.group/alfresco_images/pharma//2019/03/07/aae3f1b3-af2d-4eb0-9d05-fdd7ed4bb372/LCTC121718%20North%20America.pdf (accessed on 31 July 2019).
25. Herrington, J.S.; Myers, C.; Stidsen, G.; Kozel, S. Determination of Volatile Organic Compounds in Water (ISO 17943) with SPME Arrow. In Proceedings of the ExTech 2017, Santiago de Compostela, Spain, 27–30 June 2017.
26. Cernosek, T.; Eckert, K.E.; Carter, D.O.; Perrault, A.P. Volatile Organic Compound Profiling from Postmortem Microbes using Gas Chromatography–Mass Spectrometry. *J. Forensic Sci.* **2020**, *65*, 134–143. [CrossRef] [PubMed]
27. Hartonen, K.; Helin, A.; Parshintsev, J.; Riekkola, M.-L. Problems Caused by Moisture in Gas Chromatographic Analysis of Headspace SPME Samples of Short-Chain Amines. *Chromatographia* **2019**, *82*, 307–316. [CrossRef]
28. Gionfriddo, E.; Passarini, A.; Pawliszyn, J. A facile and fully automated on-fiber derivatization protocol for direct analysis of short-chain aliphatic amines using a matrix compatible solid-phase microextraction coating. *J. Chromatogr. A* **2016**, *1457*, 22–28. [CrossRef] [PubMed]
29. Kremser, A.; Jochmann, M.A.; Schmidt, T.C. Systematic comparison of static and dynamic headspace sampling techniques for gas chromatography. *Anal. Bioanal. Chem.* **2016**, *408*, 6567–6579. [CrossRef] [PubMed]
30. Wardencki, W.; Curyło, J.; Namieśnik, J. Trends in Solventless Sample Preparation Techniques for Environmental Analysis. *J. Biochem. Biophys. Methods* **2007**, *70*, 275–288. [CrossRef]
31. Namieśnik, J.; Zygmunt, B.; Jastrzębska, A. Application of Solid-Phase Microextraction for Determination of Organic Vapours in Gaseous Matrices. *J. Chromatogr. A* **2000**, *885*, 405–418. [CrossRef]
32. Flórez Menéndez, J.C.; Fernández Sánchez, M.L.; Sánchez Uría, J.E.; Fernández Martínez, E.; Sanz-Medel, A. Static Headspace, Solid-Phase Microextraction and Headspace Solid-Phase Microextraction for BTEX Determination in Aqueous Samples by Gas Chromatography. *Anal. Chim. Acta* **2000**, *415*, 9–20. [CrossRef]
33. Jochmann, M.A.; Yuan, X.; Schilling, B.; Schmidt, T.C. In-Tube Extraction for Enrichment of Volatile Organic Hydrocarbons from Aqueous Samples. *J. Chromatogr. A* **2008**, *1179*, 96–105. [CrossRef]
34. Laaks, J.; Jochmann, M.A.; Schilling, B.; Schmidt, T.C. In-Tube Extraction of Volatile Organic Compounds from Aqueous Samples: An Economical Alternative to Purge and Trap Enrichment. *Anal. Chem.* **2010**, *82*, 7641–7648. [CrossRef]
35. Jakubowska, N.; Polkowska, Ż.; Kujawski, W.; Konieczka, P.; Namieśnik, J. A Comparison of Three Solvent-Free Techniques Coupled with Gas Chromatography for Determining Trihalomethanes in Urine Samples. *Anal. Bioanal. Chem.* **2007**, *388*, 691–698. [CrossRef] [PubMed]
36. Meyer-Monath, M.; Beaumont, J.; Morel, I.; Rouget, F.; Tack, K.; Lestremau, F. Analysis of BTEX and Chlorinated Solvents in Meconium by Headspace-Solid-Phase Microextraction Gas Chromatography Coupled with Mass Spectrometry. *Anal. Bioanal. Chem.* **2014**, *406*, 4481–4490. [CrossRef] [PubMed]
37. Schulz, K.; Dreßler, J.; Sohnius, E.-M.; Lachenmeier, D.W. Determination of Volatile Constituents in Spirits Using Headspace Trap Technology. *J. Chromatogr. A* **2007**, *1145*, 204–209. [CrossRef] [PubMed]
38. German Standard Procedure DIN 38 407-41. German Standard Procedure NA 119-01-03-02 AK. German Standard Method Procedure 32465:2008–11. Available online: http://www.wasserchemische-

gesellschaft.de/dev/validierungsdokumente?download=32:f41-din-38407-41-2011-06&lang=de (accessed on 16 January 2020).

39. Jeleń, H.H.; Wlazły, K.; Wąsowicz, E.; Kamiński, E. Solid-Phase Microextraction for the Analysis of Some Alcohols and Esters in Beer: Comparison with Static Headspace Method. *J. Agric. Food Chem.* **1998**, *64*, 1469–1473. [CrossRef]

40. Sriseadka, T.; Wongpornchai, S.; Kitsawatpaiboon, P. Rapid Method for Quantitative Analysis of the Aroma Impact Compound, 2-Acetyl-1-Pyrroline, in Fragrant Rice Using Automated Headspace Gas Chromatography. *J. Agric. Food Chem.* **2006**, *54*, 8183–8189. [CrossRef] [PubMed]

41. Keith, L.H.; Crummett, W.; Deegan, J.; Libby, R.A.; Taylor, J.K.; Wentler, G. Principles of Environmental Analysis. *Anal. Chem.* **1983**, *55*, 2210–2218. [CrossRef]

42. Ma, J.; Lu, W.; Li, J.; Song, Z.; Liu, D.; Chen, L. Determination of Geosmin and 2-Methylisoborneol in Water by Headspace Liquid-Phase Microextraction Coupled with Gas Chromatography-Mass Spectrometry. *Anal. Lett.* **2011**, *44*, 1544–1557. [CrossRef]

43. Yu, S.; Xiao, Q.; Zhu, B.; Zhong, X.; Xu, Y.; Su, G.; Chen, M. Gas Chromatography–Mass Spectrometry Determination of Earthy–Musty Odorous Compounds in Waters by Two Phase Hollow-Fiber Liquid-Phase Microextraction Using Polyvinylidene Fluoride Fibers. *J. Chromatogr. A* **2014**, *1329*, 45–51. [CrossRef]

44. Nakamura, S.; Nakamura, N.; Ito, S. Determination of 2-Methylisoborneol and Geosmin in Water by Gas Chromatography-Mass Spectrometry Using Stir Bar Sorptive Extraction. *J. Sep. Sci.* **2001**, *24*, 674–677. [CrossRef]

45. Bagheri, H.; Salemi, A. Headspace Solvent Microextraction as a Simple and Highly Sensitive Sample Pretreatment Technique for Ultra Trace Determination of Geosmin in Aquatic Media. *J. Sep. Sci.* **2006**, *29*, 57–65. [CrossRef]

46. Wu, D.; Duirk, S.E. Quantitative Analysis of Earthy and Musty Odors in Drinking Water Sources Impacted by Wastewater and Algal Derived Contaminants. *Chemosphere* **2013**, *91*, 1495–1501. [CrossRef] [PubMed]

47. Boyaci, E.; Rodriguez-Lafuente, A.; Gorynski, K.; Mirnaghi, F.; Souza-Silva, E.A.; Hein, D.; Pawliszyn, J. Sample Preparation with Solid Phase Microextraction and Exhaustive Extraction Approaches: Comparison for Challenging Cases. *Anal Chim Acta* **2015**, *873*, 14–30. [CrossRef] [PubMed]

48. Kremser, A.; Jochmann, M.A.; Schmidt, T.C. PAL SPME Arrow—Evaluation of a Novel Solid-Phase Microextraction Device for Freely Dissolved PAHs in Water. *Anal. Bioanal. Chem.* **2016**, *408*, 943–952. [CrossRef] [PubMed]

49. Barreira, L.M.F.; Parshintsev, J.; Kärkkäinen, N.; Hartonen, K.; Jussila, M.; Kajos, M.; Kulmala, M.; Riekkola, M.-L. Field Measurements of Biogenic Volatile Organic Compounds in the Atmosphere by Dynamic Solid-Phase Microextraction and Portable Gas Chromatography-Mass Spectrometry. *Atmos. Environ.* **2015**, *115*, 214–222. [CrossRef]

50. Cheng, X.; Forsythe, J.; Peterkin, E. Some Factors Affecting SPME Analysis and PAHs in Philadelphia's Urban Waterways. *Water Res.* **2013**, *47*, 2331–2340. [CrossRef]

51. Pérez-Carrera, E.; León, V.M.L.; Parra, A.G.; González-Mazo, E. Simultaneous Determination of Pesticides, Polycyclic Aromatic Hydrocarbons and Polychlorinated Biphenyls in Seawater and Interstitial Marine Water Samples, Using Stir Bar Sorptive Extraction–Thermal Desorption–Gas Chromatography-Mass Spectrometry. *J. Chromatogr. A* **2007**, *1170*, 82–90. [CrossRef]

52. Nam, T.G.; Lee, J.-Y.; Kim, B.-K.; Song, N.-E.; Jang, H.W. Analyzing Volatiles in Brown Rice Vinegar by Headspace Solid-Phase Microextraction (SPME)–Arrow: Optimizing the Extraction Conditions and Comparisons with Conventional SPME. *Int. J. Food Prop.* **2019**, *22*, 1195–1204. [CrossRef]

53. Song, N.-E.; Lee, J.-Y.; Lee, Y.-Y.; Park, J.-D.; Jang, H.W. Comparison of Headspace–SPME and SPME-Arrow–GC–MS Methods for the Determination of Volatile Compounds in Korean Salt–Fermented Fish Sauce. *Appl. Biol. Chem.* **2019**, *62*, 16. [CrossRef]

54. Lan, H.; Rönkkö, T.; Parshintsev, J.; Hartonen, K.; Gan, N.; Sakeye, M.; Sarfraz, J.; Riekkola, M.-L. Modified Zeolitic Imidazolate Framework-8 as Solid-Phase Microextraction Arrow Coating for Sampling of Amines in Wastewater and Food Samples Followed by Gas Chromatography-Mass Spectrometry. *J. Chromatogr. A* **2017**, *1486*, 76–85. [CrossRef]

55. Yuan, Y.; Lin, X.; Li, T.; Pang, T.; Dong, Y.; Zhuo, R.; Wang, Q.; Cao, Y.; Gan, N. A Solid Phase Microextraction Arrow with Zirconium Metal–Organic Framework/Molybdenum Disulfide Coating Coupled with Gas Chromatography–Mass Spectrometer for the Determination of Polycyclic Aromatic Hydrocarbons in Fish Samples. *J. Chromatogr. A* **2019**. [CrossRef]
56. Lan, H.; Zhang, W.; Smått, J.-H.; Koivula, R.T.; Hartonen, K.; Riekkola, M.-L. Selective Extraction of Aliphatic Amines by Functionalized Mesoporous Silica-Coated Solid Phase Microextraction Arrow. *Microchim. Acta* **2019**, *186*, 412. [CrossRef] [PubMed]
57. Russo, E.B. Taming THC: Potential Cannabis Synergy and Phytocannabinoid-Terpenoid Entourage Effects. *Br. J. Pharmacol.* **2011**, *163*, 1344–1364. [CrossRef] [PubMed]
58. Halpenny, M.; Stenerson, K.K. Quantitative Determination of Terpenes in Cannabis Using Headspace Solid Phase Microextraction and GC/MS. Available online: https://www.gerstel.com/pdf/AppNote-189.pdf (accessed on 31 July 2019).
59. Huba, A.K.; Mirabelli, M.F.; Zenobi, R. High-Throughput Screening of PAHs and Polar Trace Contaminants in Water Matrices by Direct Solid-Phase Microextraction Coupled to a Dielectric Barrier Discharge Ionization Source. *Anal. Chim. Acta* **2018**, *1030*, 125–132. [CrossRef]
60. Gionfriddo, E.; Boyacı, E.; Pawliszyn, J. New Generation of Solid-Phase Microextraction Coatings for Complementary Separation Approaches: A Step toward Comprehensive Metabolomics and Multiresidue Analyses in Complex Matrices. *Anal. Chem.* **2017**, *89*, 4046–4054. [CrossRef]
61. Gomez-Rios, G.A. High Throughput Analysis for On-Site Sampling. MSc. Thesis, University of Waterloo, Waterloo, ON, Canada, November 2012.
62. Stilo, F.; Cordero, C.; Sgorbini, B.; Bicchi, C.; Liberto, E.; Stilo, F.; Cordero, C.; Sgorbini, B.; Bicchi, C.; Liberto, E. Highly Informative Fingerprinting of Extra-Virgin Olive Oil Volatiles: The Role of High Concentration-Capacity Sampling in Combination with Comprehensive Two-Dimensional Gas Chromatography. *Separations* **2019**, *6*, 34. [CrossRef]
63. Bee, M.Y.; Jastrzembski, J.A.; Sacks, G.L. Parallel Headspace Extraction onto Etched Sorbent Sheets Prior to Ambient-Ionization Mass Spectrometry for Automated, Trace-Level Volatile Analyses. *Anal. Chem.* **2018**, *90*, 13806–13813. [CrossRef]
64. Rafson, J.P.; Bee, M.Y.; Sacks, G.L. Spatially Resolved Headspace Extractions of Trace-Level Volatiles from Planar Surfaces for High-Throughput Quantitation and Mass Spectral Imaging. *J. Agric. Food Chem.* **2019**, *67*, 13840–13847. [CrossRef]
65. Vasiljevic, T.; Gómez-Ríos, G.A.; Pawliszyn, J. Single-Use Poly(Etheretherketone) Solid-Phase Microextraction—Transmission Mode Devices for Rapid Screening and Quantitation of Drugs of Abuse in Oral Fluid and Urine via Direct Analysis in Real-Time Tandem Mass Spectrometry. *Anal. Chem.* **2018**, *90*, 952–960. [CrossRef]
66. Mirabelli, M.F.; Zenobi, R. Solid-Phase Microextraction Coupled to Capillary Atmospheric Pressure Photoionization-Mass Spectrometry for Direct Analysis of Polar and Nonpolar Compounds. *Anal. Chem.* **2018**, *90*, 5015–5022. [CrossRef]
67. Wang, C.H.; Su, H.; Chou, J.H.; Huang, M.Z.; Lin, H.J.; Shiea, J. Solid Phase Microextraction Combined with Thermal-Desorption Electrospray Ionization Mass Spectrometry for High-Throughput Pharmacokinetics Assays. *Anal. Chim. Acta* **2018**, *1021*, 60–68. [CrossRef]
68. Hu, B.; Zheng, B.; Rickert, D.; Gómez-Ríos, G.A.; Bojko, B.; Pawliszyn, J.; Yao, Z.-P. Direct Coupling of Solid Phase Microextraction with Electrospray Ionization Mass Spectrometry: A Case Study for Detection of Ketamine in Urine. *Anal. Chim. Acta* **2019**, *1075*, 112–119. [CrossRef]
69. Lendor, S.; Gómez-Ríos, G.A.; Boyacı, E.; Vander Heide, H.; Pawliszyn, J. Space-Resolved Tissue Analysis by Solid-Phase Microextraction Coupled to High-Resolution Mass Spectrometry via Desorption Electrospray Ionization. *Anal. Chem.* **2019**, *91*, 10141–10148. [CrossRef] [PubMed]
70. Tascon, M.; Alam, M.N.; Gómez-Ríos, G.A.; Pawliszyn, J. Development of a Microfluidic Open Interface with Flow Isolated Desorption Volume for the Direct Coupling of SPME Devices to Mass Spectrometry. *Anal. Chem.* **2018**, *90*, 2631–2638. [CrossRef] [PubMed]
71. Gómez-Ríos, G.A.; Liu, C.; Tascon, M.; Reyes-Garcés, N.; Arnold, D.W.; Covey, T.R.; Pawliszyn, J. Open Port Probe Sampling Interface for the Direct Coupling of Biocompatible Solid-Phase Microextraction to Atmospheric Pressure Ionization Mass Spectrometry. *Anal. Chem.* **2017**, *89*. [CrossRef] [PubMed]

72. Looby, N.T.; Tascon, M.; Acquaro, V.R.; Reyes-Garcés, N.; Vasiljevic, T.; Gomez-Rios, G.A.; Wąsowicz, M.; Pawliszyn, J. Solid Phase Microextraction Coupled to Mass Spectrometry via a Microfluidic Open Interface for Rapid Therapeutic Drug Monitoring. *Analyst* **2019**, *144*, 3721–3728. [CrossRef] [PubMed]
73. Gómez-Ríos, G.A.; Reyes-Garcés, N.; Bojko, B.; Pawliszyn, J. Biocompatible Solid-Phase Microextraction Nanoelectrospray Ionization: An Unexploited Tool in Bioanalysis. *Anal. Chem.* **2016**, *88*, 1259–1265. [CrossRef]
74. Gómez-Ríos, G.A.; Vasiljevic, T.; Gionfriddo, E.; Yu, M.; Pawliszyn, J. Towards On-Site Analysis of Complex Matrices by Solid-Phase Microextraction-Transmission Mode Coupled to a Portable Mass Spectrometer via Direct Analysis in Real Time. *Analyst* **2017**, *142*, 2928–2935. [CrossRef]
75. Feider, C.L.; Krieger, A.; DeHoog, R.J.; Eberlin, L.S. Ambient Ionization Mass Spectrometry: Recent Developments and Applications. *Anal. Chem.* **2019**, *91*, 4266–4290. [CrossRef]
76. Prabhu, G.R.D.; Urban, P.L. The Dawn of Unmanned Analytical Laboratories. *TrAC Trends Anal. Chem.* **2017**, *88*, 41–52. [CrossRef]
77. Trujillo-Rodriguez, M.J.; Pino, V.; Psillakis, E.; Anderson, J.L.; Ayala, J.H.; Yiantzi, E.; Afonso, A.M. Vacuum-Assisted Headspace-Solid Phase Microextraction for Determining Volatile Free Fatty Acids and Phenols. Investigations on the Effect of Pressure on Competitive Adsorption Phenomena in a Multicomponent System. *Anal. Chim. Acta* **2017**, *962*, 41–51. [CrossRef]
78. Zhang, S.; Yang, Q.; Wang, W.; Wang, C.; Wang, Z. Covalent Bonding of Metal-Organic Framework-5/Graphene Oxide Hybrid Composite to Stainless Steel Fiber for Solid-Phase Microextraction of Triazole Fungicides from Fruit and Vegetable Samples. *J. Agric. Food Chem.* **2016**, *64*, 2792–2801. [CrossRef]
79. Song, X.-Y.; Chen, J.; Shi, Y.-P. Different Configurations of Carbon Nanotubes Reinforced Solid-Phase Microextraction Techniques and Their Applications in the Environmental Analysis. *TrAC Trends Anal. Chem.* **2017**, *86*, 263–275. [CrossRef]
80. Kalsoom, U.; Nesterenko, P.N.; Paull, B. Current and Future Impact of 3D Printing on the Separation Sciences. *TrAC Trends Anal. Chem.* **2018**, *105*, 492–502. [CrossRef]

© 2020 by the authors. Licensee MDPI, Basel, Switzerland. This article is an open access article distributed under the terms and conditions of the Creative Commons Attribution (CC BY) license (http://creativecommons.org/licenses/by/4.0/).

Review

Development, Optimization and Applications of Thin Film Solid Phase Microextraction (TF-SPME) Devices for Thermal Desorption: A Comprehensive Review

Ronald V. Emmons [1], Ramin Tajali [1] and Emanuela Gionfriddo [1,2,3,*]

1. Department of Chemistry and Biochemistry, College of Natural Sciences and Mathematics, The University of Toledo, Toledo, OH 43606, USA
2. School of Green Chemistry and Engineering, The University of Toledo, Toledo, OH 43606, USA
3. Dr. Nina McClelland Laboratories for Water Chemistry and Environmental Analysis, The University of Toledo, Toledo, OH 43606, USA
* Correspondence: emanuela.gionfriddo@utoledo.edu

Received: 5 July 2019; Accepted: 30 July 2019; Published: 5 August 2019

Abstract: Through the development of solid phase microextraction (SPME) technologies, thin film solid phase microextraction (TF-SPME) has been repeatedly validated as a novel sampling device well suited for various applications. These applications, encompassing a wide range of sampling methods such as onsite, in vivo and routine analysis, benefit greatly from the convenience and sensitivity TF-SPME offers. TF-SPME, having both an increased extraction phase volume and surface area to volume ratio compared to conventional microextraction techniques, allows high extraction rates and enhanced capacity, making it a convenient and ideal sampling tool for ultra-trace level analysis. This review provides a comprehensive discussion on the development of TF-SPME and the applications it has provided thus far. Emphasis is given on its application to thermal desorption, with method development and optimization for this desorption method discussed in detail. Moreover, a detailed outlook on the current progress of TF-SPME development and its future is also discussed with emphasis on its applications to environmental, food and fragrance analysis.

Keywords: TF-SPME; microextraction; thermal desorption; environmental analysis; flavor and fragrance; onsite sampling; in vivo analysis; ultra-trace analysis

1. Introduction

As the need for more sensitive and greener alternatives in analytical chemistry continues to grow [1–5], it is necessary for the further development of robust sample preparation technologies to meet these modern demands. Sample preparation, being the first step in any analytical procedure, is far-reaching since any and all steps in the workflow are consequently affected by the sampling and extraction method used. The sampling and clean-up of compounds before introduction into an instrument plays a critical role in the achieved sensitivity and quantitative capabilities of the method. Consequently, novel extraction techniques must be developed to enhance analytical performance, while meeting the newfound call for greener sample preparation methods. Microextraction, being characterized by a small amount of extraction phase compared to the volume of the sample [6], affords the opportunity of substantially reducing the amount of organic solvent used while still achieving similar or better results compared to more traditional extraction techniques such as solid phase extraction (SPE) [7,8] and liquid–liquid extraction (LLE) [9]. The volume of the extraction phase, being inconsequential to the overall volume of the sample [6], allows rapid non-exhaustive extraction, in some cases non-depletive, that can easily be quantitated using a variety of calibration methods [10]. Moreover, analytes from the sample matrix are extracted in their "free-form" (non-bound or free-concentration), giving the opportunity for the analysis of bio-available analytes in various matrices. Among existing microextraction techniques, solid phase microextraction (SPME) is the most widely adopted as it allows solvent-less extraction that can be easily automated and adapted for in vivo and onsite applications [11]. The conventional configuration of SPME consists of an extraction phase coated on a solid, fiber-like support composed of fused silica, stainless steel or flexible metal alloys; this geometry allows ease of use and automated extraction and analysis [11]. As a sample preparation method, the use of SPME enables sampling and pre-concentration to be performed in one simple step, making the technique more versatile in its use and able to achieve better throughput compared to more laborious exhaustive methods such as SPE and LLE.

2. From Fiber to Thin Film Format: Pros and Cons

In common with all non-exhaustive extraction techniques, the mechanism of extraction for SPME is based on the equilibrium-driven diffusion of analytes between the sample matrix and the extraction phase [6]. The amount extracted at equilibrium between these phases is described in Equation (1) and explained in Section 4, "Fundamentals of TF-SPME". This equation implies that for most SPME-based extractions, the only parameters that are consequential when optimizing extraction efficiency for a non-exhaustive extraction are the distribution coefficient between the sample and the extraction phase (K_{es}), and the volume of the extraction phase (V_e). During method development and optimization, K_{es} is maximized through changes in different physical parameters of the extraction, such as temperature, agitation, ionic strength, the amount of organic solvent in the sample (if any) and most notably extraction phase chemistry [6]. As a result, during the development of an SPME device the physiochemical properties of the coating must be carefully selected as they affect both the extraction efficiency and specificity for the targeted analyte. Additionally, the extraction phase must also be able to be efficiently desorbed, by either thermal desorption (TD) (the phase then needing to be thermally stable) or by desorption in an organic solvent (the phase not swelling in organics).

Beyond the previously mentioned parameters, an increase in extraction phase volume also contributes to the enhanced efficiency of the sampling device [12]. This phase volume allows improved capacity for the analyte, which in turn enables a more sensitive extraction, applicable to ultra-trace level analysis, doing so, however, poses practical challenges in both engineering of the device and mass transfer phenomena. For example, when optimizing the volume of the extraction phase (V_e) for fiber SPME, a simple increase in the diameter of the coating, as seen in Equation (2), drastically prolongs the equilibration time and negatively affects the desorption efficiency. An increase in phase volume for a fiber SPME also requires a redesign of the whole device assembly, as in the case of the recently introduced Arrow-SPME [13,14]. This device, while important to the overall development of SPME due to its enhanced capacity, will not be further discussed as it does not seek to maximize the device's surface-to-volume ratio as other TF-SPME devices do.

There have been multiple developments throughout the years to increase the extraction phase volume of microextraction devices to enhance their sensitivity with varying success. Still commonly used today, stir bar sorptive extraction (SBSE) seeks to increase the volume of extraction phase by coating a magnetic stir bar with polydimethylsiloxane (PDMS) for the immersive extraction of aqueous analytes, and has proven suitable for both direct immersion and headspace extraction [15]. This geometry allows higher capacities than conventional SPME fiber due to the larger extraction volume, however, it still has yet to overcome the difficulties discussed previously in timely extraction [16] (especially in regards to large sample volumes) [17] and efficient desorption. While the large capacity can grant greater sensitivity, the extraction time can take 24 h [18] and desorption can be long as well depending on the molecular weight and volatility of the analytes [17]. Thin film solid phase microextraction (TF-SPME), a novel SPME device first developed in 2003 [12], overcomes these limitations, extraction efficiency and capacity, by the use of an alternative geometry. TF-SPME consists of a large-volume thin layer of extraction phase (originally pure PDMS) for the pre-concentration of analytes. The development of TF-SPME devices differs from previous attempts as its geometry is simply a flat planar surface, effectively increasing the surface area-to-volume ratio and thus avoiding the usual caveats of increased phase volume [12]. With enhanced capacity and faster equilibration rates compared to other microextractive methods, the practicality of the first iteration of TF-SPME devices was still limited. The geometry of the earlier developments of the technique was cumbersome, its large volume required specially suited large-volume injectors which not all labs were equipped for [12]. On a more fundamental note, an increase of extraction phase volume consequently increases the amount of background and bleed from the extraction phase itself and this critical drawback has only recently been addressed by newer generations of TF-SPME devices [19]. Although the desorption of TF-SPME devices can be fully automated, their geometry still poses a barrier for online extraction and analysis, and currently an auto-sampler that can both perform extraction and desorption for conventional TF-SPME devices has yet to be developed. The suitability of TF-SPME devices, however, for ultra-trace level analysis, along with its convenience of onsite sampling, makes it a suitable and robust alternative SPME application.

3. Types of Desorption Modes for TF-SPME

There are various different approaches that can be employed toward the desorption of TF-SPME devices, and these techniques are chosen based on the characteristics of the compounds of interest and the composition of the TF-SPME device itself. Aside from TD, the second most common desorption method used for TF-SPME is liquid desorption (LD), commonly used in conjunction with liquid chromatography but also with various separation platforms [20–23]. LD utilizes an organic solvent (or a mixture of water and multiple organic solvents) to re-extract all compounds from the extraction phase before introduction of the now-analyte enriched liquid phase into an analytical instrument. In doing so, this allows the TF-SPME device to still carry out sampling and sample clean up before desorbing into a much smaller amount of organic solvent that would usually be required for LLE. This desorption mode is a necessity for liquid-phase separations; most commonly used for non-volatiles, thermally labile compounds and biomolecules [24]. Previous studies have shown promise for the direct coupling of SPME to nanoelectrospray ionization [25,26], the nanospray solvent effectively desorbing the SPME device. More recently, nanomaterial-based TF-SPME devices have been simultaneously desorbed and analyzed by total reflection X-ray fluorescence spectrometry (TXRF) [27,28] which can be applied to both the analysis of organic compounds and metals.

Desorption by Thermal Desorption Unit (TDU)

Since the advent of thermally stable extractive phases and binders, the TD of sorbents has been an attractive method for sample introduction to analytical instrumentation, as it requires no additional organic solvent as many other methods do [7–9,21]. As the science and engineering behind these thermally stable phases progress, the inherent background of newly developed phases (solid or liquid) is reduced, thus allowing the TD of appropriate sorbents to be applicable to ultra-trace level analysis [19,29,30]. Fundamentally, TD operates by heating a sorbent with hot gas to release all volatile analytes adsorbed onto the extraction phase, the increase in temperature driving the partition coefficient of the analytes to favor the gas phase thus releasing them from the sorbent. To its advantage, TD enables the introduction of all analytes to a gas chromatograph (GC), with the exception of any non-volatiles that either remain on the TF-SPME device or are desorbed and deposited in the liner/column (as can be the circumstance with liquid injection). This characteristic of TD prolongs the life of the analytical column, reducing the amount of maintenance needed due to the reduction of particulate matter introduced to the column. The ease-of-use and efficiency of TD, along with the inherent greenness of the method (when applicable to solid sorbents), allows TD to be the desorption method of choice for most volatile and semi-volatile compounds.

The efficiency of TD coupled with the convenient geometry of fiber SPME has allowed it to be easily adapted for online analysis. Taking advantage of quick extraction times and a lack of organic solvent, the TD of fiber SPME has long been used for routine analysis. As in other SPME geometries, TD is the most efficient and green desorption method for the analysis of volatiles by TF-SPME, directly desorbing into the instrument and removing the need for an organic solvent. However, as TF-SPME devices boast a larger extraction phase volume compared to the conventional fiber SPME geometry, an adapter is needed for the GC inlet to accommodate its larger size. In the development of TD adapters for larger extraction phase volumes, Wilcockson and colleagues [31] first made a custom thermal desorption unit (TDU) that employed an external heating element to accommodate a 22 mm glass disk coated with extraction phase. Only a couple of years later, Bruheim et al. [12] used a glass insert to introduce a sheet of monophasic PDMS to a commercial programmed temperature vaporizer (PTV) injector. Since then, as there is an ever-growing need for large-volume PTV injectors for various extraction techniques, the manufacturing of automated TDUs suitable for TF-SPME have become commonplace with units from companies such as GERSTEL, Inc. (Figure 1). The geometry of the TDU itself plays a large role in the development of TF-SPME technology, as the volume and overall shape of the TF-SPME device must be developed so the intended desorption unit can accommodate the device and effectively desorb it.

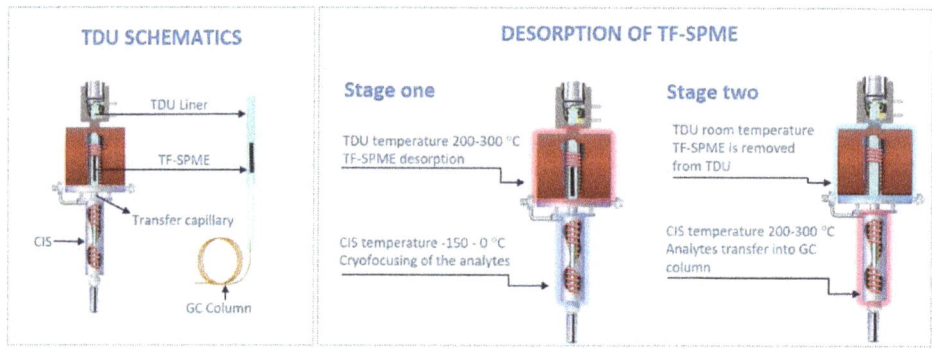

Figure 1. Two-stage thermal desorption unit for thin film solid phase microextraction (TF-SPME) (photo courtesy of GERSTEL Inc.). For the purpose of clarity, the CIS (cooled injection system) is referred to as a cryo-trap in the text. TDU, thermal desorption unit; GC, gas chromatography.

The thermal desorption of TF-SPME devices is typically carried out by a two-stage device, as opposed to the simpler one-stage TD often used for small-volume phases, such as is the case of traditional SPME fiber devices. In two-stage TD, the first stage is responsible for the release of analytes from the extraction phase into the desorption unit. This first stage is in many ways equivalent to a one-stage TDU, however, there are more parameters to carefully optimize compared to traditional one-stage desorption (an example being a typical GC inlet). Opposed to only transferring the analyte directly into the column, with a two-stages desorption the analyte must be efficiently desorbed and transferred into the cryo-trap first, and then passed into the column. This requires optimization of the different split settings and temperature programming for each stage. The first stage, the TDU, is most often run at a constant temperature to desorb all compounds enriched on the TF-SPME device, however, a temperature ramp program is possible if desired. This temperature must then be optimized depending on the volatility of the analyte (typically recommended to be 50 °C below a compound boiling point) and the thermal stability of the extraction phase, the latter usually taking priority as different extraction phases require different thermal thresholds to desorb efficiently. In most instances, the lowest operating temperature that can be used for the desorption of enriched analytes with minimal carryover is optimal, as this approach prolongs the health of the TF-SPME device and minimizes the amount of bleed from the extraction phase. As an example, the operating temperatures for the two commercially available TF-SPME devices, Carboxen®/PDMS (Car/PDMS) and divinylbenzene/PDMS (DVB/PDMS), go up to 250 °C. This parameter, like many others, needs to be optimized by trial and error at the beginning of the analytical procedure. Additionally, the split ratio must be optimized like any other injector, however, as the TDU does not directly transfer analyte into the column but the cryo-trap, the split-mode of the TDU does not necessarily reflect the overall behavior of the injector. During desorption by way of the TDU, the analyte is transferred through a heated capillary to a cryo-trap. After desorption, the TDU is cooled and the TF-SPME device removed to ensure that no residual analyte is erroneously transferred from the TDU to the cryo-trap during the second stage. Furthermore, the heated transfer capillary connecting the TDU and cryo-trap should always be at a greater temperature than the highest temperature of the TDU to ensure complete transfer of all analyte into the cryo-trap, preferably 20–30 °C higher.

Contrary to one-stage desorption, in two-stage TD, after the first stage desorbs all volatile and semi-volatile compounds from the extraction phase, the second stage is used to pre-focus analytes before their introduction into the analytical column. This pre-focusing step, while a boon to any TD method, is especially crucial when desorbing large-volume phases due to the amount of analyte desorbed and the longer desorption times typically required. During the desorption of such large-volumes, as is the case of TF-SPME, the loss of highly-volatile compounds and ultra-volatiles during a single-stage desorption method would be unavoidable. The pre-focusing step is usually achieved by the inclusion of either a sorbent tube or temperature-controlled unit (here referred to as a cryo-trap). While a sorbent-based second-stage can exhibit suitable trapping capacities of analyte, the addition of a second sorbent can complicate method optimization and thus cryo-trap stages are generally preferred. Depending on the cryo-trap device used, the trap can be cooled by liquid gas (such as nitrogen and carbon dioxide) or a solid-state device. In the use of a cryo-trapping device, while the analytes are desorbed at high temperature in the TDU, the cryo-trap is held at a low temperature, −40 to 0 °C being appropriate for most volatiles [16,32,33] with highly volatile substances requiring lower temperatures trending toward −150 °C to properly pre-focus [8,19,29,34–36]. Only after all compounds are thermally desorbed from the TF-SPME device and condensed in the cryo-trap, the temperature of the cryo-trap is increased at a high rate (usually 12 °C/s), achieving discrimination-free transfer of analytes into the analytical column with minimal sample loss. Proper method development and TF-SPME workflow (Figure 2), with the use of a cryo-trap for the focusing of volatiles, results in sharper chromatographic peaks and ensures that all compounds extracted are introduced into the analytical column, increasing the reproducibility and sensitivity of the analytical method. In case of direct immersion extraction, it is advisable to gently wipe the TF-SPME device before introduction into the desorption liner, as to avoid potential matrix contaminants such as water and oil entering the TDU.

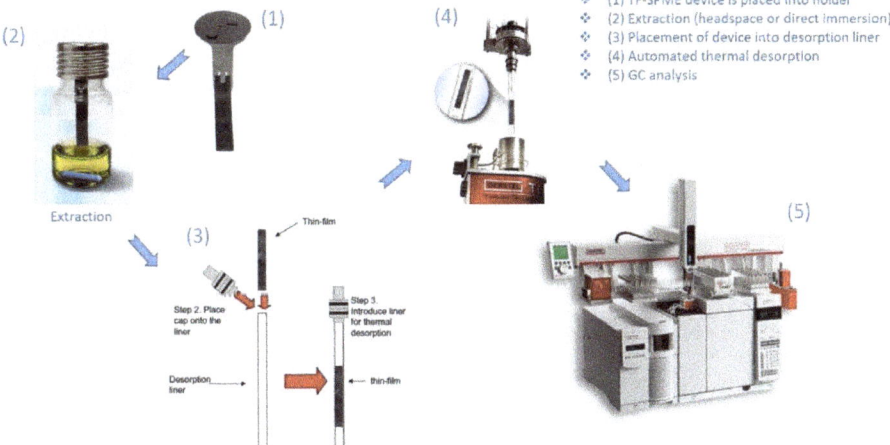

Figure 2. Optimized TF-SPME workflow (photos courtesy of GERSTEL Inc.) for extraction and thermal desorption.

Care must be taken when deciding the type of split for the cryo-trap as it is synergistic with the TDU (Table 1). It should be mentioned that the gas flow of a two-stage unit is restricted by the cryo-trap during both stages. As a result, if splitless injection is desired a solvent-venting cryo-trap is still necessary to maximize gas flow, thus allowing efficient desorption. This solvent-vent must then be closed when the cryo-trap begins to increase in temperature, ensuring a splitless injection with maximum desorption efficiency. In short, for maximum sensitivity both stages should be run in splitless mode, the cryo-trap being operated under solvent-vent conditions before desorption into the column. In the presence of solvent, the cryo-trap can typically run in a mode that vents the solvent if the boiling point differs greatly from the analyte (approximately a 150 °C difference). This allows better chromatography with no loss of analyte, provided the boiling point temperature difference is sufficient and the cryo-trap is heated after the solvent vent closes, effectively making this method a splitless injection onto the column. For unknown analysis, it is typical to run both stages in split mode, and with concentrations in the ppm level, the TDU must have an appropriate split ratio with the cryo-trap being performed in either a splitless or solvent venting fashion.

Table 1. Common parameters of two-stage thermal desorption and their applications.

Application	TDU Split	Cryo-Trap Split	Rationale
Ultra-trace analysis	Splitless	Solvent-vent [1]	Maximum amount of analyte injected
High concentrations	Splitless	Solvent-vent	Sensitivity/less column load
Unknown analysis	Split	Split	Clean cryo-trap/column
Watery/fouled devices	Split	Solvent-vent [1]	High sensitivity/clean cryo-trap

[1] Solvent-venting ends at the same time the cryo-trap begins to heat, allowing maximum desorption flow while still achieving a splitless injection.

It should be noted that alternative forms of TD for TF-SPME devices have been recently reported. These methods use "transmission mode" devices consisting of a coated mesh-like surface to enrich analytes and subsequently both desorb and ionize in one step using ambient ionization methods [37–39]. Direct analysis in real time (DART), an ambient ion source that has been shown to be well suited for onsite analysis [40], can be simply coupled to TF-SPME (coated mesh-like geometry) by positioning the device in-between the ion source and the mass spectrometer (MS). In this way, a stream of heated plasma (most often helium) desorbs the TF-S4PME transmission device while also ionizing analytes at the same time [37,38]. Another ambient ionization source, dielectric barrier discharge ionization (DBDI), has also been proven to be well suited for the analysis of illicit drugs using TF-SPME devices [39]. However, as the DBDI source does not thermally desorb compounds as the DART source can, a separate TD chamber was constructed using an aluminum body and temperature controller. This chamber was then filled with pre-humidified nitrogen to facilitate desorption, the flow of this now-analyte enriched gas entering the DBDI source [39].

4. Fundamentals of TF-SPME

SPME is one of the most attractive extraction techniques commonly used today due to its high throughput, low-cost and solvent-less extraction. In recent years, TF-SPME has been developed to better meet the demands of onsite and ultra-trace level analysis, as its high surface area-to-volume ratio enables it to achieve enhanced sensitivity and greater extraction efficiency [12]. In common with other extraction technologies, sampling using TF-SPME devices is characterized by partition equilibria between the free form of the analyte in a sample and the extraction phase constituting the extraction device. Consequently, the properties of the extraction phase play a significant role in the efficiency of the extraction process. The total amount of extracted analyte in direct immersion SPME, where the analyte equilibrates between only two phases, is described by Equation (1) [12]:

$$n^{eq} = \frac{K_{es} V_e V_s}{K_{es} V_e + V_s} C_s, \qquad (1)$$

where n^{eq} is the total amount of extracted analyte at equilibrium, K_{es} is the distribution constant of the analyte between the matrix and extraction phase, V_e is the volume of extraction phase, V_s is the volume of the sample and C_s is the initial concentration of analyte in the sample. As can be seen in Equation (1), the greater the volume of the extraction phase (V_e), the larger amount of n^{eq} is extracted.

In the following equation describing the equilibrium time for the analyte between the two phases (Equation (2)) [12],

$$t_e = t_{95\%} = \frac{3\delta(b-a)}{D_s}, \quad (2)$$

t_e is the required time for the analyte to reach equilibrium with the extraction phase, $t_{95\%}$ is the time needed to extract 95% of the equilibrium amount of an analyte on the device, δ is the thickness of the boundary layer, $(b-a)$ represents the thickness of the extraction phase and D_s is the diffusion constant of the analytes into the sample matrix. While an increase in V_e allows higher extraction efficiency and thus greater sensitivity according to Equation (1), Equation (2) demonstrates that the corresponding increase in coating thickness results in a longer equilibrium time. Hence, it is important to optimize these parameters to ensure the most efficient and practical mode of extraction [41].

Considering Equation (3) [12],

$$\frac{dn}{dt} = \left(\frac{D_s A}{\delta_s}\right) C_s, \quad (3)$$

A thin film geometry can also enhance the sampling rate due to its high surface area-to-volume ratio, reducing the time it takes to reach equilibrium and enhancing the capacity of the extraction device. In this equation, n is the amount analyte extracted over the extraction time t, A is the area of the extraction phase, D_s is the diffusion constant of the analyte into the sample matrix and δ_s is the thickness of the boundary layer [42]. In other words, employing TF-SPME devices allows rapid sampling with high extraction capacity, suitable for ultra-trace level analysis.

5. Development of the First TF-SPME Device and Improvements up to 2019

As discussed previously, the simultaneous increase of extraction phase volume and surface area for TF-SPME devices allows enhanced sensitivity with as good or better extraction rates compared to traditional fiber SPME. The development of the underlying theory of this phenomena took root in 2000 when Semenov and colleagues described the kinetics of a thin layer of extraction phase, predicting what would be the driving force for both TF-SPME and passive sampler development [43]. While not a traditional TF-SPME device, the first technique to exploit a higher surface area-to-volume ratio to increase both extraction capacity and efficiency was developed in 2001 by Wilcockson and colleagues [31]. The procedure utilized a thin film (0.05 and 0.33 μm) of ethylene-vinyl acetate as the extraction phase which was coated onto 22 mm diameter glass disks serving as the support. This passive sampling method, exhibiting a surface area to volume ratio over 1000 times higher than comparable 100 μm SPME fibers, demonstrated faster equilibration times but with the caveat of having less phase volume. Due to the decreased phase volume, sensitivity did not exceed traditional SPME fibers, however, the method has still been adopted throughout the years with success as environmental passive samplers [44–48]. While important to the development of TF-SPME, this technique does not share the same geometry or sensitivity as traditional TF-SPME devices and thus is only mentioned due to its importance in the development of parallel sampling technologies based on thin adsorbent layers.

First developed in 2003 by Bruheim and colleagues [12], the first TF-SPME device consisted of a pre-manufactured 25.4 μm sheet of polydimethylsiloxane (PDMS) as the extraction phase. This thin sheet of PDMS was attached to a stainless steel rod as support, being affixed in a "flag-like" manner during extraction and wrapped around the rod prior to manual TD in a PTV GC inlet (Figure 3). Using polycyclic aromatic hydrocarbons (PAHs) as model analytes, the study demonstrated the practical use of TF-SPME as an alternative geometry to fiber SPME in both direct immersion extraction and headspace extraction. As the extraction efficiency was up to 20 times higher when using a 1 cm × 1 cm sheet of PDMS compared to a 100 μm PDMS fiber, with the extraction rate exceeding the already-developed SBSE [15], TF-SPME devices were further developed and optimized for better sensitivity and a more convenient sampling approach.

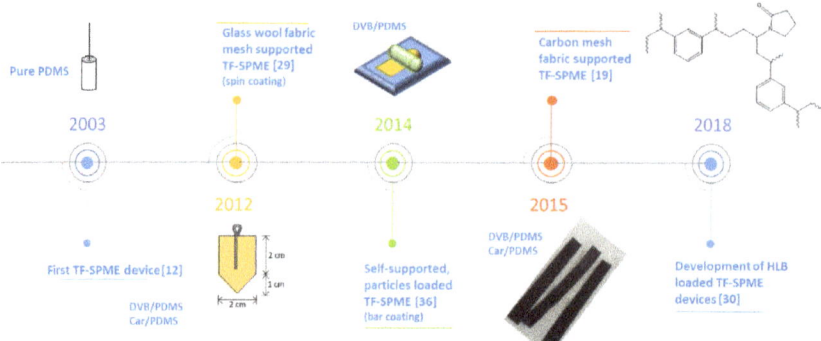

Figure 3. A timeline of pivotal moments in the evolution of TF-SPME devices. PDMS, polydimethylsiloxane; DVB, divinylbenzene; Car, Carboxen®; HLB, hydrophilic-lipophilic balance.

Being limited to PDMS as an extraction phase, due to the availability of pre-made sorbents, priority was given in the geometric optimization of TF-SPME devices. In 2006, Bragg and colleagues modified a PDMS sheet into a 127 μm thick house-like shape (Figure 3) supported by a stainless steel wire [49]. This TF-SPME device, dimensions of 2 cm × 2 cm with the triangular portion of the device being 1 cm in height, boasted an increased phase volume of 0.0635 cm^3 compared to the 0.00255 cm^3 phase volume achieved by the previous developed TF-SPME device [12]. An increase of over 20 times the volume of the extraction phase, the achievable amount of extracted analyte was greatly increased, and thus greater sensitivity was attained. Moreover, the house-like geometry of the TF-SPME device permitted an increase in surface area while still allowing the device to be easily wrapped around the support, ensuring ease of injection into the GC inlet. Additionally, the same study demonstrated the efficacy of using TF-SPME for field analysis of aqueous media, establishing TF-SPME as a convenient onsite extraction tool [12].

A distinct departure from previous developments in both chemistry and design, Rodil and colleagues [32] demonstrated the use of glass wool fabric as solid support for TF-SPME. Opposed to previous attempts, which utilized pre-manufactured sheets of PDMS supported by steel wire, the SPME devices designed consisted of a polyacrylate (PA) extraction phase bound to glass wool. During the development of these devices, an amount of glass wool fabric was saturated with a solution of PA, being cured while sandwiched between two sheets of polyethylene foil to ensure homogeneity of the applied extraction phase. Being the first composite TF-SPME device developed, the sampling device demonstrated increased mechanical stability and the final device was 6 cm × 0.3 cm after the sheets of foil were removed. These TF-SPME devices, named "PA strips", were compared to SBSE by performing their extraction with the same parameters and desorbing them into a TDU, very similar to the modern-day TD of TF-SPME devices. It was found, however, that these devices undergo thermal decomposition after multiple uses, resulting in high amounts of bleed into the GC and a lack of robustness of the device.

Incorporating the house-like geometry [49] and the glass fiber support [32] of earlier TF-SPME devices, Riazi et al. [29] developed the first mixed-mode extraction phase for TF-SPME using Car/PDMS and PDMS/divinylbenzene (PDMS/DVB). Adsorptive particles (Carboxen® or DVB) were suspended in a solution of the binding agent (PDMS) before being applied to a thin sheet of glass wool fabric. Instead of allowing the polymer solution to absorb into the supporting material as previous methods did [32], the coating procedure was performed using the spin coating method due to its ease of use and greater control over the phase thickness. After curing and cutting the material to a 2 cm × 2 cm square with a 1 cm triangle (same geometry developed by Bragg et al. [49]), the TF-SPME device was held by a cotter pin during extraction of analytes. This newly developed TF-SPME device was then desorbed into a large volume inlet (TDU-2, Gerstel GmbH, Mulheim, Germany) with a CIS-4 (Gerstel GmbH, Mulheim, Germany) cryo-trap. Accordingly, results showed a marked increase in the mechanical stability of the device during extraction, as well as improved thermal stability during the desorption process.

Building on the prior success of mixed-mode TF-SPME devices, the first self-supported particle loaded TF-SPME device was developed using DVB particles loaded onto PDMS [36]. This device was made with the intent of air-sampling using high amounts of DVB embedded into a PDMS base. It was found that when increasing the amount of DVB particles onto the PDMS membrane, up to 30% (w:w) DVB allows the membrane to achieve better mechanical stability compared to non-particle load membranes, 20% being found to be optimal for mechanical stability and extraction efficiency. As a result, the TF-SPME device boasted better sensitivity for the target analyte, benzene, than previous methods using fiber SPME or monophasic PDMS TF-SPME devices. This device, however, was still unable to appropriately be used in direct immersion extraction at high agitation rates, as the mechanical stability was not as great as previous glass-coated TF-SPME devices [29].

In 2016, Grandy and colleagues developed a sampler exhibiting far less siloxane bleed and greater robustness due to the carbon mesh support used therefore creating the first TF-SPME device well suited for untargeted analysis and onsite sampling by direct immersion extraction [19]. To achieve this, higher density PDMS was used to reduce bleed along with a slight reduction in phase volume. A mixture of two components, DVB particles embedded into PDMS, was then spread out onto a carbon mesh which primarily provided support along with some affinitive properties. In spite of the reduced phase volume of this device, the new design still afforded a highly sensitive extraction with now far lower siloxane bleed. Moreover, the carbon mesh support granted much better mechanical stability compared to previous iterations of TF-SPME, allowing the device to undergo more rigorous agitation compared to previously developed TF-SPME devices. After curing the extraction phase, the devices were then cut to different sizes well suited for TD, allowing more practical device introduction compared to previous methods. As a result of these recent developments, this version of TF-SPME is now the first commercially available TF-SPME device, currently distributed by GERSTEL Inc. (Gerstel GmbH, Mulheim, Germany). Since then, much of the development of TF-SPME devices has followed the same convenient format. Different extraction phases have been tested but not commercialized yet, including hydrophilic-lipophilic balance (HLB) phases that offer a wider range of extraction [30].

With the development of new mixed-mode extractive membranes that were both thermally and mechanically stable, new TF-SPME devices were able to overcome the limitations of previous variants of TF-SPME technology [12,16,32,49]. Initial devices, being made of pure monophasic extraction phase, exhibited poor structural rigidity making their practical application cumbersome in immersive extraction [12,49]. As a consequence of the large volumes of PDMS utilized in these devices, significant siloxane bleed was found, resulting in unacceptably high backgrounds. This was at times circumvented by use of single ion monitoring (SIM), avoiding detector saturation but nonetheless a technique not well suited for untargeted analysis. With the development of glass fiber supported TF-SPME devices [29,32], better structural rigidity was met, with the first device [32] still lacking suitable mechanical stability and thermal stability for repeated analysis. The second glass fiber supported device [29], however, not only achieved better mechanical and thermal stability compared to its earlier counterparts but also demonstrated a wider range of extraction than found previously by the use of mixed-mode extraction phases. While there is certainly a benefit of using glass fiber as a structural base for TF-SPME devices, the large amount of extraction phase and binder used is still cause of concern in terms of high siloxane backgrounds.

Further developments of TF-SPME devices have been focused on self-supported membranes with the incorporation of extractive particles, allowing them to achieve efficient extraction of a wide range of analytes. The incorporation of these extractive particles provided newfound mechanical stability [29], permitting the devices to be well suited for direct immersion extraction. Given that the qualities indispensable to the sampling method and characteristics important for the ruggedness of any GC-amenable extraction device—thermal stability, mechanical stability and extraction efficiency—were finally met, a refinement of TF-SPME as a whole was essential and further development was needed. Although it is true that the developed phases are thermally stable, a large volume of extraction phase will still cause more background and bleed compared to a smaller volume. In light of this, the newest iteration of TF-SPME devices [19], applying the extraction phase to a carbon mesh, solves this issue by retaining comparable sensitivity but with drastically reduced siloxane bleed, all the while being more mechanically stable than previous TF-SPME devices. As much as the geometry is far more convenient than previous iterations, the planar surface still poses a practical obstacle in the engineering of automated sampling and desorption. The major caveat in the new design is that currently only two phases currently are offered (DVB/PDMS and Car/PDMS).

6. TF-SPME Coating Methods

The various extraction phases and coating methods used in the development of TF-SPME devices (Figure 4) are chosen based on the chemical properties of the analyte, the surrounding matrix and the adopted desorption method. There are many techniques for the coating of TF-SPME devices, including dip coating [39,50], spin coating [51] bar coating [34,36,52], electrospinning [53] and spray coating [54]. Among these different coating methods, bar and spin coating are the most common coating methods for the production of TF-SPME devices amendable for TD. The most common coating method discussed in this review is bar coating in which the liquid extraction phase is set on a substrate and then this extraction phase is spread by a bar to develop the device [24]. The first TF-SPME device made by bar coating was prepared using DVB particles impregnated on PDMS and this device was later used for air sampling with good results [36]. Moving forward, Grandy et al. prepared bar coated DVB/PDMS onto a carbon-based mesh support coupled with a portable GC-MS for the quantitation of volatile and semi-volatile organic compounds [19]. The bar coating procedure is needed to be repeated on both sides of the carbon mesh support to ensure an even coating for the final device [9]. In 2017, Piri-Moghadam et al. used bar coating for the preparation of DVB/PDMS TF-SPME devices for the analysis of common pesticides in surface water samples. They found that the use of TF-SPME, in comparison to LLE, provided enhanced selectivity, reproducibility and faster rates of extraction [9]. In another study by Piri-Moghadam and colleagues in 2018, bar coating was utilized in the development of novel TF-SPME devices in an effort to analyze various pesticides found in river water, demonstrating a greener and more sensitive alternative than LLE [55]. The use of these developed DVB/PDMS TF-SPME devices for the quantification of pesticides demonstrated enhanced sensitivity and extraction efficiency for both onsite and bench-top analysis. In the spin coating method, in similar fashion to bar coating, a layer of extraction phase is placed on the substrate, after which by spinning the substrate, a thin layer of extraction phase is homogenously produced [24]. In both bar coating and spin coating, the thickness of the extraction phase can easily be controlled by the pressure of the bar onto the substrate and the intensity of spinning, respectively. Spray coating, one of the simplest methods for the preparation of TF-SPME devices, utilizes a dissolved mixture of extractive phase in a suitable solvent to be sprayed onto a stage until the formation of a uniform film [24]. An example of this technique, Mirnaghi et al. prepared polyacrylonitrile-polystyrene (PAN-PS)-DVB and polyacrylonitrile–phenylboronic acid (PAN-PBA) TF-SPME devices for the analysis of a variety of pharmaceuticals from human plasma [56]. Through the development of these two new TF-SPME devices for consequent analysis by LC-MS/MS, the analysis of a wide spectrum of polar compounds in human plasma with high efficiency and rapid throughput was achieved. Finally, another coating method which is often used for TD is electrospinning or electrospray coating. In this method, a mixture containing a polymer is sprayed by electrical energy on the surface of substrate [24]. In 2015, TF-SPME devices were prepared by the electrospinning method using polyimide nanofibers for the investigation and quantification of phenol compounds in environmental water using GC-MS [53]. Extraction devices were first activated by acetone, increasing hydrophilicity, resulting in greatly enhanced extraction efficiency with results demonstrating limits of quantification (LOQs) in the ppt level.

Figure 4. The different coating methods for preparation of TF-SPME devices.

7. Applications

7.1. Environmental Analysis

The applications of TF-SPME have been traditionally environmental in nature, as much of their development has been expedited by the need for rugged and high capacity samplers, both passive and active. Initial progress toward highly efficient samplers with greater surface-to-volume ratios were developed to meet the need for the trace level extraction of contaminants found in complex environmental matrices, but these matrices proved to be challenging due to a variety of parameters such as large volumes for air analysis or the incredible sensitivity needed for persistent contaminants in aqueous media. Under these circumstances, the enhanced sensitivity of TF-SPME coupled with its convenient geometry for both extraction and introduction to onsite and benchtop instrumentation affords it the opportunity to outperform other SPME technologies in trace level environmental analysis.

Initial developments of membrane-based TF-SPME were evaluated by the extraction of polycyclic aromatic hydrocarbons (PAHs) from aqueous samples [12]. PAHs, a class of hydrophobic semivolatiles that are commonly released into the environment through the combustion of hydrocarbons and other organic matter, continue to rise in their environmental significance as they are readily distributed throughout biosystems and are a known cause of cancer among other mutagenic effects [57]. Since then, many studies have analyzed PAHs from aqueous samples using TF-SPME in various different modes, utilizing PDMS film [16,49,58,59], TF-SPME membrane [20] and an alternative thin film-based sampler using polymer-coated aluminum [45]. In addition, glass fiber reinforced TF-SPME has also been used for the analysis of PAHs in aqueous samples, using polyacrylate coatings to extract PAHs along with organochlorous and organophosphorus pesticides, demonstrating greatly enhanced efficiency with the partition coefficients for the TF-SPME device being up to 15 times higher than the SBSE device [32]. Less conventional extraction phases have also been introduced, as the recent success of carbonaceous nanomaterials being suitable extraction phases for SPME have led the path toward the development of mixed-mode carbonaceous TF-SPME devices. These carbonaceous TF-SPME devices have in turn been proven to be effective in the extraction of different organic compound classes, an example again being the analysis of PAHs in aqueous samples [20]. Furthermore, the analysis of PAHs have also been performed using other "parallel" thin film technologies, that is, extraction which utilizes a large surface area but does not exemplify the large phase volume needed for enhanced sensitivity. This is accomplished using PDMS-coated vials for the determination of PAHs in soil [46]. Similar alternative geometries have also been tested for their uses as passive samplers, such as ethylene-vinyl acetate (EVA) coated glass fibers for the aqueous extraction of pesticides [48] and EVA coated glass cylinders for the air sampling of volatile PCBs [44].

TF-SPME devices made of pure monophasic extraction phase, such as the first generation of TF-SPME devices [12], are still used in some studies as passive samplers. In the case of pyrethroids, a class of significantly hydrophobic insecticides that are known to cause damage to beneficial insects and fish, the suitability of TF-SPME has been just recently studied using thin films (25–500 μm) of different materials (silicon, polyethylene, polymethylmethacrylate, polyoxymethylene and polyurethane) as passive samplers [60]. Another alternative geometry of passive sampling, sorptive tape extraction (STE) [61], has also been effectively used for the direct sampling of plant volatiles [33,62] by both headspace and direct application to the plant surface. This geometry uses a tape-like PDMS thin film as a sampling device, providing an easy method of application for the non-invasive sampling of environmental and biological [61] matrices, recently being demonstrated by Boggia et al. [33] to be suitable for the analysis of herbivory-induced plant volatiles. While viewed currently as a passive sampling device, similar technology could be implemented in the development of new, more convenient TF-SPME devices.

During the development of particle-loaded TF-SPME devices, in this instance DVB particles in PDMS, Jiang et al. demonstrated the greatly enhanced sensitivity TF-SPME offers in trace air sampling and monitoring [36]. This particle loaded TF-SPME device was able to achieve extraction of a wide range of analytes with differing volatilities at high capacity. Furthermore, the study quantitatively samples benzene as a model analyte, a known carcinogen that is found commonly in fuels and smoking devices or from polluted air near a high-traffic road. Other more recent studies have confirmed the use of TF-SPME as a validated onsite sampling tool by its use in the analysis of biocides and UV blockers in sunscreen found in rivers, utilizing HLB/PAN and octadecyl (C_{18})/PAN TF-SPME devices [54]. Moreover, the efficacy of TF-SPME for the analysis of trace and ultra-trace level analysis in environmental matrices has been repeatedly tested and validated since its inception. In more recent times, since the development of carbon mesh-based TF-SPME devices, TF-SPME has been proven to be even more of a convenient sampling tool for onsite analysis due to the structural robustness of the device and its ease of introduction to portable instrumentation [19]. Since then, an interlaboratory study comparing these newly improved TF-SPME devices (DVB/PDMS on carbon mesh) to EPA-validated LLE methods demonstrated the efficacy of TF-SPME as a greener and more sensitive technique for the routine analysis of pesticides in water [55]. This comparative study by Piri-Moghadam and colleagues compared the suitability of different TF-SPME approaches to LLE. These approaches comprised of an in-bottle TF-SPME method using benchtop GC/MS, an onsite drill-assisted sampling that later used benchtop GC/MS and a procedure that used the same drill-assisted sampling as the previous but instead used a portable GC/MS to achieve both onsite sampling and onsite analysis. Results demonstrated (Table 2) the robustness of all TF-SPME-based methods, with the onsite extraction and analysis being the most environmentally friendly of all methods in the study [55]. In a similar fashion, the performance of TF-SPME has further been compared to other extraction methods, namely fiber SPME and SPE, in the analysis of harmful coal frothing agents that have been reportedly released into environmental waters and, by consequence, contaminated drinking water supplies. This study, the first to use a Car/PDMS TF-SPME device with carbon mesh support, was able to reliably extract multiple components of the coal frothing agent, crude (4-methylcyclohexyl)methanol (MCHM), along with a tentative metabolite with minimal sample manipulation. Results again showed TF-SPME devices to be more efficient at trace-level analysis than other methods (Table 2) [8]. Furthermore, with the development of novel HLB TF-SPME devices, Grandy and colleagues demonstrated the wide range of analytes that can be extracted by the analysis of chlorination byproducts in residential hot tubs [30].

Applications of TF-SPME devices using other modes of desorption have also been applied in the environmental sector. Recently, de la Calle and colleagues have utilized graphene TF-SPME devices coupled with chelating agents to sample various different metals in aqueous samples before analysis using total reflection X-ray fluorescence (TXRF) [28]. Similar methods have also employed similar TF-SPME-TXRF protocols with the use of nanomaterial TF-SPME devices [27]. In recent years, nanomaterial-based TF-SPME devices have been successfully applied to TD [63]. Mohammadi et al. demonstrated the use of a self-supported TF-SPME device composed of a zeolitic imidazolate framework, which extracts the organophosphorus pesticide ethion with subsequent TD for the analysis of environmental water samples [63]. In other respects, there are examples of more conventional TF-SPME being used in conjunction with LD for environmental analysis [64], ranging from the analysis of fluorinated benzoic acids in aqueous samples [21] to the analysis of polychlorinated biphenyls (PCBs) by direct application of a TF-SPME device into fish tissue [65].

7.2. Flavors and Fragrance Analysis

TF-SPME has various applications for the analysis of flavors and fragrances in a variety of different matrices. In this regard, Stuff and colleagues applied carbon mesh-supported DVB/PDMS as an extraction phase for the sampling of various volatile compounds, such as alcohols and ethyl esters [66]. The improved sensitivity for these various classes of compounds, afforded by TF-SPME, is essential for the rigorous demands of quality control in the beverage industry. A similar study by Vernarelli et al. investigated the efficacy of TF-SPME technology for the analysis of foodstuffs using DVB/PDMS TF-SPME devices and analyzed dark chocolate, cheeses and Caesar dressing [34]. These studies [34,66] compared the performance of TF-SPME to fiber SPME with the same stationary phase, revealing enhanced extraction efficiency and capacity for the TF-SPME devices. In 2016, another study investigated the fragrance of various essential volatile compounds from grapes, including linalool and 3-isobutyl-2-methoxypyrazine (IBMP), which were measured using PDMS-based TF-SPME devices [38]. This novel solid-phase mesh-enhanced sorption from headspace (SPMESH) method developed by Jastrzembski and colleagues resulted in greater throughput and enhanced limits of quantitation (ppb) during the analysis of volatile compounds (such as odorants) compared to other traditional methods. Sol-gel coating of a stainless steel mesh substrate with a thin film of PDMS provided great thermal stability and high sensitivity for the direct analysis with SPME-DART [38]. In 2007, Bicchi and coworkers compared the results of headspace and direct contact sorptive tape extraction (STE), all the while employing fiber HS-SPME as a reference standard [62]. The analytes extracted, volatile compounds found in various solid biological matrices such as apple, perfume on human skin, rosemary and spearmint, all demonstrated increased sensitivity and extraction efficiency compared to the more conventional fiber HS-SPME method.

7.3. Other Applications of TF-SPME

In addition to the diverse applications TF-SPME provides in the environmental and flavor/fragrance fields, many other procedures have been validated with a variety of matrices. Of these applications, one of the most striking uses of TF-SPME has been the analysis of sebum, developed by Sisalli and colleagues for the non-invasive detection of sebum in vivo [61]. This novel TF-SPME device, utilizing an adhesive thin layer of PDMS, demonstrated good performance in the extraction of sebum and other constituents of skin by a simple placement of the device on human skin, later being thermally desorbed in a TDU/cryo-trap system. This rapid and non-invasive sampling of human skin is crucial for the further development of sampling methods for the cosmetic and pharmaceutical industries. In a similar fashion, Jiang et al. introduced a novel in vivo sampling method for the analysis of human skin constituents using a thin layer of PDMS [35]. To prevent saturation of the device from sebum and other common skin oils, the PDMS-TF was emplaced between two pieces of stainless steel mesh and then placed on the surface of skin for sampling (Figure 5). As a result, this developed approach was able to demonstrate great promise for applications in clinical settings due to its high reproducibility and non-invasive sampling procedure. In another example of TF-SPME being well suited for both in vivo and ex vivo sampling in clinical settings, Bessonneau et al. prepared HLB/PDMS and (C_{18})/PDMS TF-SPME devices for the investigation of prohibited substances in saliva, analyzed with both LC-MS/MS and GC-MS [52]. Furthermore, this study demonstrated increased analytical precision compared to their similarly developed ex vivo method, confirming the need for more rugged in vivo sampling devices for clinical settings. In 2019, Shigeyama and colleagues reported another application of TF-SPME by use of zeolite-based devices for the extraction of volatile organic compounds (VOCs) in saliva, a class of compounds that is used to determine if a patient has oral cancer [67]. This study again demonstrated the robustness of TF-SPME devices as a non-invasive method. In a more recent study, Mirabelli et al. [39] utilized self-supported TF-SPME prepared according to the method proposed by Jiang et al. [36] to extract illicit drugs in both beverages and biofluids. In this study, for the first time, it was demonstrated that the DVB/PDMS TF-SPME devices were suitable for ultrasound-assisted extraction, with a consequent drastic reduction of extraction time prior to direct coupling to a DBDI source. Moreover, this approach allowed rapid quantitative desorption, reducing the likelihood of the thermal degradation of sensitive analytes. As a result, this newly developed method demonstrated several advantageous aspects, including rapid analytical throughput and enhanced sensitivity. With the simplicity for TF-SPME devices to be used as onsite samplers, along with their incredible sensitivity, they have been proven to be suitable tools for environmental analysis, while their fast and non-invasive sampling affords them great potential for both the pharmaceutical and clinical industries (Table 3).

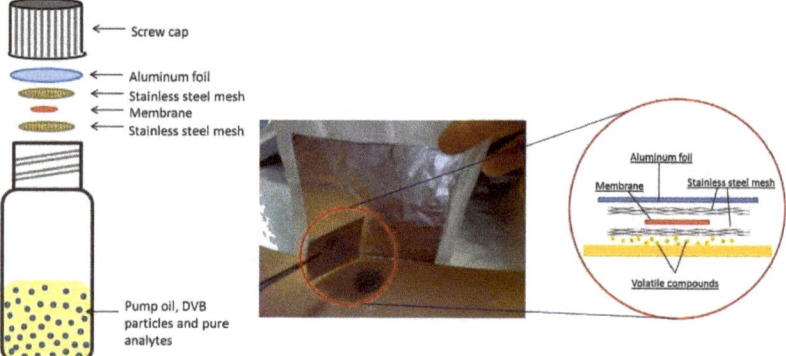

Figure 5. The development of polydimethylsiloxane (PDMS) TF-SPME devices for the analysis of skin volatiles. Reproduced from [35], with permission from Elsevier, 2013.

Table 2. Comparative studies demonstrating the efficacy of TF-SPME over similar extraction methods. Data are in limits of quantitation (LOQs) if not otherwise noted.

	Piri-Moghadam et al. [9]			Rodriguez-Lafuente et al. [68]		Piri-Moghadam et al. [55]				Emmons et al. [8]		
Compounds	TF-SPME Carbon Mesh Supported (μg L^{-1})	TF-SPME Non-Supported (μg L^{-1})	LLE (μg L^{-1}) LOD	Compounds	Fiber SPME (μg L^{-1}) RDL	Compounds	TF-SPME (ng L^{-1})	TF-SPME Drill Agitation (ng L^{-1})	LLE (ng L^{-1})	Compounds	TF-SPME (μg L^{-1})	SPME (μg L^{-1})	SPE (μg L^{-1})
2,4,6-TCP	0.050	0.025	0.50	2,4,6-TCP	0.20	2,4,6-TCP	10	100	500	MCHM	0.10	1.0	500
2,3,4,6-TeCP	0.025	0.025	0.50	2,3,4,6-TeCP	0.20	2,3,4,6-TeCP	3.0	250	500	MMCHC	0.10	2.5	250
Trifluralin	0.050	0.050	1.0	Trifluralin	0.050	Trifluralin	3.0	50	1000	4MMCH	1.0	2.5	500
Diazinon	0.050	0.050	1.0	Diazinon	0.50	Diazinon	100	1000	1000	1-4CHDM	0.50	0.25	5000
Triallate	0.050	0.050	1.0	Triallate	0.050	Triallate	3.0	50	1000	DM-1-4CHC	0.10	0.50	500
Methyl parathion	0.50	0.50	1.0	Methyl parathion	0.20	Methyl parathion	100	1000	1000	MCHCA	2.0	2.50	25,000
Alachlor	0.50	0.050	0.50	Alachlor	0.10	Alachlor	10	100	500				
Metalachlor	0.050	0.025	5.0	Metalachlor	0.10	Metalachlor	3.0	250	500				
Chlorpyrifos	0.25	0.10	1.0	Chlorpyrifos	0.20	Chlorpyrifos	10	1000	1000				
Cyanazine	0.10	0.10	1.0	Cyanazine	1.0	Cyanazine	10	100	1000				
2,4-DCP	0.10	0.050	0.25	2,4-DCP	0.10								
Bendiocarb	0.050	0.025	2.0	Bendiocarb	0.05								
Phorate	0.25	0.25	0.5	Phorate	0.25								
Carbofurane	0.10	0.10	5.0	Carbofurane	0.20								
Simazine	0.25	0.25	1.0	Simazine	0.050								
Atrazine	0.075	0.25	0.50	Atrazine	0.050								
PCP	0.075	0.10	0.50	PCP	0.20								
Terbufos	0.25	0.25	0.50	Terbufos	0.50								
Metribuzine	0.25	0.10	5.0	Metribuzine	0.20								
Carbaryl	0.10	0.10	5.0	Carbaryl	0.50								
Prometryn	0.075	0.075	0.25	Prometryn	0.20								
Malathion	0.50	0.50	5.0	Malathion	0.20								

LLE, liquid-liquid extraction; SPE, solid phase extraction; LOD, limit of detection; RDL, reported detection limits.

Table 3. Applications of TF-SPME and closely related devices.

Extraction Phase Chemistry	Sample	Targeted Analytes	Instrumentation	Coating Method	Ref.
Environmental Analysis					
PDMS	Environmental waters	PAHs	GC-MS	N/A	[12]
PDMS	Environmental waters	PAHs	GC-MS	N/A	[16]
PDMS	Environmental waters	PAHs	GC-MS	N/A	[49]
PDMS	Deionized water	PAHs	GC-MS	N/A	[58]
PDMS	Environmental waters	PAHs	GC-MS	N/A	[59]
Carbonaceous Nanomaterials	Wastewater	PAHs	GC-MS	Evaporation	[20]
EVA and LDPE	Air	PAHs	GC-MS	Dipping	[45]
Polyacrylate	Environmental waters	PAHs	GC-MS	Dispersion	[32]
PDMS	Soil	PAHs	GC-MS	Dispersion	[46]
EVA	Environmental waters	Pesticides and PCBs	GC-MS	Dip coating	[48]
EVA	Air	PCBs	HPLC-FLD	Dispersion	[44]
Various monophasic polymers	Soil	Pyrethroids	GC-MS	N/A	[60]
PDMS	Human skin	Sebum constituents (squalene/fatty acids)	GC-MS	N/A	[61]
PDMS	Plant Tissue	Plant volatiles	GC-MS	N/A	[33]
PDMS	Plants/Perfumes	Plant volatiles/perfumes	GC-MS	N/A	[62]
DVB/PDMS	Air	Benzene	GC-MS	Bar coating	[36]
Polyacrylonitrile	Environmental waters	UV filters/biocides	LC-MS	Spray coating	[54]
DVB/PDMS	Environmental waters	Toluene/xylene/pesticides	GC-MS	Bar coating	[19]
DVB/PDMS	Environmental waters	Pesticides	GC-MS	Bar coating	[55]
Car/PDMS	Environmental waters	Crude MCHM constituents	GC-MS	Bar coating	[8]
HLB/PDMS	Hot tub water	Chlorination byproducts	GC-MS	Bar coating	[30]
Graphene	Various aqueous samples	Trace metals	TXRF	Drop-casting	[28]
Graphene	Aqueous	Methylcyclopentadienyl manganese tricarbonyl	TXRF	Drop-casting	[27]
Zeolitic imidazolate framework	Agricultural Wastewater	Ethion (insecticide)	IMS	Electrospinning	[63]
PDMS	Lake water	Organochlorine pesticides	GC-MS	Bar coating	[64]
(HLB, C18, SAX, PS-DVB-WAX, HLB-SAX)	Synthetic Seawater	Fluorinated benzoic acids	LC-MS, GC-MS	Spray coating	[21]
PDMS	Fish tissue	Polychlorinated biphenyls	GC-MS	N/A	[65]
Flavors and fragrance					
DVB/PDMS	Beverage	Aroma and flavor components	GC-MS	Bar coating	[66]
DVB/PDMS	Grapes	Linalool/IBMP	DART-MS	Dip coating	[69]
PDMS	Perfumes and foodstuffs	Volatile compounds	GC-MS	N/A	[62]
DVB/PDMS	Food	Aroma and flavor components	GC-MS	Bar coating	[34]
PDMS	Grapes	Linalool/IBMP	DART-MS	Dip coating	[38]

Table 3. *Cont.*

		Other			
PDMS	Human Skin	Volatile organic compounds	GC-MS	N/A	[35]
PDMS	Human Skin	Sebum	GC-MS	N/A	[61]
PDMS	Human Saliva	Prohibited substances and endogenous steroids	GC-MS, LC-MS	Bar coating	[52]
PDMS/DVB	Beverages and biological fluids	Drugs	GC-MS	Dip coating	[39]
ZSM-5/PDMS	Human saliva	Volatile metabolites	GC-MS	Deposition	[67]

EVA, ethylene-vinyl acetate; LDPE, low–density polyethylene; HLB, hydrophobic lipophilic balance; SAX, strong anion exchange; PAHs, polycyclic aromatic hydrocarbons; PCBs, polychlorinated biphenyls; PS-DVB-WAX, polystyrene divinylbenzene weak anion exchange; TXRF, total reflection X-ray fluorescence; IMS, ion mobility spectrometry.

8. Concluding Remarks and Future Directions

According to the principles of green analytical chemistry (GAC), the environmental impact of analytical methodologies should be minimized by reducing the amount of solvents used for sample pre-treatment and the use of toxic reagents, as well as developing alternative methodologies not requiring solvents and reagents [70]. As broadly discussed in this review article, TF-SPME, as an alternative geometry of SPME, is able to satisfy the requirements for greener sampling strategies yet providing ease of use, high-throughput workflows, extraction capability for trace analysis, robustness for onsite sampling, suitability for in vivo analysis and easy coupling to various separation platforms and direct MS analysis. When an analytical eco-scale for assessing the greenness of analytical procedures is performed [2,9] on conventional LLE methods, currently in use by regulatory agencies, versus newly developed TF-SPME-based approaches, the minimized use of organic solvents and production of laboratory waste allows TF-SPME to collect at least 50% less penalty points (based on parameters of an analytical process that are not in agreement with the ideal green analysis) compared to LLE. The versatility of the TF-SPME geometry both in the self-supported or supported devices enable this technique to fit various analytical needs for environmental, food and bioanalysis. As an example, TF-SPME can be used as wearable devices for passive sampling of skin emissions or as extractive probes for remote sampling by use of drones, when sampling sites are not easily reachable or their contamination levels could pose hazards to the analyst.

Due to the recent commercialization of carbon mesh supported TF-SPME devices, we envision that their use in academic and industrial premises will increase quickly in the upcoming years. Further work can be envisioned to test the ruggedness of these devices in complex fluids and evaluate their capabilities of performing multiple extraction/desorption cycles in these matrices without significant loss of extraction efficiency. Moreover, further development of automation strategies for TF-SPME could aim for the complete automation of extraction/desorption cycles when needed, achieving the same throughput capabilities of fiber SPME.

Author Contributions: Conceptualization, E.G.; writing—original draft preparation, R.V.E. and R.T.; writing—review and editing, E.G. and R.V.E.; visualization, E.G., R.V.E. and R.T.; supervision, E.G.; funding acquisition, E.G.

Funding: This research received no external funding.

Acknowledgments: E.G., R.V.E. and R.T. thank The University of Toledo for funding. E.G., R.V.E. and R.T. thank John Stuff from GERSTEL Inc. for providing some of the graphics for Figures 1 and 2.

Conflicts of Interest: The authors declare no conflicts of interest.

References

1. Tobiszewski, M.; Mechlińska, A.; Namieśnik, J. Green analytical chemistry—Theory and practice. *Chem. Soc. Rev.* **2010**, *39*, 2869–2878. [CrossRef] [PubMed]
2. Gałuszka, A.; Migaszewski, Z.M.; Konieczka, P.; Namieśnik, J. Analytical Eco-Scale for assessing the greenness of analytical procedures. *TrAC—Trends Anal. Chem.* **2012**, *37*, 61–72. [CrossRef]
3. Gałuszka, A.; Migaszewski, Z.; Namieśnik, J. The 12 principles of green analytical chemistry and the SIGNIFICANCE mnemonic of green analytical practices. *TrAC—Trends Anal. Chem.* **2013**, *50*, 78–84. [CrossRef]
4. Spietelun, A.; Marcinkowski, Ł.; De La Guardia, M.; Namieśnik, J. Green aspects, developments and perspectives of liquid phase microextraction techniques. *Talanta* **2014**, *119*, 34–45. [CrossRef] [PubMed]
5. Płotka-Wasylka, J. A new tool for the evaluation of the analytical procedure: Green Analytical Procedure Index. *Talanta* **2018**, *181*, 204–209. [CrossRef] [PubMed]
6. Pawliszyn, J. Sample preparation: Quo vadis? *Anal. Chem.* **2003**, *75*, 2543–2558. [CrossRef] [PubMed]
7. Mirnaghi, F.S.; Goryński, K.; Rodriguez-Lafuente, A.; Boyaci, E.; Bojko, B.; Pawliszyn, J. Microextraction versus exhaustive extraction approaches for simultaneous analysis of compounds in wide range of polarity. *J. Chromatogr. A* **2013**, *1316*, 37–43. [CrossRef] [PubMed]

8. Emmons, R.; Devasurendra, A.; Godage, N.; Gionfriddo, E. Exploring the Efficiency of Various Extraction Approaches for Determination of Crude (4-methylcyclohexyl)methanol (MCHM) Constituents in Environmental Samples. *LC-GC N. Am.* **2019**, *37*, 27–34.
9. Piri-Moghadam, H.; Gionfriddo, E.; Rodriguez-Lafuente, A.; Grandy, J.J.; Lord, H.L.; Obal, T.; Pawliszyn, J. Inter-laboratory validation of a thin film microextraction technique for determination of pesticides in surface water samples. *Anal. Chim. Acta* **2017**, *964*, 74–84. [CrossRef]
10. Ouyang, G.; Pawliszyn, J. A critical review in calibration methods for solid-phase microextraction. *Anal. Chim. Acta* **2008**, *627*, 184–197. [CrossRef]
11. Reyes-Garcés, N.; Gionfriddo, E.; Gómez-Ríos, G.A.; Alam, M.N.; Boyaci, E.; Bojko, B.; Singh, V.; Grandy, J.; Pawliszyn, J. Advances in Solid Phase Microextraction and Perspective on Future Directions. *Anal. Chem.* **2018**, *90*, 302–360. [CrossRef]
12. Bruheim, I.; Liu, X.; Pawliszyn, J. Thin-film microextraction. *Anal. Chem.* **2003**, *75*, 1002–1010. [CrossRef]
13. Kremser, A.; Jochmann, M.A.; Schmidt, T.C. PAL SPME Arrow—Evaluation of a novel solid-phase microextraction device for freely dissolved PAHs in water. *Anal. Bioanal. Chem.* **2016**, *408*, 943–952. [CrossRef]
14. Yuan, Y.; Lin, X.; Li, T.; Pang, T.; Dong, Y.; Zhuo, R.; Wang, Q.; Cao, Y.; Gan, N. A solid phase microextraction Arrow with zirconium metal–organic framework/molybdenum disulfide coating coupled with gas chromatography–mass spectrometer for the determination of polycyclic aromatic hydrocarbons in fish samples. *J. Chromatogr. A* **2019**, *1592*, 9–18. [CrossRef]
15. Baltussen, E.; Sandra, P.; David, F.; Cramers, C. Stir Bar Sorptive Extraction (SBSE), a Novel Extraction Technique for Aqueous Samples: Theory and Principles. *J. Microcolumn Sep.* **1999**, 737–747. [CrossRef]
16. Qin, Z.; Bragg, L.; Ouyang, G.; Pawliszyn, J. Comparison of thin-film microextraction and stir bar sorptive extraction for the analysis of polycyclic aromatic hydrocarbons in aqueous samples with controlled agitation conditions. *J. Chromatogr. A* **2008**, *1196–1197*, 89–95. [CrossRef]
17. David, F.; Ochiai, N.; Sandra, P. Two decades of stir bar sorptive extraction: A retrospective and future outlook. *TrAC—Trends Anal. Chem.* **2019**, *112*, 102–111. [CrossRef]
18. Yamaguchi, M.S.; McCartney, M.M.; Linderholm, A.L.; Ebeler, S.E.; Schivo, M.; Davis, C.E. Headspace sorptive extraction-gas chromatography–mass spectrometry method to measure volatile emissions from human airway cell cultures. *J. Chromatogr. B Anal. Technol. Biomed. Life Sci.* **2018**, *1090*, 36–42. [CrossRef]
19. Grandy, J.J.; Boyaci, E.; Pawliszyn, J. Development of a Carbon Mesh Supported Thin Film Microextraction Membrane As a Means to Lower the Detection Limits of Benchtop and Portable GC/MS Instrumentation. *Anal. Chem.* **2016**, *88*, 1760–1767. [CrossRef]
20. Mukhtar, N.H.; See, H.H. Carbonaceous nanomaterials immobilised mixed matrix membrane microextraction for the determination of polycyclic aromatic hydrocarbons in sewage pond water samples. *Anal. Chim. Acta* **2016**, *931*, 57–63. [CrossRef]
21. Boyaci, E.; Goryński, K.; Viteri, C.R.; Pawliszyn, J. A study of thin film solid phase microextraction methods for analysis of fluorinated benzoic acids in seawater. *J. Chromatogr. A* **2016**, *1436*, 51–58. [CrossRef]
22. Gionfriddo, E.; Boyacl, E.; Pawliszyn, J. New Generation of Solid-Phase Microextraction Coatings for Complementary Separation Approaches: A Step toward Comprehensive Metabolomics and Multiresidue Analyses in Complex Matrices. *Anal. Chem.* **2017**, *89*, 4046–4054. [CrossRef]
23. Ghani, M.; Ghoreishi, S.M.; Salehinia, S.; Mousavi, N.; Ansarinejad, H. Electrochemically decorated network-like cobalt oxide nanosheets on nickel oxide nanoworms substrate as a sorbent for the thin film microextraction of diclofenac. *Microchem. J.* **2019**, *146*, 149–156. [CrossRef]
24. Olcer, Y.A.; Tascon, M.; Eroglu, A.E.; Boyaci, E. Thin film microextraction: Towards faster and more sensitive microextraction. *TrAC Trends Anal. Chem.* **2019**. [CrossRef]
25. Walles, M.; Gu, Y.; Dartiguenave, C.; Musteata, F.M.; Waldron, K.; Lubda, D.; Pawliszyn, J. Approaches for coupling solid-phase microextraction to nanospray. *J. Chromatogr. A* **2005**, *1067*, 197–205. [CrossRef]
26. Gómez-Ríos, G.A.; Reyes-Garcés, N.; Bojko, B.; Pawliszyn, J. Biocompatible Solid-Phase Microextraction Nanoelectrospray Ionization: An Unexploited Tool in Bioanalysis. *Anal. Chem.* **2016**, *88*, 1259–1265. [CrossRef]
27. Romero, V.; Costas-Mora, I.; Lavilla, I.; Bendicho, C. Headspace thin-film microextraction onto graphene membranes for specific detection of methyl(cyclopentadienyl)-tricarbonyl manganese in water samples by total reflection X-ray fluorescence. *Spectrochim. Acta—Part B At. Spectrosc.* **2016**, *126*, 65–70. [CrossRef]

28. De la Calle, I.; Ruibal, T.; Lavilla, I.; Bendicho, C. Direct immersion thin-film microextraction method based on the sorption of pyrrolidine dithiocarbamate metal chelates onto graphene membranes followed by total reflection X-ray fluorescence analysis. *Spectrochim. Acta—Part B At. Spectrosc.* **2019**, *152*, 14–24. [CrossRef]
29. Riazi Kermani, F.; Pawliszyn, J. Sorbent coated glass wool fabric as a thin film microextraction device. *Anal. Chem.* **2012**, *84*, 8990–8995. [CrossRef]
30. Grandy, J.J.; Singh, V.; Lashgari, M.; Gauthier, M.; Pawliszyn, J. Development of a Hydrophilic Lipophilic Balanced Thin Film Solid Phase Microextraction Device for Balanced Determination of Volatile Organic Compounds. *Anal. Chem.* **2018**, *90*, 14072–14080. [CrossRef]
31. Wilcockson, J.B.; Gobas, F.A.P.C. Thin-film solid-phase extraction to measure fugacities of organic chemicals with low volatility in biological samples. *Environ. Sci. Technol.* **2001**, *35*, 1425–1431. [CrossRef]
32. Rodil, R.; von Sonntag, J.; Montero, L.; Popp, P.; Buchmeiser, M.R. Glass-fiber reinforced poly(acrylate)-based sorptive materials for the enrichment of organic micropollutants from aqueous samples. *J. Chromatogr. A* **2007**, *1138*, 1–9. [CrossRef]
33. Boggia, L.; Sgorbini, B.; Bertea, C.M.; Cagliero, C.; Bicchi, C.; Maffei, M.E.; Rubiolo, P. Direct Contact—Sorptive Tape Extraction coupled with Gas Chromatography—Mass Spectrometry to reveal volatile topographical dynamics of lima bean (*Phaseolus lunatus* L.) upon herbivory by Spodoptera littoralis Boisd. *BMC Plant Biol.* **2015**, *15*. [CrossRef]
34. Vernarelli, L.; Whitecavage, J.A.; Stuff, J. Analysis of Food Samples using Thin Film Solid Phase Microextraction (TF-SPME) and Thermal Desorption GC/MS. **2019**, 1–7. Available online: http://www.gerstel.com/pdf/AppNote-202.pdf (accessed on 30 July 2019).
35. Jiang, R.; Cudjoe, E.; Bojko, B.; Abaffy, T.; Pawliszyn, J. A non-invasive method for in vivo skin volatile compounds sampling. *Anal. Chim. Acta* **2013**, *804*, 111–119. [CrossRef]
36. Jiang, R.; Pawliszyn, J. Preparation of a particle-loaded membrane for trace gas sampling. *Anal. Chem.* **2014**, *86*, 403–410. [CrossRef]
37. Gómez-Ríos, G.A.; Gionfriddo, E.; Poole, J.; Pawliszyn, J. Ultrafast Screening and Quantitation of Pesticides in Food and Environmental Matrices by Solid-Phase Microextraction-Transmission Mode (SPME-TM) and Direct Analysis in Real Time (DART). *Anal. Chem.* **2017**, *89*, 7240–7248. [CrossRef]
38. Jastrzembski, J.A.; Sacks, G.L. Solid Phase Mesh Enhanced Sorption from Headspace (SPMESH) Coupled to DART-MS for Rapid Quantification of Trace-Level Volatiles. *Anal. Chem.* **2016**, *88*, 8617–8623. [CrossRef]
39. Mirabelli, M.F.; Gionfriddo, E.; Pawliszyn, J.; Zenobi, R. Fast screening of illicit drugs in beverages and biological fluids by direct coupling of thin film microextraction to dielectric barrier discharge ionization-mass spectrometry. *Analyst* **2019**, *144*, 2788–2796. [CrossRef]
40. Cho, D.S.; Gibson, S.C.; Bhandari, D.; McNally, M.E.; Hoffman, R.M.; Cook, K.D.; Song, L. Evaluation of direct analysis in real time mass spectrometry for onsite monitoring of batch slurry reactions. *Rapid Commun. Mass Spectrom.* **2011**, *25*, 3575–3580. [CrossRef]
41. Piri-Moghadam, H.; Alam, M.N.; Pawliszyn, J. Review of geometries and coating materials in solid phase microextraction: Opportunities, limitations, and future perspectives. *Anal. Chim. Acta* **2017**, *984*, 42–65. [CrossRef]
42. Jiang, R.; Pawliszyn, J. Thin-film microextraction offers another geometry for solid-phase microextraction. *TrAC—Trends Anal. Chem.* **2012**, *39*, 245–253. [CrossRef]
43. Semenov, S.N.; Koziel, J.A.; Pawliszyn, J. Kinetics of solid-phase extraction and solid-phase microextraction in thin adsorbent layer with saturation sorption isotherm. *J. Chromatogr. A* **2000**, *873*, 39–51. [CrossRef]
44. Harner, T.; Farrar, N.J.; Shoeib, M.; Jones, K.C.; Gobas, F.A.P.C. Characterization of polymer-coated glass as a passive air sampler for persistent organic pollutants. *Environ. Sci. Technol.* **2003**, *37*, 2486–2493. [CrossRef]
45. Kennedy, K.E.; Hawker, D.W.; Müller, J.F.; Bartkow, M.E.; Truss, R.W. A field comparison of ethylene vinyl acetate and low-density polyethylene thin films for equilibrium phase passive air sampling of polycyclic aromatic hydrocarbons. *Atmos. Environ.* **2007**, *41*, 5778–5787. [CrossRef]
46. Reichenberg, F.; Smedes, F.; Jönsson, J.A.; Mayer, P. Determining the chemical activity of hydrophobic organic compounds in soil using polymer coated vials. *Chem. Cent. J.* **2008**, *2*, 1–10. [CrossRef]
47. Victoria Otton, S.; deBruyn, A.M.H.; Meloche, L.M.; Gobas, F.A.P.C.; Ikonomou, M.G. Assessing Exposure of Sediment Biota To Organic Contaminants By Thin-Film Solid Phase Extraction. *Environ. Toxicol. Chem.* **2008**, *28*, 247.

48. St. George, T.; Vlahos, P.; Harner, T.; Helm, P.; Wilford, B. A rapidly equilibrating, thin film, passive water sampler for organic contaminants; Characterization and field testing. *Environ. Pollut.* **2011**, *159*, 481–486. [CrossRef]
49. Bragg, L.; Qin, Z.; Alaee, M.; Pawliszyn, J. Field sampling with a polydimethylsiloxane thin-film. *J. Chromatogr. Sci.* **2006**, *44*, 317–323. [CrossRef]
50. Godage, N.H.; Gionfriddo, E. A critical outlook on recent developments and applications of matrix compatible coatings for solid phase microextraction. *TrAC—Trends Anal. Chem.* **2019**, *111*, 220–228. [CrossRef]
51. Guerra, P.; Lai, H.; Almirall, J.R. Analysis of the volatile chemical markers of explosives using novel solid phase microextraction coupled to ion mobility spectrometry. *J. Sep. Sci.* **2008**, *31*, 2891–2898. [CrossRef]
52. Bessonneau, V.; Boyaci, E.; Maciazek-Jurczyk, M.; Pawliszyn, J. In vivo solid phase microextraction sampling of human saliva for non-invasive and on-site monitoring. *Anal. Chim. Acta* **2015**, *856*, 35–45. [CrossRef]
53. Li, S.; Wu, D.; Yan, X.; Guan, Y. Acetone-activated polyimide electrospun nanofiber membrane for thin-film microextraction and thermal desorption-gas chromatography-mass spectrometric analysis of phenols in environmental water. *J. Chromatogr. A* **2015**, *1411*, 1–8. [CrossRef]
54. Ahmadi, F.; Sparham, C.; Boyaci, E.; Pawliszyn, J. Time Weighted Average Concentration Monitoring Based on Thin Film Solid Phase Microextraction. *Environ. Sci. Technol.* **2017**, *51*, 3929–3937. [CrossRef]
55. Piri-Moghadam, H.; Gionfriddo, E.; Grandy, J.J.; Alam, M.N.; Pawliszyn, J. Development and validation of eco-friendly strategies based on thin film microextraction for water analysis. *J. Chromatogr. A* **2018**, *1579*, 20–30. [CrossRef]
56. Mirnaghi, F.S.; Pawliszyn, J. Development of coatings for automated 96-blade solid phase microextraction-liquid chromatography-tandem mass spectrometry system, capable of extracting a wide polarity range of analytes from biological fluids. *J. Chromatogr. A* **2012**, *1261*, 91–98. [CrossRef]
57. Kim, K.H.; Jahan, S.A.; Kabir, E.; Brown, R.J.C. A review of airborne polycyclic aromatic hydrocarbons (PAHs) and their human health effects. *Environ. Int.* **2013**, *60*, 71–80. [CrossRef]
58. Qin, Z.; Mok, S.; Ouyang, G.; Dixon, D.G.; Pawliszyn, J. Partitioning and accumulation rates of polycyclic aromatic hydrocarbons into polydimethylsiloxane thin films and black worms from aqueous samples. *Anal. Chim. Acta* **2010**, *667*, 71–76. [CrossRef]
59. Qin, Z.; Bragg, L.; Ouyang, G.; Niri, V.H.; Pawliszyn, J. Solid-phase microextraction under controlled agitation conditions for rapid on-site sampling of organic pollutants in water. *J. Chromatogr. A* **2009**, *1216*, 6979–6985. [CrossRef]
60. Xu, C.; Wang, J.; Richards, J.; Xu, T.; Liu, W.; Gan, J. Development of film-based passive samplers for in situ monitoring of trace levels of pyrethroids in sediment. *Environ. Pollut.* **2018**, *242*, 1684–1692. [CrossRef]
61. Sisalli, S.; Adao, A.; Lebel, M.; Le Fur, I.; Sandra, P. Sorptive Tape Extraction—A Novel Sampling Method for the in vivo Study of Skin. *LC-GC Eur.* **2006**, *19*, 33–39.
62. Bicchi, C.; Cordero, C.; Liberto, E.; Rubiolo, P.; Sgorbini, B.; Sandra, P. Sorptive tape extraction in the analysis of the volatile fraction emitted from biological solid matrices. *J. Chromatogr. A* **2007**, *1148*, 137–144. [CrossRef]
63. Mohammadi, V.; Jafari, M.T.; Saraji, M. Flexible/self-supported zeolitic imidazolate framework-67 film as an adsorbent for thin-film microextraction. *Microchem. J.* **2019**, *146*, 98–105. [CrossRef]
64. Wei, F.; Zhang, F.F.; Liao, H.; Dong, X.Y.; Li, Y.H.; Chen, H. Preparation of novel polydimethylsiloxane solid-phase microextraction film and its application in liquid sample pretreatment. *J. Sep. Sci.* **2011**, *34*, 331–339. [CrossRef]
65. Jahnke, A.; Mayer, P.; Broman, D.; McLachlan, M.S. Possibilities and limitations of equilibrium sampling using polydimethylsiloxane in fish tissue. *Chemosphere* **2009**, *77*, 764–770. [CrossRef]
66. Stuff, J.R.; Whitecavage, J.A.; Grandy, J.J.; Pawliszyn, J. Analysis of Beverage Samples using Thin Film Solid Phase Microextraction (TF-SPME) and Thermal Desorption GC/MS. **2018**, 1–10. Available online: http://www.gerstel.com/pdf/AppNote-200.pdf (accessed on 30 July 2019).
67. Shigeyama, H.; Wang, T.; Ichinose, M.; Ansai, T.; Lee, S.W. Identification of volatile metabolites in human saliva from patients with oral squamous cell carcinoma via zeolite-based thin-film microextraction coupled with GC–MS. *J. Chromatogr. B Anal. Technol. Biomed. Life Sci.* **2019**, *1104*, 49–58. [CrossRef]
68. Rodriguez-Lafuente, A.; Piri-Moghadam, H.; Lord, H.L.; Obal, T.; Pawliszyn, J. Inter-laboratory validation of automated SPME-GC/MS for determination of pesticides in surface and ground water samples: Sensitive and green alternative to liquid–liquid extraction. *Water Qual. Res. J. Canada* **2016**, *51*, 331–343. [CrossRef]

69. Jastrzembski, J.A.; Bee, M.Y.; Sacks, G.L. Trace-Level Volatile Quantitation by Direct Analysis in Real Time Mass Spectrometry following Headspace Extraction: Optimization and Validation in Grapes. *J. Agric. Food Chem.* **2017**, *65*, 9353–9359. [CrossRef]
70. Armenta, S.; Garrigues, S.; de la Guardia, M. Green Analytical Chemistry. *TrAC Trends Anal. Chem.* **2008**, *27*, 497–511. [CrossRef]

© 2019 by the authors. Licensee MDPI, Basel, Switzerland. This article is an open access article distributed under the terms and conditions of the Creative Commons Attribution (CC BY) license (http://creativecommons.org/licenses/by/4.0/).

Review

Advancements in Non-Invasive Biological Surface Sampling and Emerging Applications

Atakan Arda Nalbant and Ezel Boyacı *

Department of Chemistry, Middle East Technical University, 06800 Ankara, Turkey; nalbant.atakan@metu.edu.tr
* Correspondence: ezel@metu.edu.tr; Tel.: +90-312-210-3208

Received: 8 July 2019; Accepted: 11 October 2019; Published: 4 November 2019

Abstract: Biological surfaces such as skin and ocular surface provide a plethora of information about the underlying biological activity of living organisms. However, they pose unique problems arising from their innate complexity, constant exposure of the surface to the surrounding elements, and the general requirement of any sampling method to be as minimally invasive as possible. Therefore, it is challenging but also rewarding to develop novel analytical tools that are suitable for in vivo and in situ sampling from biological surfaces. In this context, wearable extraction devices including passive samplers, extractive patches, and different microextraction technologies come forward as versatile, low-invasive, fast, and reliable sampling and sample preparation tools that are applicable for in vivo and in situ sampling. This review aims to address recent developments in non-invasive in vivo and in situ sampling methods from biological surfaces that introduce new ways and improve upon existing ones. Directions for the development of future technology and potential areas of applications such as clinical, bioanalytical, and doping analyses will also be discussed. These advancements include various types of passive samplers, hydrogels, and polydimethylsiloxane (PDMS) patches/microarrays, and other wearable extraction devices used mainly in skin sampling, among other novel techniques developed for ocular surface and oral tissue/fluid sampling.

Keywords: non-invasive sampling; wearable devices; extractive patches; skin sampling; passive sampling; green sampling technologies

1. Introduction

Biological surfaces provide easy-to-access information about the underlying metabolism and workings of the human body. In broad terms, there are two approaches that can be followed to gather information from biological surfaces. The first approach is the development of wearable sensors, where a direct sensing mechanism is used as a wearable device. Wearable sensors have complex designs that can track a relatively small number of compounds from various biological surfaces, or monitor the specific features of the medical condition of the wearer in different ways [1–5]. While these wearable sensors are quite intricate in their design and mechanisms, they usually provide very basic information in terms of their ability to detect only a limited number of biologically relevant information. Another approach is the development of wearable extractive sampling devices examined in this review, which can be described as being on the opposite side of the coin in the sense that they are quite simple in their design but with the help of advanced detection methods, i.e., high sensitivity mass spectrometry (MS) coupled with the myriad of ionization techniques available, provide a vast amount of information.

2. Skin Sampling

Sampling and profiling skin metabolites has been a topic of interest over the last several years [6–9]. The majority of the biological information is gathered from sweat, which is secreted from eccrine and apocrine glands, and sebum which is secreted by sebaceous glands. The sebum contains many distinctive compounds that result from the breakdown of proteins and enzymes in the outer layer of the skin, known as the epidermis [10]. The underlying biological pathways that produce these compounds in sweat, sebum, and epidermis are quite varied and the compounds of interest have a wide range of polarity, size, and structure as a result. These diverging compounds are of interest to clinical research fields such as forensics and toxicology [11–13], not to mention the obvious applications in medicinal diagnosis [14–16], doping tests [17,18], and environmental/occupational exposomics [19–21].

Some of the major difficulties in sampling skin secretions are the challenges of collecting large volumes of sweat and the low concentrations [22,23] of metabolites and other compounds of interest. Therefore, initial studies in the mid-20th century required patients to sweat copious amounts into plastic skin bags, a rather exhaustive method of sample collection [24]. A more commonly employed method nowadays is to use small vessels filled with organic solvents [25] to extract metabolites from a relatively small area, but this method leaves the skin irritated and many of the organic solvents are not biocompatible nor environmentally friendly. Less invasive methods such as Macroduct® [26] and cotton pads [27] for sampling from skin surfaces as well as micro dialysis [28] and subcutaneous solid phase microextraction (SPME) [29] for sampling from inner skin layers have been developed but have some limitations. For instance, some of these methods can only be applied for one type of sample collection, e.g., only for sweat collection or only for sebum sampling. Some methods require additional sample preparation steps to be performed which are time consuming and may introduce additional errors into the analysis. There are substantial efforts to address the limits of the methods listed above and to pave the way for non-invasive, biocompatible in vivo and in situ skin sampling techniques and to integrate them with state-of-art analysis instruments.

2.1. Direct Contact Type Sampling

In this type of sampling approach, the sampling device, or patch, is placed directly on the skin without leaving any gap between the sampler and the surface. The advantage to this approach is that it provides good extraction capability for both volatile and non-volatile analytes. The main drawback is that many unwanted compounds can adhere to the sampling device by smearing through the skin surface resulting in a considerable amount of matrix contamination. These contaminations might affect sampler performance, resulting in absolute matrix effects, including ionization suppression/enhancement in electrospray ionization (ESI), and may cause frequent instrumental maintenance.

Patch type sampling employs sorbent materials embedded on scaffold or directly placed onto the sampling area, removed after a satisfactory sampling duration, and can either be analyzed by various analytical instrumentations allowing direct analysis (i.e., direct mass spectrometry) or be subjected to a desorption step before analysis by suitable instrumentation. Materials used in patch type sampling should be elastic enough to conform to the shape of the sampling area but resilient enough to sustain the expected wear and tear of normal daily activity if the patch is required to be left on the skin of the subject for an extended period of time. The biocompatibility of the material is another important criterion to prevent irritation of the skin for the same reasons as above.

2.1.1. Polydimethylsiloxane (PDMS)

PDMS has been commonly used as an extractive phase for thin film microextraction (TFME) [30,31] as well as solid phase microextraction (SPME) [32,33] for various different analytes, matrices, and applications. Controlled polymerization of PDMS allows for adjustable elasticity and molding with relatively inert and biocompatible features which are useful for extended sampling periods. Thermal and chemical stability of PDMS, particularly against commonly used cleaning and surface activation solvents, further reinforces its applicability as an extractive phase for patch type sampling purposes.

PDMS can be used as an extractive phase in two different modes: first is the headspace sampling method where the material is not in direct contact with the sampling area and volatile compounds that emanate from skin are captured. The second mode is the direct contact mode, where the PDMS patch is directly placed on top of the sampling area. In this mode, both volatile and semi-volatile compounds can be extracted. However, the skin is usually contaminated by dust particles and many other chemicals, such as cosmetic products; therefore, the patch might convey these unwanted matrix components to the instrument. These matrix components could potentially contaminate the analytical instrument, which would require clean-up before and after each use. In an effort to prevent such compounds from adhering to the PDMS surface, Jiang et al. developed a sampling method which allowed the patch to be used either as a headspace sampling device or for direct contact type sampling from skin [34]. Jiang and colleagues employed two layers of stainless-steel mesh to separate the PDMS film from coming into direct contact with the skin, as shown in Figure 1a. In order to evaluate the reproducibility of the so-called "membrane sandwich" method, an in-vial experiment was conducted in parallel, and intra-membrane irreproducibility was found to be 9.8% ($n = 6$) and inter-membrane reproducibility was determined to be 8.2%. Inter-membrane reproducibility is particularly important for potential clinical applications, where membranes would be used only once. When detected quantities of volatile organic compounds (VOCs) were investigated, such as octanal, nonanal, and decanal, similar intensities were observed for both headspace and direct contact methods. However, for semi- and low-volatile compounds, such as 1-tetradecanol and 1-octadecanaol, higher peak intensities were observed for direct contact sampling. Some of the heavier compounds such as squalene were not detected at all by headspace sampling. These results indicate that, although the chromatograms for the headspace sampling were less contaminated by matrix related background peaks, direct sampling from the surface could be considered as a compromise between detection sensitivity and instrumental contamination. The same study also reported monitoring of food metabolites and alcohol intake directly from skin could be a potential application for such patches.

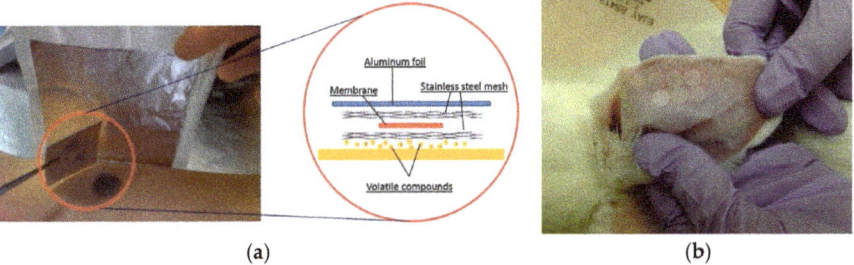

Figure 1. (a) Schematic diagram of the "membrane sandwich" headspace sampling for skin volatiles sampling. (b) Polydimethylsiloxane (PDMS) patches placed over the ventral ear surface for direct sampling. These were then covered with PTFE (polytetrafluoroethylene) and Tegaderm® (not shown). Reproduced from [34,35], with permission from Elsevier and IOP Publishing, 2019.

Direct sampling mode for PDMS were employed in many other studies showing the potential of PDMS patches for clinical applications, all highlighting the simplicity of using the patches in direct sampling mode. For instance, a rabbit model study was conducted by Schivo et al. that employed PDMS patches (Figure 1b) placed inside the ears of subjects to investigate histological evidence of early-stage ulcer formation by metabolomic screening [35]. The study results identified that the patches made of PDMS can be used for monitoring the differences between healthy and diseased skin metabolites. Approximately 150 unique skin related compounds were characterized, and 12 biologically relevant compounds were detected abundantly in the ulcer group. Most of the skin metabolites observed in the study were waxy compounds, although a large number of VOCs were detected as well. Results of the study underline that the abundance of non-volatile compounds makes direct contact sampling with PDMS a good fit and emphasize that the distribution of volatile and non-volatile compounds can be compared to human skin profile. In another study, Martin et al. employed PDMS patches placed onto the axilla area of female participants for VOC sampling. The study aimed to investigate skin metabolites responsible for body odor with an emphasis on the effect of a single nucleotide polymorphism 538G → A in the ABCC11 gene on the concentrations of apocrine derived axillary odor molecules, especially 3-methy-2-hexenoic acid [36]. The investigation was concluded with four volatile fatty acid (VFA) target compounds being identified with limits of detection approximately around 100 s pg cm^{-2}, with authors noting that a typical skin patch has an area of 0.5 cm^2. It was also noted that the ABCC11 gene, other than controlling body odor, is studied for its role in breast cancer and drug resistance in cancer cells [37]. Based on the study results, combining the sampling patches with a thermal desorption unit coupled to a secondary electrospray ionization and mass spectrometry (TD–SESI–MS) setup is a promising approach for unattended automation of analysis and could be employed in non-invasive fast screening of metabolites for personalized medicine.

In a pilot study [38], Martin et al. investigated changes in VOC profiles for stress-related biomarkers using PDMS devices placed on the foreheads of volunteer subjects for sample collection. The effect of stress on VOCs exhaled in breath was investigated previously [39], identifying six potential marker compounds. Similarly, the results obtained by Martin et al. suggested that some of the compounds identified in breath exhale in the previous study [39] also show altered levels in skin, namely terpenes (3-carene is suggested as a likely candidate), benzoic acid, and N-decanoic acid seem to be involved in the metabolic response to stress, which the authors believe a potential increase in oxidative metabolic pathways could be the cause of.

Moreover, a clinical study by Stevens et al. have explored the spatial distribution of bacterial populations and related VOCs using various sampling techniques, with PDMS patch type sampling being one of them [40]. In this study, PDMS patches were used for VOC sampling. Different areas of the foot, including toe clefts, and several dorsal and plantar surfaces, were sampled for microbiological and VOC analyses. Then, a spatial map of the foot was constructed for microbiological and VOC concentrations which showed significant differences between different regions. Specific volatiles such as acetic, butyric, valeric, and isovaleric acids were found in considerably high concentrations in the sole region of the foot compared to the dorsal region, and, in the case of isovaleric acid, it was never found in dorsal regions while it was readily detected on the plantar surface. Their findings indicate that key volatiles responsible for malodors are abundantly found on the sole of the foot rather than in other regions sampled in the study. This study clearly indicates that sampling from different regions of the body might give different levels of particular metabolite which could be simply related to differences in body temperature, excretion rates, or skin thickness or might be indicative of biologically relevant information. Sampling conditions of such results has to be carefully designed in order to avoid any misinterpretation.

As it has been remarked above, PDMS based patches are versatile tools and easy to adopt for numerous investigations. In addition, PDMS films can be purchased or prepared in the laboratory using well defined protocols and can be easily shaped in any size. In addition, the thermal stability of PDMS enables the thermal desorption of extracted analytes from patches using large thermal desorption units connected to gas chromatography mass spectrometry (GC–MS) [35,36,38]. However, it is worth mentioning that before use PDMS patches require extensive cleaning and activation steps that might involve a considerable amount of preparation before sampling (4–24 h). Therefore, it is crucial to prevent contamination at each step of the workflow including cleaning/pre-treatment and sampling. Moreover, during the sampling, PDMS patches should be protected from environmental wear and tear and possible contaminations by a placing a durable cover over them, as shown by Schivo et al. in their pilot study [35].

2.1.2. Agarose Hydrogel

Agarose hydrogel is another commonly used material for direct contact type sampling. Agarose hydrogel lacks the robustness of PDMS patches, but it is more biocompatible than PDMS, as it is a naturally occurring homopolymer derived from red algae [41]. Agarose has completely different extractive properties suitable for hydrophilic skin metabolites owing to its complex polar structure and adjustable water content. These properties lead to agarose patches being mostly used for extraction of relatively polar metabolites as can be seen from the examples below.

For instance, Dutkiewicz et al. have developed agarose micropatches to detect several low-molecular-weight skin metabolites as a proof-of-concept study [42]. In this study, the lack of robustness of agarose hydrogels was overcome by embedding them inside cavities on a PTFE chip. Various prevalent skin metabolites were detected, and some were identified, as given in Table 1. The authors stated that high signal-to-noise ratios were achieved for clinically relevant analytes. In another study, Dutkiewicz et al. also employed micropatch-arrayed pads (MAPAs) to profile topically applied drugs. In this study, five rows of five cavities, a total of 25, containing agarose hydrogels were applied to both in vivo human skin and ex vivo porcine skin by self-adhesive 3-D printed PTFE scaffolds [43]. In a later study by the same group, MAPAs were implemented to investigate psoriasis-related skin metabolites and an automated diagnostic method was developed with satisfactory results that imply clinical use of MAPAs is a reasonable possibility [44]. A number of metabolites, namely choline, citrulline, glutamic acid, urocanic acid, lactic acid, and phenylalanine, were discovered to have altered concentrations in diseased skin areas, indicating they are related to psoriasis metabolism. One of the key advantages of MAPAs is the direct coupling with nano desorption electrospray ionization (nanoDESI). In this approach, a solvent mixture can be continuously pumped through a capillary to one of the hydrogel micropatches while another capillary delivers the solvent to the nanoDESI loading dock. This approach eliminates further steps (e.g., desorption, pre-concentration, etc.) and provides a quick and simple online method for direct mass spectrometric analysis.

Table 1. Identification of peaks in the mass spectra of sweat [44].

Peak No	m/z^* (IT)	m/z^* (FT-ICR)	Metabolite Formula	Metabolite Name	Predicted [b] m/z^*	MS/MS	Compared with Standard
1	89.0	89.02438	$C_3H_6O_3$	lactic acid	89.02442	+	+
2	93.0	93.04578	$C_5H_6N_2$	fragment of uronic acid	93.04582	+	+
3	104.0	104.03530	$C_3H_7NO_3$	serine	104.03532	+	+
4	118.0	118.05089	$C_4H_9NO_3$	threonine	118.05097	+	+
5	128.0	128.03530	$C_5H_7NO_3$	pyroglutamic acid	128.03532	+	+
6	131.0	131.08259	$C_5H_{12}N_2O_2$	ornithine [a]	131.08260	+	-
7	137.0	137.03561	$C_6H_6N_2O_2$	urocanic acid	137.03565	+	+
8	154.0	154.06218	$C_6H_9N_3O_2$	histidine [a]	154.06220	+	-
9	179.0	179.05731	$C_7H_8N_4O_2$	paraxanthine [a]	179.05745	-	-

[a] Putative formula and name of the metabolite. [b] Values calculated for [M − H]⁻ ions (mass of an electron is included). *m/z: mass-to-charge ratio. IT: ion trap; FT–ICR: Fourier transform ion cyclotron resonance. Reproduced from [44], with permission from ACS Publications, 2019.

As it has been pointed out several times by now in the studies shown above, agarose hydrogels are limited for extraction of relatively polar analytes. This limitation narrows the biologically relevant information that can be obtained from the studied system. Other biocompatible hydrogels, such as gelatin and polyvinyl pyrrolidone, could potentially offer the same, or even greater, advantage for skin sampling and further functionalization of agarose or other hydrogels could very well be a prospective area of research in the future. Selective sampling can be also be achieved by using copolymers of different hydrogels with different functional groups to target compounds with varying polarities.

2.1.3. Microneedle Arrays

In recent years there has been considerable effort to employ low-invasive microneedle structures (<1-mm long, hollow structures) in drug delivery through skin [45,46]. More recently, this technique was modified to be employed in capturing circulating biomarkers [47–52], combining microneedle structures with biomarker-based immunoassay approach. This was achieved by functionalizing the tips of the microneedles with specific antibodies. Although functionalized microarrays provide high selectivity and could benefit many clinical diagnostic applications, only a limited number of studies showing the potential of the approach have been conducted. In one of the studies, Ng et al.

developed a multiplex microneedle device coupled with a blotting method for fast and selective detection of multiple selected biomarkers from skin surface [53]. Microneedle devices were produced by micromoulding: polylactic acid (PLA) was melted in vacuo into the template micromould which was produced from the Sylgard® 184 (PDMS) elastomer and surface activated using chemical methods. Surface characterization of the PLA microneedle device was performed using a scanning electron microscope (SEM) at various production stages, which can be seen in Figure 2. For immobilization of different antibodies on the multiplex microneedle device, each microneedle array was individually dipped into the desired antibody solution 10 times. Microneedle arrays were then placed for 1 h on mice skin doped with mouse (Interleukin 6 and 1) as well as human (Tumor Necrosis Factor alpha) antigens. After successful antigen capture, two detection methods were employed: UV/Vis spectrophotometry for quantitative detection and scanning the images of blotting patterns left by the microneedles for qualitative visualization. Blotting technique employed by the authors should be further underlined as it requires no specialized equipment and allows rapid and selective visualization of specific antigens, albeit qualitatively. It should be kept in mind that employing the detection of antigens with this method has its disadvantages such as having a limited number of antibodies that can be loaded into the microneedle arrays. Therefore, scanning for a large number of different antigens would require several runs which would increase the invasiveness of the method since different areas have to be sampled.

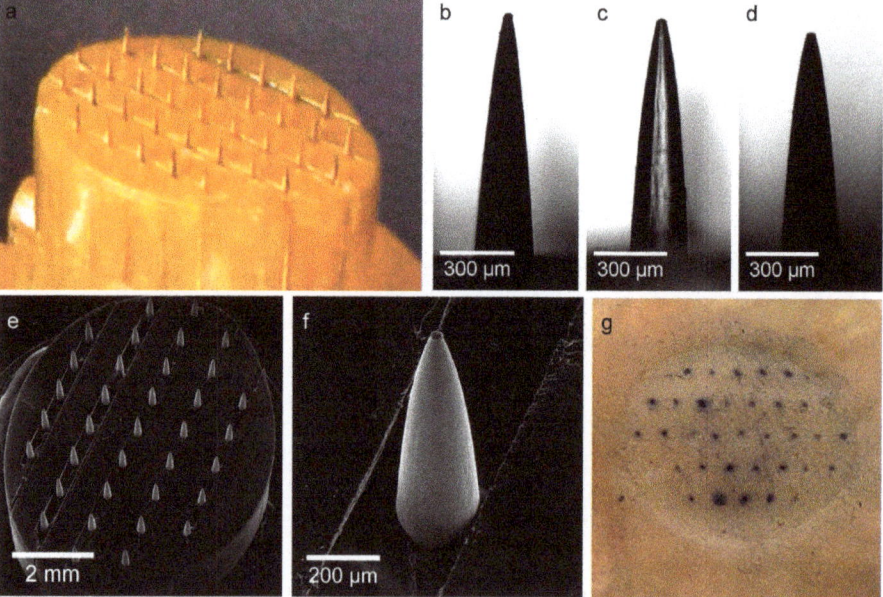

Figure 2. (a) A photograph of the template microneedle array. (b–d) Light micrographs showing: (b) a template microneedle, (c) a polylactic acid (PLA) microneedle prior to surface activation, and (d) a surface-activated PLA microneedle. (e) A surface-activated PLA microneedle array under SEM; (f) a single microneedle on this array is enlarged and shown. (g) Perforation marks on hairless mouse skin, visualized by methylene blue staining. Reproduced from [53], with permission from Springer, 2019.

Microneedle arrays, as indicated by this study, show promise as a low-invasive method that can be employed in diagnosis of skin diseases. The highly selective nature of using antibodies further promotes their diagnostic potential since extensively studied diseases can be targeted by specifically targeting selected antigens. Creative approaches to the geometric structures of the arrays may allow the number of microneedles in the same area to be increased, enabling more antigens to be detected.

2.2. Headspace Sampling

2.2.1. Conventional SPME Fibers

Patch type or other contact-based sampling methods [6,34–36,38,54–56] may not always be preferable for skin VOC analysis as contamination and introduction of non-volatile compounds is always a possibility. Headspace SPME is a commonly used sampling approach preferred for extracting volatiles present in various samples including complex matrices [57–60]. Several studies have scrutinized their applicability for analysis of skin VOCs [61,62] albeit not without its limitations due to their fragility, which is one of the main reasons why in situ headspace sampling is practiced in controlled environments. A simple solution for this problem is employing a housing for the SPME fibers during the sampling. This approach creates a controlled environment and ensures that extracted analytes on the fiber are only associated with the sampling area. Similar to how PDMS and agarose patches can be protected by durable covers or by being placed on PTFE scaffolds, Duffy et al. have shown that headspace SPME fibers can be protected by using wearable housing vials, seen in Figure 3a [63]. In this study, researchers compared skin VOCs before and after the acute barrier disruption of the sampling area by tape stripping to simulate impaired skin. After tape stripping, commercial divinylbenzene/carboxen/polydimethylsiloxane Stableflex fibers were used to sample skin VOCs. A total of 37 compounds were identified that were significantly altered after barrier disruption, mainly consisting of aldehydes (hexanal, nonanal, decanal), acids (nonanoic, decanoic, dodecanoic, tetradecanoic and pentadecanoic acids), and hydrocarbons (squalene). Duffy et al. have also utilized their wearable headspace SPME device in another study to investigate fragrance longevity and scent profiles of participants. In this study, skin VOCs of volunteers were analyzed before and after fragrance-derived compounds were applied to their skin to compare the difference and fragrance permanence [64]. Following the instrumental analysis, 32 fragrance-derived and 19 endogenous compounds were identified. Several endogenous VOCs were found to be suppressed by fragrance application, most apparent ones being 2-decanal, 2-undecenal, 1-dodecanol, pentadecanal, and octadecanoic acid. The most noticeable decrease in endogenous VOCs was observed immediately after fragrance application where several skin gland secretions and their oxidation products were not detected including acids, aldehydes, ketones, and hydrocarbons. Temporal and spatial profiling of fragrance-derived and endogenous compounds can prove unique insights to improve fragrance longevity and can lead to niche personalized fragrance production in the cosmetics industry.

As can be seen from the outlined studies, although such housings are not as practical as wearable patches described in previous section, they provide unique advantage for creating a well-controlled environment during sampling. Moreover, various commercial SPME fibers are readily available for purchase with different extractive properties and homemade fibers can be coated with materials that can be specialized for unique sampling purposes. Another important advantage of the SPME fibers is that they can be directly coupled with ambient ionization techniques such as nanoelectrospray ionization (nanoESI) [65], dielectric barrier discharge ionization (DBDI) [66], and open port probe (OPP) interfaces [67]. Direct coupling to MS instruments eliminates desorption and chromatographic separation steps from the analytical workflow, speeding up the overall process.

2.2.2. Passive Flux Samplers

Passive flux samplers (PFSs) are commonly employed in environmental studies to measure toxic or greenhouse gases emitted from animal slaughterhouses [68–70], greenhouses [71], and building materials [72], which was later adapted to sample VOCs emanating from skin [62]. So-called passive sampling devices (PSDs) consist of a small stainless-steel plate, a trapping filter that contains an extractive phase to extract relatively volatile analytes, a PTFE O-ring creating a headspace between the extractive phase and skin surface, and a back-up plate. Its schematics can be seen in Figure 3b. PFSs can be considered more practical and robust than wearable SPME fibers with integrated housing in terms of ease-of-use. PFSs, however, are more susceptible to the environment and there is always a possibility of environmental background to be considered when using this technique in sampling unless completely isolated from the environment. Due to their ease-of-use feature these devices can be adapted for sampling from various parts of the body and track metabolic changes associated with particular conditions. For example, a recent study by Kimura et al. investigated the causes of the characteristic "elderly body odor" by sampling from different areas of the skin from participants in three different age groups (young-, middle- and old-aged) using a passive sampling device [73]. MonoTrap® DCC18 containing octadecylsilane functionalized monolithic silicate and activated carbon as extractive media providing with a large surface area and high trapping capacity (O.D. 10 mm × 1 mm thick, >150 m^2 g^{-1}) was chosen for the study. Two compounds, 2-nonenal and diacetyl, were chosen as likely candidates to study as to the cause of the elderly body odor phenomenon. The emission fluxes of the compounds with respect to sampling position (left forearm, left thigh, left calf, forehead, nape of the neck and abdomen), the effect of diffusion length on the dermal emission flux, and distribution of the emission fluxes, and changes by age and sex were investigated. Analysis of the collected samples illustrated that emission flux of 2-nonenal increases significantly with age while diacetyl was found to have the highest emission flux for middle aged participants. Moreover, male participants were found to have higher emission fluxes compared to female participants of similar age. Interestingly, out of all sampling areas, nape of the neck was found to be the most reliable sampling spot, since the emission fluxes of both compounds altered significantly in other sampling areas but remained relatively stable in this area. Due to the fact that eccrine and sebaceous glands are believed to be potential sources for both compounds, their abundant presence in the area is reasonable.

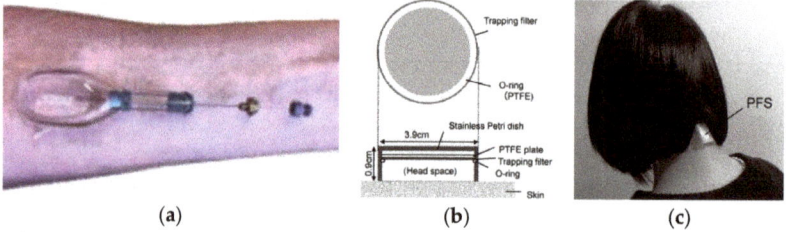

Figure 3. (a) Headspace sampling volatiles on the volar forearm using a wearable housing integrating solid phase microextraction (SPME) fiber, affixed to the skin with surgical tape. (b) Schematic view of the passive flux sampler for human skin gas. (c) Sampling of 2-nonenal and diacetyl at the nape of the neck. The PFS (passive flux sampler) was fixed by a piece of medical tape. Reproduced from [63,73], with permission from Wiley and Elsevier, 2019.

In another study, a homemade PSD (Figure 3c) was employed by Furukawa et al. to determine the whole body dermal emission rate of ammonia by taking simultaneous measurements from a total of 13 chosen spots around the body in an attempt to cover all regions of the body [74]. In order to prepare homemade PSD for ammonia, a commercial cellulose filter paper first was dipped in methanol solution containing 2% phosphoric acid and 1% glycerol and then dried under vacuum. After sampling from selected body parts, the trapped ammonia was back extracted with pure water and its concentration was determined by ion chromatography. Results revealed that the total rates of ammonia emission ranged from 2.9 to 12 mg h^{-1} with an average emission rate of 5.9 ± 3.2 mg h^{-1} for each person. Partial emission rates obtained in the study per body part can be found below in Figure 4. Higher emission flux of ammonia was found at the feet, back, and lumbar, where large sweat glands are located, and lower flux was observed at upper arms, buttock, thighs, and legs. Based on these results the authors speculated that there are two routes for ammonia emission in the body. The first route follows blood ammonia, produced by metabolic reactions that involve proteins, which can directly rise to the skin surface from blood capillaries. The second proposed route is mixing of blood ammonia into sweat, which explains the increased emission flux levels in areas where sweat glands are concentrated. As can be inferred, such ease-of-use devices are very useful for tracking metabolic pathways in the human body.

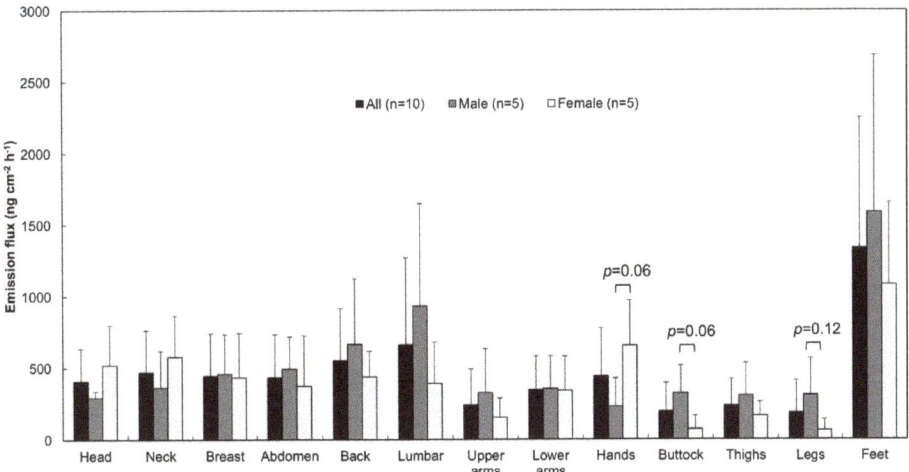

Figure 4. Comparison of emission flux of ammonia emanating from 13 sampling positions of ten volunteers. Error bars show standard deviation of measured values of tested volunteers ($n = 10$ for all, $n = 5$ for male, and $n = 5$ for female volunteers). Reproduced from [74], with permission from Elsevier, 2019.

In addition to metabolic profiling of natural skin metabolites summarized above, there are also some preliminary studies performed using PSDs for exposomic monitoring showing the potential of wearable devices for exposure monitoring. For instance, Sekine et al. have investigated VOCs emanating from the skin of smokers and volunteers who were exposed to second hand smoke using PSDs [75]. The device employed was identical to the ones used in the group's previous studies (MonoTrap® DCC18 trapping medium). Six volunteers (four smokers and two non-smokers) participated in three different experiments to investigate the effects of cigarette smoking (for smokers) and second-hand smoke (for non-smokers) on dermal VOC emissions. PSDs were placed for 1 h on the forearm and the back of the hand (Figure 5a) of the volunteers. In the first test, a smoker and non-smoker volunteer stayed in the same room for 15 min while the smoking volunteer smoked a single cigarette. Based on the results of the first test, acetaldehyde, toluene, 3-methyl furan (3-MF), 2,5-dimethyl furan(2,5-DMF), 3-ethenyl pyridine(3-EP), and nicotine were chosen as the dermal VOCs to be investigated in further analyses. Following tests investigated the concentrations of these compounds in smokers and non-smokers (through second-hand smoking) in different conditions. Results revealed that for smokers, there is an initial spike in nicotine and its derivative 3-EP immediately after smoking which decreases to its initial value after 2 h. In the case of secondhand smokers, the maximum dermal emission for the same compound was observed at the 2-h mark indicating that cigarette smoke has different routes of entrance to body. Another interesting result obtained in this study revealed that there is higher skin emanation of 3-MF, 2,5-DMF, 3-EP, and nicotine when a non-smoking volunteer was exposed to direct smoke emanating from a burning cigarette compared to skin emanation when the same volunteer was exposed to environmental second hand smoke. Another recent study investigating the biotic and abiotic exposome was able to successfully use a home-made wearable passive sampling device containing 3D printed cartridge filled with molecular sieves (zeolite), capable of adsorbing both relatively polar and non-polar compounds [76]. The cartridge was attached to the particle free air flow apparatus producing a portable device collecting both biotic and abiotic species from the exposed environment with a constant rate of sampling at 0.5 L min^{-1}. In the study, 15 participants were monitored for a time span covering up to two years in over more than 60 different geographical locations. The sampling was performed by wearing the device when possible or keeping it in two-meter proximity to the participant when it was not possible to wear the device. The study results revealed that the human exposome is varied and dynamic and closely effected by parameters including lifestyle, environment, geographic location, as well as that it can be unique for each person even for people living under similar environmental conditions. Both studies clearly highlight the importance of wearable extractive devices for monitoring public health, microbiome, and environmental exposure to chemicals just to name a few.

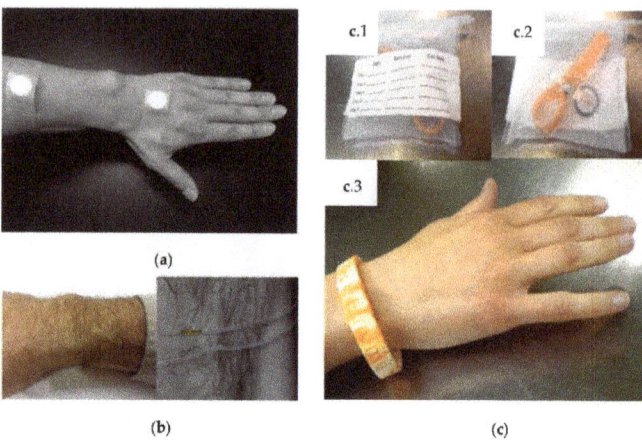

Figure 5. (a) Sampling of volatile chemicals on the forearm and back of the hand before and after cigarette smoking or exposure to second hand smoke (SHS). (b) Wrist (left) and ankle (right) sampling using a PDMS loop passive sampling device. (c) c.1–2 bags used for transport that were attached to track participant ID and exposure time in the occupational deployments; c.3 single wristband deployment. Reproduced from [75,77,78], with permission from Elsevier and ACS Publishing, 2019.

2.2.3. Other Wearable Headspace Extractive Samplers

Unconventional headspace extractive samplers including PDMS tubing sampling loops and silicone wristbands have been suggested in several studies for qualitative monitoring purposes. For instance, Roodt et al. used PDMS tubing in a study investigating the relationship between skin microbiome and anthropophilic mosquito disease vectors [77]. PDMS tubing was cut (180 and 240 mm tubing length, 0.25 mm inner diameter) to manufacture sampler loops (Figure 5b). The flexible nature of PDMS tubing allowed samplers to be employed in non-conventional sampling areas, in the wrists and ankles of the participants. Among the wide range of extracted analytes, 88 were identified in total, several of which, most notably cyclic ketones, were not previously reported in skin volatile literature. As an exemplary application area, the correlation between human skin microbiome and the attractiveness of participants to anthropophilic blood host seeking mosquitoes were investigated. Identification of previously unreported skin VOCs indicates that such passive sampling devices have not yet reached their full potential and can be explored to sample from unconventional skin areas. Another recent study using silicone wristbands was able to demonstrate the applicability of wearable extractive devices for monitoring of occupational exposure of roofers to chemical compounds released during hot asphalt application [78]. The obtained results indicated that the silicone wristbands were able to absorb 25 polycyclic aromatic hydrocarbons (PAHs) during 8 h of exposure under working conditions as well as differentiate the variations in the amounts of the PAH collected in divergent environmental condition, suggesting its applicability for sensitive monitoring. The silicone bands can be seen above in Figure 5c. PDMS tubing sampling loops and silicone wristbands represent the quintessential wearable sampling devices in the sense that they are simple, practical, robust devices that have the capability to be fine-tuned by surface or bulk material functionalization. The possibility of functionalization could allow these types of devices to either be used to target a specific range of analytes in terms of polarity, size, and volatility or to increase their extractive range to be used in untargeted studies. However, it should be kept in mind that these devices are completely open to the environment; therefore, a high background signal should be expected.

A summary of recent non-invasive skin sampling techniques, devices, and their applications can be found below in Table 2.

Table 2. Summary of reviewed skin sampling methods.

Method	Materials	Body Part	Analytes	Sampling Time	Instrumentations	Comments
Patch type [34]	PDMS	Upper back, forearms, back thigh	VOCs	1 h	GC–MS	"sandwich membrane", minimal contamination
Patch type [36]	PDMS	Armpit	Fatty acid metabolites, VOCs	30 min	TD–SESI–MS/TD–GC–MS	suitable for automation
Patch type [38]	PDMS	Forehead	VOCs	30 min	TD–GC–MS	complementary to breath analysis
Patch type [35]	PDMS	Ear	Rabbit skin metabolites, ulcer metabolites	30 min	GC–MS	rabbit model study
Patch type [40]	PDMS	Foot	VOCs	30 min	TD–GC–MS	complementary to bacterial mapping
Patch type [42]	Agarose hydrogel	Lower arm	Skin metabolites	1 min–3 h	nanoDESI–MS	direct mass spectrometry
Patch type [44]	Agarose hydrogel	Upper and lower limbs, abdomen, back	Psoriatic skin metabolites	20 min	nanoDESI–MS	direct mass spectrometry
Patch type [43]	Agarose hydrogels	Lower arm	Topical drug metabolites, nicotine and scopolamine metabolites	10 min	nanoDESI–MS	direct mass spectrometry
Microneedle [53]	Polylactic acid	Mouse skin	Skin biomarkers	1 h	microplate UV/VIS spectrophotometry, densitometric analysis	limited to biomarkers with known antibodies, can only sample from specific skin depth
Headspace [63]	DVB/carboxen/PDMS (Stableflex)	Volar forearm	VOCs	15 min	GC–MS	glass housing
Headspace [64]	DVB/carboxen/PDMS (Stableflex)	Volar forearm	Skin and fragrance-derived VOCs	5–40 min	GC–MS	glass housing
Passive sampling [77]	PDMS	Wrist, ankle	Skin VOCs, mosquito semiochemicals	4 h	GC × GC–TOFMS	controlled environment is recommended for good repeatability

Table 2. Cont.

Method	Materials	Body Part	Analytes	Sampling Time	Instrumentations	Comments
Passive flux sampling [73]	MonoTrap® DCC18	Forearm, thigh, calf, forehead, neck, abdomen	2-nonenal, diacetyl	7 h	GC–MS	flux flow knowledge is required for quantitation
Passive flux sampling [74]	Conditioned cellulose paper	see Figure 4	Ammonia	1 h	Ion chromatography	flux flow knowledge is required for quantitation
Passive flux sampling [75]	MonoTrap® DCC18	Forearm, back of the hand	VOCs	1 h	GC–MS	flux flow knowledge is required for quantitation
Passive sampling [78]	Silicone	Wrist	PAHs, environmental chemicals	2–24 h	GC–MS	low repeatability

3. Oral Fluid and Ocular Surface Sampling

As summarized above, most of the research including wearable extractive devices has focused on detection of skin related analytes. Studies discussed above were mostly focused on implementing already existing tools/extractive materials directly for skin sampling, neglecting other possible matrices that can be used to gather biological information. In fact, saliva and ocular surfaces can be sampled easily and could provide unique information about a biological system. However, as it has been shown below, the potential of wearable devices for those matrices have not been thoroughly explored yet.

3.1. Saliva Sampling

Saliva sampling has been gathering interest as an alternative sampling matrix to blood and urine for forensic applications, disease biomarkers, drug and doping control, and flavor studies amongst other niche areas. One of the main reasons why saliva is an attractive alternative is that the samples can be collected relatively easy without privacy concerns which are especially important in doping tests. Saliva is commonly sampled through draining, spitting, and contact sampling with commercial products such as Salivette® or Drugwipe®, and by chewing inert materials [79–82]. Saliva sampling is inherently less invasive compared to some other biological specimens like blood or urine, and there is still potential for developing alternative methods applicable for in vivo sampling.

One advancement towards in vivo saliva analysis was realized by Bessonneau et al. employing thin film microextraction (TFME) with GC–MS and LC–MS techniques to evaluate and compare the ability of in vivo and ex vivo sampling, as well as the ability to determine prohibited substances in saliva [83]. In order to obtain maximum metabolite coverage two different type of extractive phases were used for salivary sampling. Hydrophilic lipophilic balanced (HLB) particles were embedded in PDMS to be used as the extractive phase for GC–MS analysis and HLB particles were embedded in polyacrylonitrile (PAN) to be used as extractive phase for LC–MS analysis. For in vivo sampling, TFME devices were placed under the tongue of participants for 5 min and then extracted analytes on the HLB–PDMS phase were desorbed directly to GC–MS using a high volume thermal desorption unit while extracted analytes on HLB–PAN phase were first desorbed into a suitable solvent prior to their LC–MS analysis. The same desorption protocol was followed for 5 min ex vivo sampling from 1 mL of collected saliva. Comparison of ex vivo to in vivo studies revealed similar results for hydrophilic compounds while higher peak intensities were observed for in vivo sampling for hydrophobic compounds. This is a reasonable result since hydrophobic compounds tend to have secondary interactions with labware. This study highlights the advantage of employing in vivo sampling if hydrophobic compounds are of interest and secondary interactions can potentially disrupt sampling. Moreover, in vivo sampling may provide the unique advantage of capturing short lived metabolites that are susceptible to decomposition which can be difficult to be determined with other methods.

3.2. Oral Tissue Sampling

Chen et al. developed a different approach for oral sampling by using a moving string sampling probe for in situ endoscopic MS of a living mouse and to take samples from oral surfaces [84]. The device consists of a disposable cotton sampling string, which moves through the sampling area. The string is smeared and carries a small amount of tissue samples that adhere to its surface directly to an ionization source to be subsequently analyzed by MS. Figure 6 depicts the schematic of the sampling device. In the study, the sampling probe was attached to an industrial endoscope with a camera and miniature super-bright LED. A metallic tube was bent into a V-shape and partially cut to expose the sampling strip, with the authors noting that the camera allows the sampling process to be monitored in real-time which can be recorded if desired. In this study, the samples were taken from the surface of the tongue of a volunteer who had consumed a caffeinated beverage. Both atmospheric pressure chemical ionization (APCI) and ESI mass spectra showed a strong peak at m/z 195.09 ([Caffeine + H]$^+$) and two peaks at m/z values 116.07 ([$C_5H_9NO_2$ + H]$^+$) and 118.09 ([$C_5H_{11}NO_2$ + H]$^+$) which were tentatively identified as proline and valine respectively. This further affirms the capability of this method to be used for sampling small molecules such as amino acids. The proof-of-concept setup designed by the investigators employed a cotton string and APCI/ESI–MS. The moving string approach can be realized with any type string, as the authors mention, and with any type of ionization setup after necessary changes are made. Although not used in the study for such purpose, employing a camera to record the sampling process makes it possible to generate a spatial metabolite map of the sampling area to provide an additional layer of information.

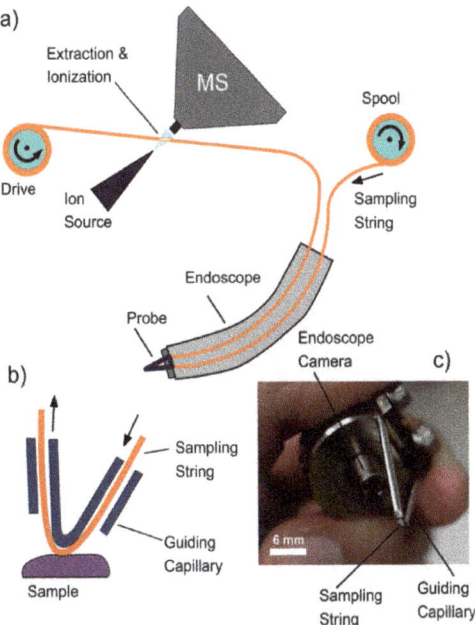

Figure 6. (a) Simplified schematic of the endoscopic mass spectrometry system that uses a moving string as the sampling and transportation material. (b) Configuration of the sampling probe. The sample is wiped off from the surface by the moving string and carried to the extraction and ionization region near the mass spectrometry (MS) inlet. (c) Photograph of the probe tip. Reproduced from [84], with permission from RSC Publishing, 2019.

3.3. Ocular Surface Sampling

The human eye is an extremely intricate and complex organ that consists of many distinct parts with distinct characteristics and qualities that are unique in the human body. Seemingly small disruptions to this delicate system can cause visual impairment and discomfort. It has been an ongoing effort to identify and investigate the non-redundant proteins in the human eye, the most notable effort being The Human Eye Proteome Project launched in 2013 [85]. Diagnosis of ocular infections caused by bacteria, fungi, parasites, or viruses is another area of research where there is an interest in developing new clinical tools [86]. A recent study in 2018 conducted by the association for Research in Vision and Ophthalmology (AVRO) reported that 43% of researchers find it "difficult" or "very difficult" to obtain human eye tissue, and 43% of the participants stated that they regularly limit the scope of their work due to inability of finding eye tissue that meets the needs of their research [87]. Therefore, it is of paramount importance to develop new tools to sample compounds from the human eye to tackle the problems researchers face in this area, which can greatly benefit from the advancements in low- and non-invasive sampling techniques.

Such an attempt was made by Liou et al. in a study to sample from the eyelids of volunteers using a watercolor brush to profile caffeine and its metabolites [88]. The ethanol-soaked tip of the watercolor brush was simply rubbed against eyelids of 4 volunteers who had consumed three cups of coffee with 3 h interval between each cup. After sampling, the brush was moved in front of the inlet of the MS instrument. The analytes evaporated from the brush surface by the aid of the ethanol and then subsequently ionized by charged ESI plum before finally being detected by the instrument. The intensity of the ions and the distance of the brush to the ESI needle were investigated, with the findings summarized in Figure 7. Although the authors chose to sample from the eyelid because of the high concentration of blood vessels in the area and the natural trapping properties of eyelashes and not because of any intention to sample eye-related metabolites or compounds, the sampling concept can still be applied to detect eye related metabolites and compounds as the detection of caffeine metabolites indicates. It should also be noted that ethanol which was used to soak the tip of the brush could extract many unwanted skin matrix components and environmental contaminants resulting in high matrix effect during such direct-to-MS approaches.

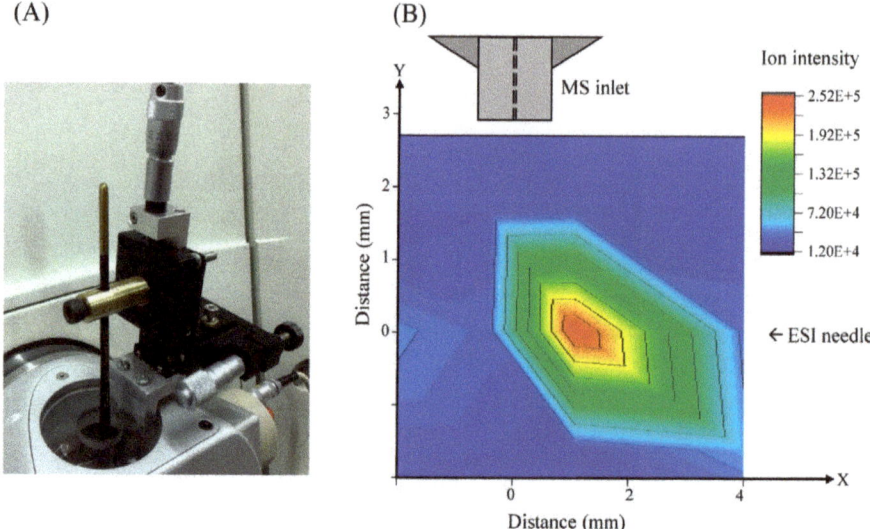

Figure 7. Photo (**A**), a picture of the sampling brush–spray ionization/mass spectrometry set up used in this study. The sampling brush was located between the mass inlet and the ESI needle (not shown in the picture). (**B**) relationship between ion intensities and the location of the brush. In this case, caffeine was used as the test sample (concentration levels, 10 µg mL^{-1}) and the ESI voltage was +4.5 kV. The position, indicated in red, denotes the optimized location. Reproduced from [88], with permission from Elsevier, 2019.

An interesting study, which could act as a lighthouse to steer future research, was conducted by Lopez et al. where a hydrogel contact lens was developed to slow the progression of corneal blindness caused by overexpression of zinc-dependent matrix metalloproteinases (MMPs) [89]. Monomer of poly(2-hydroxyetyl methacrylate) (pHEMA) were modified before polymerization with dipicolylamine (DPA) to synthesize the pDPA–HEMA hydrogel. DPA has a selective binding affinity toward zinc ions (up to $K_d = 1 \times 10^{-11}$ M [90]) which is essential for MMPs to function. Removal of zinc by the pDPA–HEMA hydrogel contact lens deactivates MMPs and slows down the degradation of the cornea. As summarized in Figure 8, ex vivo studies using porcine cornea to simulate human cornea were conducted by the authors which demonstrated advantageous results compared to conventional treatments. Because there is no systemic circulation of zinc targeting drugs in the body, the treatment is targeted and limited to the cornea only. In addition, DPA only selectively binds to zinc out of all biologically active metal ions excluding the possibility of side effects. Although this study is not directly related to the ocular surface sampling, it is vital to show the potential of using such biocompatible functionalized materials for extracting biological information directly in vivo from the ocular surface.

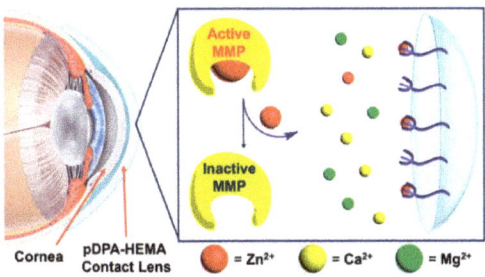

Figure 8. Schema of deactivation of matrix metalloproteinases (MMPs) by the dipicolylamine–poly(2-hydroxyetyl methacrylate) (pDPA–HEMA)-based contact lens. Reproduced from [89], with permission from ACS Publications, 2019.

4. Extractive Patches for Imaging Applications

Recently, ambient mass spectrometric (AMS) imaging techniques have gained popularity in many areas of research providing insights for metabolomics and lipidomic differences between samples, the discovery of biomarkers creating these differences, and associated clinical applications. However, as can be predicted, these techniques when applied directly to the samples may experience high background, or ionization suppression which decreases the sensitivity of the analysis. Therefore, involvement of suitable sample preparation methods for surface imaging using direct-to-MS approaches are becoming more and more crucial. With this intention, Hemalatha et al. imprinted patterns of printing inks, plant parts, and fungal growth on electrospun nanofiber mats and employed desorption electrospray ionization mass spectrometry (DESI–MS) to rapidly analyze and capture images of the imprinted patterns and analyte droplets [91]. Nanofiber mats were electrospun from a solution of Nylon-6 dissolved in formic acid and characterized by SEM. Patterns were produced on the electrospun mats by either imprinting in the case of plant slices and ink-printed patterns or by dispersing a single drop of selected analyte directly on top of the mat. Imprinted mats were then "scanned" using a 2D moving stage with a 250 µm step distance. In a typical experiment, a spray of charged solvent was pointed to the surface of the imprinted mat. The charged solvent dissolves the compounds and ionizes them similar to electrospray ionization (ESI). Subsequently, the generated molecular ions were transferred to a MS for detection. The images of the samples were obtained by raster scanning the mats under the DESI spray and detection of the molecular ions at each spot. Major metabolites of several plants, analytes, and constituents of dyes were detected and their imprints were imaged, some of the results can be seen in Figure 9.

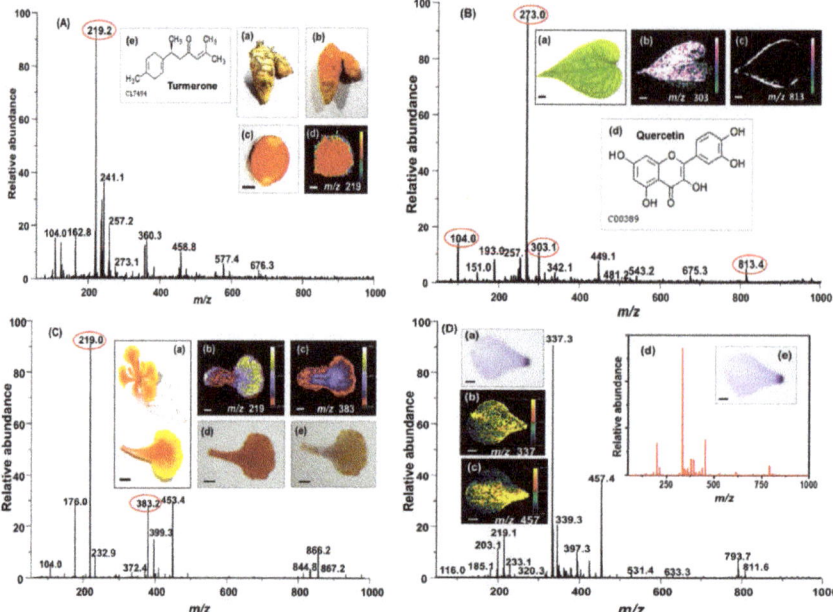

Figure 9. (a) Desorption electrospray ionization mass spectrometry (DESI–MS) spectrum from a turmeric rhizome slice of an imprinted nylon nanofiber mat. (a,b) Optical images of a whole and unskinned turmeric rhizome. (c) Imprinted slice on a nylon nanofiber mat. (d) DESI–MS image at m/z 219 due to α-turmerone shown. Reproduced from [91], with permission from ACS Publications, 2019.

Li et al. utilized a different imprinting (extractive) media using a similar approach by developing a solid-phase extraction device employing micro-funnels to scan the image of an imprinted strawberry as a proof of concept [92]. In order to functionalize the micro-funnel membrane Parafilm M® and Teflon tape with silicone adhesive was used to fix a mask with a chosen pattern on the surface, and C18 functionalized silicate powder was used to create grooves on the membrane with extractive micro-funnels. Figure 10 shows the workflow of the preparation and analysis of the micro-funnel based SPE device as well as the mapping of major metabolites found in the imprinted strawberry slice. Although both approaches are good candidates to decrease background signal in direct-to-MS studies, the effect of imprinting in chemical image resolution is not discussed thoroughly. In fact, limited image resolution should be expected, mainly due to diffusion phenomenon. The diffusion of analytes in the extractive phase during the chemical imprinting and further diffusion of analyte during the desorption process will be the main reasons for limited resolution. Also, it should be kept in mind that none of these imaging studies were performed directly in situ on living systems. However, they show the potential of the current progress in imprint imaging with MS towards the applicability of wearable extractive devices for non-destructive in vivo and in situ small molecule imaging.

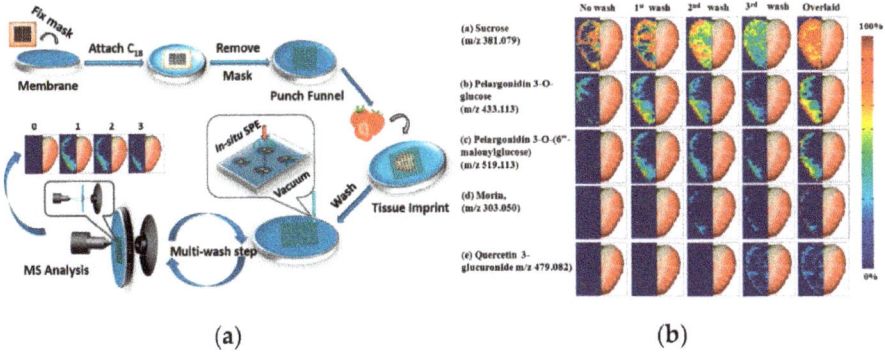

Figure 10. (a) Flow chart of C18-mounted solid phase extraction (SPE) micro-funnel-based spray for Ion-mobility spectrometry (IMS). (b) Chemical images of major components of a strawberry using C18-mounted SPE micro-funnel based spray MS with 0–3 washing steps. Reproduced from [92], with permission from SAGE, 2019.

5. Perspectives in Future Directions and Concluding Remarks

The goal of this review was to summarize the current technology that has been paving the way for novel applications for non-invasive in situ sampling devices and highlighting potential methods that can be used for wearable extraction-based devices. There are two routes of progress in this area. The first is focused on collection methods of the representative specimen in a non-invasive way from the investigated system, with very little technological improvement in collection devices. The second and more important aspect is the development of wearable extractive devices which are designed to allow free movement of the object during the sampling period with the advantage of integrating the sampling and sample preparation into a single step. This approach in general allows short sampling times with information relevant to the current state of the system, and long-term monitoring providing insightful information about time weighted average (TWA) concentration of the species in the studied system. In this review we focused only on non-invasive extractive sampling tools and wearable devices that can be used in various areas, including clinical application, exposure monitoring, food quality control, and cosmetics. A brief overview of the methods and their advantages as well as their shortcomings is shown in Table 3. The current progress in the area is promising; however, only a fraction of the full potential of such devices is currently being developed and investigated. It is clear that with current direction of the development in noninvasive wearable technologies, soon many different areas will start to use the advantage of these technologies.

Table 3. Comparison of reviewed sampling methods (biocompatibility, non-invasiveness, and ease-of-use: ☆ low, ☆☆ medium, ☆☆☆ high).

Method	Biocompatibility	Non-Invasiveness	Ease-of-Use	Commercial Availability	Sampler to Instrument Coupling	Real Time Monitoring
PDMS patches	☆☆☆	☆☆☆	☆☆☆	Yes [c]	Yes [f]	No
Hydrogel patches	☆☆☆	☆☆☆	☆☆☆	No	Yes	No
Microneedle arrays	☆☆	☆	☆☆	No	No	No
SPME headspace sampler	☆☆ [a]	☆☆☆	☆	Yes [d]	Yes	No

Table 3. *Cont.*

Method	Biocompatibility	Non-Invasiveness	Ease-of-Use	Commercial Availability	Sampler to Instrument Coupling	Real Time Monitoring
Passive flux sampler	☆☆ [a]	☆☆☆	☆☆☆	Yes [e]	No	No
Wearable headspace sampling	☆☆☆	☆☆☆	☆☆☆	No	Yes [f]	No
TFME sampling	☆☆☆	☆☆☆	☆☆☆	Yes [e]	Yes	No
String type sampler	☆☆☆	☆☆☆	☆	No	Yes	Yes
Brush type sampler	☆☆ [b]	☆☆	☆☆	Not applicable	Yes	No

[a] Depends on the extractive coating. [b] Ethanol used as solvent reduces biocompatibility. [c] PDMS sheets are commercially available. [d] SPME fibers are commercially sold, however the housing is home-made. [e] There are some commercially available. [f] Can be coupled to GC via large volume thermal desorbers.

For instance, doping control could be one of the areas that could benefit from the wearable extractive patches. Such patches will protect athletes' privacy during the sample collection process as it does not require any assistance for the sampling. Currently, there are some wearable patches that are being considered for sweat sampling; however, their main target is to collect the sweat itself rather than to focus on extraction of analytes that may be present in the sweat, giving the information about drug abuse or other type of doping.

Another area that such patches will benefit is roadside sampling. Particularly if it is combined with a portable device it might benefit opioid and cannabinoid analysis. Taking in to account that marijuana has been legalized in some countries, it is necessary to have tools that can be employed by law enforcement to perform fast roadside tests compatible with portable detection systems.

As can be seen from this review all current sampling devices are focused on skin and sweat sampling. However, alternative matrices such as ocular surface and oral fluids/surface have high potential for information as many sensor-based wearable devices have been developed.

Exposomics is another area that has a great potential to use such devises, especially for evaluation of hazards arising from occupational and environmental exposure to various chemicals that are detrimental to human health. There are already some early studies in this area using extractive wearable devices to investigate human exposure to environmental chemicals. However, more research is needed, and wearable extractive devices will be one of the main tools in the analytical toolbox for exposomic studies.

Moreover, such devices will facilitate the development of protocols for hazard analysis at critical control points. One of these areas would be pesticide control in agricultural products. Because of their frequent uncontrolled use in agricultural products, pesticide residues accumulate in food which often is associated with various health problems. There are many methods developed for pesticide analysis; however, there is still a high demand for novel and reliable methods applicable to multi-residue analyses on-site. Such extractive devices can be used for pesticide profiling on fruits and vegetables during the growth which will allow regulation of the safe amount of pesticides used in the production.

There is also a reasonable demand for wearable devices with different selectivity targeting specific needs. In the near future, the current developments in nanotechnology and nanostructured materials and the synthesis of novel extractive phase will allow the properties of materials to be tuned towards enhancing the selectivity towards targeted analytes. Potentially promising smart selective materials can be summarized as, but not limited to, molecularly imprinted polymeric materials, aptamer-based surface modifications, and metal-organic frameworks.

Finally, more work will be done in future for coupling these extractive wearable devices with portable instrumentation for direct sampling and monitoring on site.

Author Contributions: A.A.N.: Literature review, writing and revisions. E.B.: Editing, review and revisions.

Funding: This research received no external funding.

Acknowledgments: The authors would like to thank Daniel Rickert for his help in the proofreading and editing of the manuscript.

Conflicts of Interest: The authors declare no conflict of interest.

References

1. Shim, B.S.; Chen, W.; Doty, C.; Xu, C.; Kotov, N.A. Smart Electronic Yarns and Wearable Fabrics for Human Biomonitoring made by Carbon Nanotube Coating with Polyelectrolytes. *Nano Lett.* **2008**, *8*, 4151–4157. [CrossRef] [PubMed]
2. Amjadi, M.; Pichitpajongkit, A.; Lee, S.; Ryu, S.; Park, I. Highly Stretchable and Sensitive Strain Sensor Based on Silver Nanowire–Elastomer Nanocomposite. *ACS Nano* **2014**, *8*, 5154–5163. [CrossRef] [PubMed]
3. Gao, W.; Emaminejad, S.; Nyein, H.Y.Y.; Challa, S.; Chen, K.; Peck, A.; Fahad, H.M.; Ota, H.; Shiraki, H.; Kiriya, D.; et al. Fully integrated wearable sensor arrays for multiplexed in situ perspiration analysis. *Nature* **2016**, *529*, 509–514. [CrossRef]
4. Amjadi, M.; Kyung, K.-U.; Park, I.; Sitti, M. Stretchable, Skin-Mountable, and Wearable Strain Sensors and Their Potential Applications: A Review. *Adv. Funct. Mater.* **2016**, *26*, 1678–1698. [CrossRef]
5. Trung, T.Q.; Lee, N.-E. Flexible and Stretchable Physical Sensor Integrated Platforms for Wearable Human-Activity Monitoringand Personal Healthcare. *Adv. Mater.* **2016**, *28*, 4338–4372. [CrossRef] [PubMed]
6. Penn, D.J.; Oberzaucher, E.; Grammer, K.; Fischer, G.; Soini, H.A.; Wiesler, D.; Novotny, M.V.; Dixon, S.J.; Xu, Y.; Brereton, R.G. Individual and gender fingerprints in human body odour. *J. R. Soc. Interface* **2007**, *4*, 331–340. [CrossRef]
7. Nemes, P.; Vertes, A. Ambient mass spectrometry for in vivo local analysis and in situ molecular tissue imaging. *Trac.-Trends Anal. Chem.* **2012**, *34*, 22–34. [CrossRef]
8. Dormont, L.; Bessière, J.-M.; Cohuet, A. Human Skin Volatiles: A Review. *J. Chem. Ecol.* **2013**, *39*, 569–578. [CrossRef]
9. Andrisic, L.; Dudzik, D.; Barbas, C.; Milkovic, L.; Grune, T.; Zarkovic, N. Short overview on metabolomics approach to study pathophysiology of oxidative stress in cancer. *Redox Biol.* **2017**, *14*, 47–58. [CrossRef]
10. Noël, F.; Piérard-Franchimont, C.; Piérard, G.E.; Quatresooz, P. Sweaty skin, background and assessments. *Int. J. Dermatol.* **2012**, *51*, 647–655. [CrossRef]
11. Prada, A.; Furton, K.G. Recent Advances in Solid-Phase Microextraction for Forensic Applications. In *Comprehensive Sampling and Sample Preparation*; Pawliszyn, J., Ed.; Academic Press: Oxford, UK, 2012; pp. 877–891.
12. Curran, A.M.; Prada, P.A.; Furton, K.G. Canine human scent identifications with post-blast debris collected from improvised explosive devices. *Forensic Sci. Int.* **2010**, *199*, 103–108. [CrossRef] [PubMed]
13. DeGreeff, L.E.; Furton, K.G. Collection and identification of human remains volatiles by non-contact, dynamic airflow sampling and SPME-GC/MS using various sorbent materials. *Anal. Bioanal. Chem.* **2011**, *401*, 1295–1307. [CrossRef] [PubMed]
14. Prugnolle, F.; Lefèvre, T.; Renaud, F.; Møller, A.P.; Missé, D.; Thomas, F. Infection and body odours: Evolutionary and medical perspectives. *Infect. Genet. Evol.* **2009**, *9*, 1006–1009. [CrossRef] [PubMed]
15. Kim, K.-H.; Jahan, S.A.; Kabir, E. A review of breath analysis for diagnosis of human health. *TrAC Trends Anal. Chem.* **2012**, *33*, 1–8. [CrossRef]
16. Santonico, M.; Lucantoni, G.; Pennazza, G.; Capuano, R.; Galluccio, G.; Roscioni, C.; La Delfa, G.; Consoli, D.; Martinelli, E.; Paolesse, R.; et al. In situ detection of lung cancer volatile fingerprints using bronchoscopic air-sampling. *Lung Cancer* **2012**, *77*, 46–50. [CrossRef] [PubMed]
17. Saudan, C.; Baume, N.; Robinson, N.; Avois, L.; Mangin, P.; Saugy, M. Testosterone and doping control. *Br. J. Sports Med.* **2006**, *40*, i21–i24. [CrossRef] [PubMed]

18. Geyer, H.; Schänzer, W.; Thevis, M. Anabolic agents: Recent strategies for their detection and protection from inadvertent doping. *Br. J. Sports Med.* **2014**, *48*, 820–826. [CrossRef]
19. Xu, X.; Weisel, C.P. Dermal uptake of chloroform and haloketones during bathing. *J. Expo. Anal. Environ. Epidemiol.* **2005**, *15*, 289–296. [CrossRef]
20. Jiřík, V.; Machaczka, O.; Miturová, H.; Tomášek, I.; Šlachtová, H.; Janoutová, J.; Veliká, H.; Janout, V. Air Pollution and Potential Health Risk in Ostrava Region—A Review. *Cent. Eur. J. Public Health* **2016**, *24*, S4–S17. [CrossRef]
21. Vineis, P.; Chadeau-Hyam, M.; Gmuender, H.; Gulliver, J.; Herceg, Z.; Kleinjans, J.; Kogevinas, M.; Kyrtopoulos, S.; Nieuwenhuijsen, M.; Phillips, D.H.; et al. The exposome in practice: Design of the EXPOsOMICS project. *Int. J. Hyg. Environ. Health* **2017**, *220*, 142–151. [CrossRef]
22. Taylor, R.P.; Polliack, A.A.; Bader, D.L. The Analysis of Metabolites in Human Sweat: Analytical Methods and Potential Application to Investigation of Pressure Ischaemia of Soft Tissues. *Ann. Clin. Biochem. Int. J. Biochem. Lab. Med.* **1994**, *31*, 18–24. [CrossRef] [PubMed]
23. Kutyshenko, V.P.; Molchanov, M.; Beskaravayny, P.; Uversky, V.N.; Timchenko, M.A. Analyzing and Mapping Sweat Metabolomics by High-Resolution NMR Spectroscopy. *PLoS ONE* **2011**, *6*, e28824. [CrossRef] [PubMed]
24. Shirreffs, S.M.; Maughan, R.J. Whole body sweat collection in humans: An improved method with preliminary data on electrolyte content. *J. Appl. Physiol.* **1997**, *82*, 336–341. [CrossRef] [PubMed]
25. Gallagher, M.; Wysocki, C.J.; Leyden, J.J.; Spielman, A.I.; Sun, X.; Preti, G. Analyses of volatile organic compounds from human skin. *Br. J. Dermatol.* **2008**, *159*, 780–791. [CrossRef]
26. Calderón-Santiago, M.; Priego-Capote, F.; Jurado-Gámez, B.; Luque de Castro, M.D. Optimization study for metabolomics analysis of human sweat by liquid chromatography–tandem mass spectrometry in high resolution mode. *J. Chromatogr. A* **2014**, *1333*, 70–78. [CrossRef]
27. Kintz, P.; Cirimele, V.; Ludes, B. Detection of Cannabis in Oral Fluid (Saliva) and Forehead Wipes (Sweat) from Impaired Drivers. *J. Anal. Toxicol.* **2000**, *24*, 557–561. [CrossRef]
28. Wang, L.; Pi, Z.; Liu, S.; Liu, Z.; Song, F. Targeted metabolome profiling by dual-probe microdialysis sampling and treatment using Gardenia jasminoides for rats with type 2 diabetes. *Sci. Rep.* **2017**, *7*, 10105. [CrossRef]
29. Poole, J.J.; Grandy, J.J.; Yu, M.; Boyaci, E.; Gómez-Ríos, G.A.; Reyes-Garcés, N.; Bojko, B.; Heide, H.V.; Pawliszyn, J. Deposition of a Sorbent into a Recession on a Solid Support to Provide a New, Mechanically Robust Solid-Phase Microextraction Device. *Anal. Chem.* **2017**, *89*, 8021–8026. [CrossRef]
30. Bruheim, I.; Liu, X.; Pawliszyn, J. Thin-Film Microextraction. *Anal. Chem.* **2003**, *75*, 1002–1010. [CrossRef]
31. Jiang, R.; Pawliszyn, J. Thin-film microextraction offers another geometry for solid-phase microextraction. *TrAC Trends Anal. Chem.* **2012**, *39*, 245–253. [CrossRef]
32. Souza-Silva, É.A.; Gionfriddo, E.; Shirey, R.; Sidisky, L.; Pawliszyn, J. Methodical evaluation and improvement of matrix compatible PDMS-overcoated coating for direct immersion solid phase microextraction gas chromatography (DI-SPME-GC)-based applications. *Anal. Chim. Acta* **2016**, *920*, 54–62. [CrossRef] [PubMed]
33. Risticevic, S.; Pawliszyn, J. Solid-Phase Microextraction in Targeted and Nontargeted Analysis: Displacement and Desorption Effects. *Anal. Chem.* **2013**, *85*, 8987–8995. [CrossRef] [PubMed]
34. Jiang, R.; Cudjoe, E.; Bojko, B.; Abaffy, T.; Pawliszyn, J. A non-invasive method for in vivo skin volatile compounds sampling. *Anal. Chim. Acta* **2013**, *804*, 111–119. [CrossRef] [PubMed]
35. Schivo, M.; Aksenov, A.A.; Pasamontes, A.; Cumeras, R.; Weisker, S.; Oberbauer, A.M.; Davis, C.E. A rabbit model for assessment of volatile metabolite changes observed from skin: A pressure ulcer case study. *J. Breath Res.* **2017**, *11*, 016007. [CrossRef]
36. Martin, H.J.; Reynolds, J.C.; Riazanskaia, S.; Thomas, C.L.P. High throughput volatile fatty acid skin metabolite profiling by thermal desorption secondary electrospray ionisation mass spectrometry. *Analyst* **2014**, *139*, 4279–4286. [CrossRef]
37. Lang, T.; Justenhoven, C.; Winter, S.; Baisch, C.; Hamann, U.; Harth, V.; Ko, Y.-D.; Rabstein, S.; Spickenheuer, A.; Pesch, B.; et al. The earwax-associated SNP c.538G>A (G180R) in ABCC11 is not associated with breast cancer risk in Europeans. *Breast Cancer Res. Treat* **2011**, *129*, 993–999. [CrossRef]
38. Martin, H.J.; Turner, M.A.; Bandelow, S.; Edwards, L.; Riazanskaia, S.; Thomas, C.L.P. Volatile organic compound markers of psychological stress in skin: A pilot study. *J. Breath Res.* **2016**, *10*, 046012. [CrossRef]

39. Turner, M.A.; Bandelow, S.; Edwards, L.; Patel, P.; Martin, H.J.; Wilson, I.D.; Thomas, C.L.P. The effect of a paced auditory serial addition test (PASAT) intervention on the profile of volatile organic compounds in human breath: A pilot study. *J. Breath Res.* **2013**, *7*, 017102. [CrossRef]
40. Stevens, D.; Cornmell, R.; Taylor, D.; Grimshaw, S.G.; Riazanskaia, S.; Arnold, D.S.; Fernstad, S.J.; Smith, A.M.; Heaney, L.M.; Reynolds, J.C.; et al. Spatial variations in the microbial community structure and diversity of the human foot is associated with the production of odorous volatiles. *FEMS Microbiol. Ecol.* **2015**, *91*, 1–11. [CrossRef]
41. McHugh, D.J. Issue 288 of FAO fisheries technical paper. In *Production and Utilization of Products from Commercial Seaweeds*; Food and Agriculture Organization of the United Nations: Rome, Italy, 1987; p. 189.
42. Dutkiewicz, E.P.; Lin, J.-D.; Tseng, T.-W.; Wang, Y.-S.; Urban, P.L. Hydrogel Micropatches for Sampling and Profiling Skin Metabolites. *Anal. Chem.* **2014**, *86*, 2337–2344. [CrossRef]
43. Dutkiewicz, E.P.; Chiu, H.-Y.; Urban, P.L. Micropatch-arrayed pads for non-invasive spatial and temporal profiling of topical drugs on skin surface: Skin analysis. *J. Mass Spectrom.* **2015**, *50*, 1321–1325. [CrossRef] [PubMed]
44. Dutkiewicz, E.P.; Hsieh, K.-T.; Wang, Y.-S.; Chiu, H.-Y.; Urban, P.L. Hydrogel Micropatch and Mass Spectrometry-Assisted Screening for Psoriasis-Related Skin Metabolites. *Clin. Chem.* **2016**, *62*, 1120–1128. [CrossRef] [PubMed]
45. Kim, Y.-C.; Park, J.-H.; Prausnitz, M.R. Microneedles for drug and vaccine delivery. *Adv. Drug Deliv. Rev.* **2012**, *64*, 1547–1568. [CrossRef] [PubMed]
46. Tuan-Mahmood, T.-M.; McCrudden, M.T.C.; Torrisi, B.M.; McAlister, E.; Garland, M.J.; Singh, T.R.R.; Donnelly, R.F. Microneedles for intradermal and transdermal drug delivery. *Eur. J. Pharm. Sci.* **2013**, *50*, 623–637. [CrossRef] [PubMed]
47. Bhargav, A.; Muller, D.A.; Kendall, M.A.F.; Corrie, S.R. Surface Modifications of Microprojection Arrays for Improved Biomarker Capture in the Skin of Live Mice. *ACS Appl. Mater. Interfaces* **2012**, *4*, 2483–2489. [CrossRef]
48. Coffey, J.W.; Corrie, S.R.; Kendall, M.A.F. Early circulating biomarker detection using a wearable microprojection array skin patch. *Biomaterials* **2013**, *34*, 9572–9583. [CrossRef]
49. Lee, K.T.; Muller, D.A.; Coffey, J.W.; Robinson, K.J.; McCarthy, J.S.; Kendall, M.A.F.; Corrie, S.R. Capture of the Circulating Plasmodium falciparum Biomarker HRP2 in a Multiplexed Format, via a Wearable Skin Patch. *Anal. Chem.* **2014**, *86*, 10474–10483. [CrossRef]
50. Yeow, B.; Coffey, J.W.; Muller, D.A.; Grøndahl, L.; Kendall, M.A.F.; Corrie, S.R. Surface Modification and Characterization of Polycarbonate Microdevices for Capture of Circulating Biomarkers, Both in Vitro and in Vivo. *Anal. Chem.* **2013**, *85*, 10196–10204. [CrossRef]
51. Muller, D.A.; Corrie, S.R.; Coffey, J.; Young, P.R.; Kendall, M.A. Surface Modified Microprojection Arrays for the Selective Extraction of the Dengue Virus NS1 Protein as a Marker for Disease. *Anal. Chem.* **2012**, *84*, 3262–3268. [CrossRef]
52. Corrie, S.R.; Fernando, G.J.P.; Crichton, M.L.; Brunck, M.E.G.; Anderson, C.D.; Kendall, M.A.F. Surface-modified microprojection arrays for intradermal biomarker capture, with low non-specific protein binding. *Lab Chip* **2010**, *10*, 2655–2658. [CrossRef]
53. Ng, K.W.; Lau, W.M.; Williams, A.C. Towards pain-free diagnosis of skin diseases through multiplexed microneedles: Biomarker extraction and detection using a highly sensitive blotting method. *Drug Deliv. Transl. Res.* **2015**, *5*, 387–396. [CrossRef] [PubMed]
54. Bernier, U.R.; Kline, D.L.; Barnard, D.R.; Schreck, C.E.; Yost, R.A. Analysis of Human Skin Emanations by Gas Chromatography/Mass Spectrometry. 2. Identification of Volatile Compounds That Are Candidate Attractants for the Yellow Fever Mosquito (*Aedes aegypti*). *Anal. Chem.* **2000**, *72*, 747–756. [CrossRef] [PubMed]
55. Soini, H.A.; Bruce, K.E.; Klouckova, I.; Brereton, R.G.; Penn, D.J.; Novotny, M.V. In Situ Surface Sampling of Biological Objects and Preconcentration of Their Volatiles for Chromatographic Analysis. *Anal. Chem.* **2006**, *78*, 7161–7168. [CrossRef] [PubMed]
56. Riazanskaia, S.; Blackburn, G.; Harker, M.; Taylor, D.; Thomas, C.L.P. The analytical utility of thermally desorbed polydimethylsilicone membranes for in-vivo sampling of volatile organic compounds in and on human skin. *Analyst* **2008**, *133*, 1020–1027. [CrossRef] [PubMed]
57. Zhang, Z.; Pawliszyn, J. Headspace solid-phase microextraction. *Anal. Chem.* **1993**, *65*, 1843–1852. [CrossRef]

58. Camarasu, C.C. Headspace SPME method development for the analysis of volatile polar residual solvents by GC-MS. *J. Pharm. Biomed. Anal.* **2000**, *23*, 197–210. [CrossRef]
59. Rocha, S.M.; Ramalheira, V.; Barros, A.A.C.; Delgadillo, I.; Coimbra, M.A. Headspace solid phase microextraction (SPME) analysis of flavor compounds in wines. Effect of the matrix volatile composition in the relative response factors in a wine model. *J. Agric. Food Chem.* **2001**, *49*, 5142–5151. [CrossRef]
60. Pini, G.F.; de Brito, E.S.; García, N.H.P.; Valente, A.L.P.; Augusto, F. A Headspace Solid Phase Microextraction (HS-SPME) method for the chromatographic determination of alkylpyrazines in cocoa samples. *J. Braz. Chem. Soc.* **2004**, *15*, 267–271. [CrossRef]
61. Zhang, Z.-M.; Cai, J.-J.; Ruan, G.-H.; Li, G.-K. The study of fingerprint characteristics of the emanations from human arm skin using the original sampling system by SPME-GC/MS. *J. Chromatogr. B* **2005**, *822*, 244–252. [CrossRef]
62. Sekine, Y.; Toyooka, S.; Watts, S.F. Determination of acetaldehyde and acetone emanating from human skin using a passive flux sampler—HPLC system. *J. Chromatogr. B* **2007**, *859*, 201–207. [CrossRef]
63. Duffy, E.; Jacobs, M.R.; Kirby, B.; Morrin, A. Probing skin physiology through the volatile footprint: Discriminating volatile emissions before and after acute barrier disruption. *Exp. Dermatol.* **2017**, *26*, 919–925. [CrossRef] [PubMed]
64. Duffy, E.; Albero, G.; Morrin, A. Headspace Solid-Phase Microextraction Gas Chromatography-Mass Spectrometry Analysis of Scent Profiles from Human Skin. *Cosmetics* **2018**, *5*, 62. [CrossRef]
65. Gómez-Ríos, G.A.; Reyes-Garcés, N.; Bojko, B.; Pawliszyn, J. Biocompatible Solid-Phase Microextraction Nanoelectrospray Ionization: An Unexploited Tool in Bioanalysis. *Anal. Chem.* **2016**, *88*, 1259–1265. [CrossRef] [PubMed]
66. Mirabelli, M.F.; Wolf, J.-C.; Zenobi, R. Direct Coupling of Solid-Phase Microextraction with Mass Spectrometry: Sub-pg/g Sensitivity Achieved Using a Dielectric Barrier Discharge Ionization Source. *Anal. Chem.* **2016**, *88*, 7252–7258. [CrossRef]
67. Gómez-Ríos, G.A.; Liu, C.; Tascon, M.; Reyes-Garcés, N.; Arnold, D.W.; Covey, T.R.; Pawliszyn, J. Open Port Probe Sampling Interface for the Direct Coupling of Biocompatible Solid-Phase Microextraction to Atmospheric Pressure Ionization Mass Spectrometry. *Anal. Chem.* **2017**, *89*, 3805–3809. [CrossRef]
68. Mosquera, J.; Scholtens, R.; Ogink, N. Using Passive Flux Samplers to determine the ammonia emission from mechanically ventilated animal houses. In Proceedings of the 2003 ASAE Annual Meeting, Las Vegas, NV, USA, 27–30 July 2003; American Society of Agricultural and Biological Engineers: San Jose, MI, USA, 2003.
69. Larios, A.D.; Brar, S.K.; Ramírez, A.A.; Godbout, S.; Sandoval-Salas, F.; Palacios, J.H.; Dubé, P.; Delgado, B.; Giroir-Fendler, A. Parameters determining the use of zeolite 5A as collector medium in passive flux samplers to estimate N2O emissions from livestock sources. *Environ. Sci. Pollut. Res.* **2017**, *24*, 12136–12143. [CrossRef]
70. Larios, A.D.; Kaur Brar, S.; Avalos Ramírez, A.; Godbout, S.; Sandoval-Salas, F.; Palacios, J.H. Challenges in the measurement of emissions of nitrous oxide and methane from livestock sector. *Rev. Environ. Sci. Biotechnol.* **2016**, *15*, 285–297. [CrossRef]
71. Debbagh, M.; Adamchuk, V.; Madramootoo, C.; Whalen, J. Development of a Wireless Sensor Network for Passive in situ Measurement of Soil CO_2 Gas Emissions in the Agriculture Landscape. In Proceedings of the 14th International Conference on Precision Agriculture, Montreal, QC, Canada, 24–27 June 2018.
72. Fujii, M.; Shinohara, N.; Lim, A.; Otake, T.; Kumagai, K.; Yanagisawa, Y. A study on emission of phthalate esters from plastic materials using a passive flux sampler. *Atmos. Environ.* **2003**, *37*, 5495–5504. [CrossRef]
73. Kimura, K.; Sekine, Y.; Furukawa, S.; Takahashi, M.; Oikawa, D. Measurement of 2-nonenal and diacetyl emanating from human skin surface employing passive flux sampler—GCMS system. *J. Chromatogr. B* **2016**, *1028*, 181–185. [CrossRef]
74. Furukawa, S.; Sekine, Y.; Kimura, K.; Umezawa, K.; Asai, S.; Miyachi, H. Simultaneous and multi-point measurement of ammonia emanating from human skin surface for the estimation of whole body dermal emission rate. *J. Chromatogr. B* **2017**, *1053*, 60–64. [CrossRef]
75. Sekine, Y.; Sato, S.; Kimura, K.; Sato, H.; Nakai, S.; Yanagisawa, Y. Detection of tobacco smoke emanating from human skin surface of smokers employing passive flux sampler—GCMS system. *J. Chromatogr. B* **2018**, *1092*, 394–401. [CrossRef] [PubMed]
76. Jiang, C.; Wang, X.; Li, X.; Inlora, J.; Wang, T.; Liu, Q.; Snyder, M. Dynamic Human Environmental Exposome Revealed by Longitudinal Personal Monitoring. *Cell* **2018**, *175*, 277–291. [CrossRef] [PubMed]

77. Roodt, A.P.; Naudé, Y.; Stoltz, A.; Rohwer, E. Human skin volatiles: Passive sampling and GC × GC-ToFMS analysis as a tool to investigate the skin microbiome and interactions with anthropophilic mosquito disease vectors. *J. Chromatogr. B* **2018**, *1097*, 83–93. [CrossRef] [PubMed]
78. O'Connell, S.G.; Kincl, L.D.; Anderson, K.A. Silicone Wristbands as Personal Passive Samplers. *Environ. Sci. Technol.* **2014**, *48*, 3327–3335. [CrossRef] [PubMed]
79. Gallardo, E.; Queiroz, J.A. The role of alternative specimens in toxicological analysis. *Biomed. Chromatogr.* **2008**, *22*, 795–821. [CrossRef]
80. De Almeida, P.D.V.; Grégio, A.M.T.; Machado, M.A.N.; De Lima, A.A.S.; Azevedo, L.R. Saliva composition and functions: A comprehensive review. *J. Contemp. Dent. Pract.* **2008**, *9*, 72–80.
81. Gröschl, M.; Köhler, H.; Topf, H.-G.; Rupprecht, T.; Rauh, M. Evaluation of saliva collection devices for the analysis of steroids, peptides and therapeutic drugs. *J. Pharm. Biomed. Anal.* **2008**, *47*, 478–486. [CrossRef]
82. Higashi, T. Salivary Hormone Measurement Using LC/MS/MS: Specific and Patient-Friendly Tool for Assessment of Endocrine Function. *Biol. Pharm. Bull.* **2012**, *35*, 1401–1408. [CrossRef]
83. Bessonneau, V.; Boyaci, E.; Maciazek-Jurczyk, M.; Pawliszyn, J. In vivo solid phase microextraction sampling of human saliva for non-invasive and on-site monitoring. *Anal. Chim. Acta* **2015**, *856*, 35–45. [CrossRef]
84. Chen, L.C.; Naito, T.; Tsutsui, S.; Yamada, Y.; Ninomiya, S.; Yoshimura, K.; Takeda, S.; Hiraoka, K. In vivo endoscopic mass spectrometry using a moving string sampling probe. *Analyst* **2017**, *142*, 2735–2740. [CrossRef]
85. Semba, R.D.; Enghild, J.J.; Venkatraman, V.; Dyrlund, T.F.; Van Eyk, J.E. The Human Eye Proteome Project: Perspectives on an emerging proteome. *Proteomics* **2013**, *13*, 2500–2511. [CrossRef] [PubMed]
86. Sharma, S. Diagnosis of infectious diseases of the eye. *Eye* **2012**, *26*, 177–184. [CrossRef] [PubMed]
87. Stamer, W.D.; Williams, A.M.; Pflugfelder, S.; Coupland, S.E. Accessibility to and Quality of Human Eye Tissue for Research: A Cross-Sectional Survey of ARVO Members. *Invest. Ophthalmol. Vis. Sci.* **2018**, *59*, 4783–4792. [CrossRef] [PubMed]
88. Liou, Y.-W.; Chang, K.-Y.; Lin, C.-H. Sampling and profiling caffeine and its metabolites from an eyelid using a watercolor pen based on electrospray ionization/mass spectrometry. *Int. J. Mass Spectrom.* **2017**, *422*, 51–55. [CrossRef]
89. Lopez, C.; Park, S.; Edwards, S.; Vong, S.; Hou, S.; Lee, M.; Sauerland, H.; Lee, J.-J.; Jeong, K.J. Matrix Metalloproteinase-Deactivating Contact Lens for Corneal Melting. *ACS Biomater. Sci. Eng.* **2019**, *5*, 1195–1199. [CrossRef]
90. Maruyama, S.; Kikuchi, K.; Hirano, T.; Urano, Y.; Nagano, T. A Novel, Cell-Permeable, Fluorescent Probe for Ratiometric Imaging of Zinc Ion. *J. Am. Chem. Soc.* **2002**, *124*, 10650–10651. [CrossRef]
91. Hemalatha, R.G.; Ganayee, M.A.; Pradeep, T. Electrospun Nanofiber Mats as "Smart Surfaces" for Desorption Electrospray Ionization Mass Spectrometry (DESI MS)-Based Analysis and Imprint Imaging. *Anal. Chem.* **2016**, *88*, 5710–5717. [CrossRef]
92. Li, W.; Chen, X.; Wang, Z.; Wong, Y.E.; Wu, R.; Hung, Y.-L.W.; Chan, T.-W.D. Tissue imaging with in situ solid-phase extraction micro-funnel based spray ionization mass spectrometry. *Eur. J. Mass Spectrom.* **2018**, *24*, 66–73. [CrossRef]

© 2019 by the authors. Licensee MDPI, Basel, Switzerland. This article is an open access article distributed under the terms and conditions of the Creative Commons Attribution (CC BY) license (http://creativecommons.org/licenses/by/4.0/).

Review

Modern Approaches to Preparation of Body Fluids for Determination of Bioactive Compounds

Katarzyna Madej [1,*] and Wojciech Piekoszewski [1,2]

1. Department of Analytical Chemistry, Faculty of Chemistry, Jagiellonian University, 30-387 Krakow, Poland; wpiekosz@tlen.pl
2. Department of Food Science and Technology, School of Biomedicine, Far Eastern Federal University, Vladivostok 690950, Russia
* Correspondence: madejk@chemia.uj.edu.pl

Received: 3 July 2019; Accepted: 11 October 2019; Published: 5 November 2019

Abstract: The current clinical and forensic toxicological analysis of body fluids requires a modern approach to sample preparation characterized by high selectivity and enrichment capability, suitability for micro-samples, simplicity and speed, and the possibility of automation and miniaturization, as well as the use of small amounts of reagents, especially toxic solvents. Most of the abovementioned features may be realized using so-called microextraction techniques which cover liquid-phase techniques (e.g., single-drop microextraction, SDME; dispersive liquid–liquid microextraction, DLLME; hollow-fiber liquid-phase microextraction, HF-LPME) and solid-phase extraction techniques (solid-phase microextraction, SPME; microextraction in packed syringes, MEPS; disposable pipette tip extraction, DPX; stir bar sorption extraction, SBSE). Some other extraction methodologies like dispersive solid-phase extraction (d-SPE) or magnetic solid-phase extraction (MSPE) can also be easily miniaturized. This review briefly describes and characterizes the abovementioned extraction methods, and then presents their current applications to the preparation of body fluids analyzed for bioactive compounds in combination with appropriate analytical methods, mainly chromatographic and related techniques. The perspectives of the analytical area we are interested in are also indicated.

Keywords: microextraction techniques; body fluids; bioactive compounds; clinical and forensic analysis

---

1. Introduction

Body fluids belong to the most often analyzed samples in clinical investigations, and in related fields such as forensic toxicology. Plasma/serum, whole blood, urine, and saliva constitute biological materials of special interest.

Due to complexity of the matrix, body fluids are a challenge for those who are interested in bioanalysis. In most cases of such analyses, the selection of an appropriate sample preparation method is required because of main factors such as the chemical nature of the analyte and type of sample matrix, the analytical interferences of the target compounds with their metabolites and/or with constituents of the matrix, the low concentration levels of analytes, which are not detectable by analytical instruments, and the contamination and subsequent shortening of life of the used instrument or its accessories (e.g., due to interaction of matrix constituents with sorbents of chromatographic columns).

At present, conventional extractions techniques, i.e., liquid–liquid extraction (LLE) and solid-phase extraction (SPE), are still extensively applied to the analysis of body fluids for determination of bioactive compounds [1]. Although they demonstrate many advantageous features, they also possess some drawbacks, e.g., time consumption, large consumption of toxic organic solvents and possible formation of emulsion (LLE), or the requirement of relatively expensive extraction columns and multi-step processes (SPEs) [1]. Therefore, the development of alternative sample preparation methods is desired. The development of novel sample preparation approaches is strongly stimulated by the following requirements of modern clinical or forensic analysis: (a) high selectivity and enrichment capability, (b) simplicity and rapidity, (c) suitability for micro-amount biological samples, (d) automation, and (e) miniaturization [2]. The other key factor affecting the dynamic development of extraction methods is the introduction of new materials, which may be used as potential sorbents in bioanalysis. There are MOFs (metal–organic frameworks), [3] ionic liquids [4], fabric phases [5], graphene oxide tablets [6], cork [7], bract [8], or coated papers [9].

This review focuses on modern approaches to the preparation of body fluid samples. The principles of microextraction techniques are briefly described and then exemplified using selected applications for the preparation of body fluids analyzed for medicines, drugs of abuse, and other bioactive compounds.

2. Modern and Novel Sample Preparation Techniques

There are several sample treatment methodologies that may be applied to the preparation of body fluids analyzed for bioactive compounds. The selection of the below sample preparation approaches was mainly based on the consideration of efficient and cost-effective miniaturized techniques, so-called microextraction methods. Generally, microextraction methods may be divided into two groups: Solid-phase extractions and Liquid-phase extractions.

2.1. Solid-Phase Extractions

2.1.1. Solid-Phase Microextraction and Derived Techniques

A solvent-free solid-phase microextraction (SPME) was proposed by Arthur and Pawliszyn in 1990 [10], and then, in 1993, Supelco introduced the first commercial version of an SPME device [11]. The SPME technique, like conventional SPE, involves the partitioning of the target compound between an organic phase coating a fiber (usually made of fused silica) and the sample matrix. This extraction methodology may be conducted into two ways: directly by placing a small-diameter fiber coated with a stationary phase film in an aqueous sample or in headspace mode, by placing the fiber on the headspace of the sample. The adsorbed/absorbed analytes are then thermally desorbed from the stationary phase in the injector of a gas chromatograph. However, a new trend in developing faster analytical procedures involving SPME, i.e., direct coupling of this extraction approach with mass spectrometry, was noted [12]. Although, the SPME technique is known to have some drawbacks (fragile coating layers, degradation of the fibers with multi-use, carryover problems, batch-to-batch variations of fiber coatings, and relatively high cost), currently, it is dynamically developing, mainly through the introduction of new SPME fiber materials and selective coatings [2].

The SPME approach may be also performed "in-tube", where an open tubular fused-silica capillary column is exploited as the SPME device instead of an SPME fiber [13]. A comparison of fiber and in-tube SPME techniques in combination with liquid chromatography was also provided [13]. In-tube SPME is superior to fiber SPME mainly due to the shorter equilibrium time and better suitability for automation, as well as a wider diversity of stationary phases coating commercial capillary columns.

Thin-film microextraction (TFME) constitutes another geometry for SPME, which was proposed by Jiang and Pawliszyn [14]. In TFME, a sheet of flat film, employed as the extraction phase, is reinforced by an extra support such as a stainless-steel rod, stainless-steel mesh, or blade-shaped substrate [15]. Due to the large surface area-to-volume ratio, TFME is superior to SPME in terms of adsorption capability and extraction rate. This technique, in combination with novel biocompatible matrix membranes, may be especially useful in the medicine field as a non-invasive diagnostic tool in the analysis of breath, skin, and saliva.

2.1.2. Microextraction in Packed Syringe

Microextraction in a packed syringe (MEPS) is a relatively new technique because it was first applied in a fully automated procedure by Abdel-Rehim in 2004 [16] to extract local anesthetics from human plasma. The MEPS technique is a miniaturized version of conventional solid-phase extraction

(SPE), which can be connected online to GC or LC, without modification [16]. In MEPS, 1 mg of the sorbent material is usually placed into a syringe (100–250 µL) as a plug. A sample is drawn through the syringe (e.g., by an autosampler) before passing through the solid material where analytes are adsorbed. The sorbent is washed to remove the biological matrix, and then the analytes are eluted with an appropriate organic solvent (e.g., directly into the instrument's injector). In comparison with conventional SPE, the main advantages of MEPS are congruence for full automation (which minimizes the sample preparation time to a minute), as well as the possibility to reuse a packed syringe (more than 100 times with body fluid samples), whereas an SPE column usually can only be used once. Compared with SPME, the MEPS technique is more powerful for the preparation of biological samples with complex matrices. Moreover, much higher extraction yields can be obtained (60–90%) compared to standard SPME (1–10%), and smaller sample volumes can be handled (10 µL) compared to SPME (>1000 µL) [16].

2.1.3. Disposable Pipette Extraction

Disposable pipette extraction (DPX, TIPS, PT-SPE) is a miniaturization of conventional SPE, which was developed by Brewer [17]. In DPX, the sorbent is placed in a pipette tip and it is mixed well with the sample. This reduces the sorption material needed to retain analytes, and leads to faster and more efficient extraction process [18]. Similarly to conventional SPE or MEPS, a four-step extraction process takes place: (1) conditioning for activation of the sorbent sites, (2) aspiration of the sample, (3) removing sample matrix interferences, and (4) elution of analytes. In the steps 1, 3, and 4, an appropriate solvent (or solvents) is aspirated with air and removed one or more times. After conditioning (step 1), the sample is suctioned with air (step 2) for mixing with the adsorbent. The mixing/contact time must be controlled to reach a dynamic equilibrium during the interaction of analytes with the sorbent, and then the sample is removed from the tip. DPX is a simple and fast sample preparation approach, which minimizes the consumption of the sample and organic solvents. Recently, some miniaturized concepts for chromatographic analysis involving a pipette tip and spin column were reviewed [19]. However, until now, their application to routine analysis is limited mainly due to the small number of commercially available extracting materials and the higher cost compared to traditional SPE columns [20].

2.1.4. Dispersive Micro-Solid-Phase Extraction and Magnetic Solid-Phase Extraction

Dispersive micro-solid-phase extraction (D-µ-SPE) is a miniaturized dispersive solid-phase extraction (d-SPE) consisting of the dispersion of micro- or nanosorbents in the sample solution, followed by separation of the solid sorbent from the extracted analytes by centrifugation and filtration [21]. Among the advantageous features in favor of the D-µ-SPE technique in relation to conventional SPE are rapidity (due to the dispersion phenomenon which increases the sorbent surface for interaction with analyte molecules), possibility to use a large spectrum of sorbents (which are able to disperse in the sample solution), simplicity, and reducing costs in terms of the used sorbent amounts and laboratory equipment. One of the d-SPE variants is magnetic solid-phase extraction (MSPE), where magnetic nanoparticles (MNPs) are used as sorbents in the sample preparation process. MNPs attract the consideration of many analytical chemists because they considerably simplify and accelerate the extraction process due to their capacity to be easily isolated from the sample solution through an external magnet [2]. Regarding the trend of MNPs to form agglomerates and the loss of magnetism due to their chemical activity, these materials are modified into core–shell composites by coating them with an organic (e.g., surfactant) or an inorganic (e.g., graphene or carbon nanotubes) layer. The magnetic composites created, which are used as sorbents, may also have a higher extraction ability than simple magnetic nanoparticles (e.g., Fe_3O_4) when samples with a complex matrix are analyzed. A recent general review of D-µ-SPE [22] focused mainly on the dispersion strategies and sorbents used, and also indicated several trends in the near future for the development of this extraction technique.

2.1.5. Stir Bar Sorption Extraction

Stir bar sorption extraction (SBSE) was introduced by Baltussen with his co-workers in 1999 [23]. In this approach, a stir bar (magnetic element) is coated with a sorbent and immersed in a sample solution. The sample is stirred with appropriate speed for a specified time to reach equilibrium. After adsorption of the analyte on the sorption material, the magnetic element is transferred to a small amount of a selected organic solvent to desorb the analyte into it. Although SBSE is theoretically similar to SPME, its capacity is greater than SPME. This results from the fact that more sorbent mass is usually present in SBSE than in SPME and more analyte is transferred to the sorbent in SBSE. Currently, polydimethylsiloxane (PDME) is mainly employed as the sorbent coating, and the SBSE technique is not used very often. However, the current trend is to employ this technique in combination with many novel materials, including restricted access materials, carbon adsorbents, molecularly imprinted polymers, ionic liquids, microporous monoliths, sol–gel prepared coatings, and dual-phase materials [24]. The main advances in various "extraction/stirring integrated techniques", including the source SBSE technique, emphasizing their analytical potential, were also previously reviewed and compared [25].

2.2. Liquid-Phase Extractions

2.2.1. Single-Drop Microextraction

The term "liquid-phase microextraction (LPME)" means that the extraction technique consists of a microsyringe playing two roles: a funnel for extraction and a syringe for injection into a GC port. Single-drop microextraction (SDME), the simplest mode of LPME, was introduced by Liu and Dasgupta [26], as well as by Jeannot and Cantwell in 1996 [27]. In the SDME technique, analytes are isolated from an aqueous sample (being stirred) into a small (ca. one µL) drop of a water-immiscible organic solvent hanging on the needle of a microsyringe [28]. After the extraction process, the drop is withdrawn into the syringe and usually injected directly into a GC instrument. The SDME technique may be realized via three ways: through direct immersion, in the headspace mode, and as three-phase SDME. In contrast to the direct immersion mode, where the extracting solvent drop is submerged into the aqueous sample, in the headspace mode, the drop hangs in the headspace of the sample. In three-phase SDME, the droplet contains two immiscible solvents, i.e., the polar acceptor solution and the polar donor solution. These two solvents are separated by a non-polar solvent. The major problem of the SDME technique is drop instability, but attempts are being made to solve this issue via the creation of new devices for SDME. Current trends in developments in SDME, such as the use of non-conventional solvents, novel materials (for combining sorbent and liquid-phase extractions), and the creation of more suitable SDME devices, as well as their automation and implementation in microfluidic chip technologies, were presented in a recent review paper [29].

2.2.2. Dispersive Liquid–Liquid Microextraction

Dispersive liquid–liquid microextraction (DLLME) was proposed by Rezaee and co-workers in 2006 [30]. This technique is based on a ternary-component solvent system consisting of an extraction solvent, a dispenser, and an aqueous sample. In DLLME, a mixture of extraction and dispenser solvents is rapidly injected into an aqueous sample via a syringe. The extraction solvent is dispersed into the aqueous phase and forms a cloudy solution of fine droplets, which interact with an analyte. The centrifugation allows the separation of two phases, whereby the sediment phase is enriched with the analyte. The extraction solvent must be a high-density water-immiscible solvent, whereas the dispenser solvent must be a polar water-miscible one. Carbon tetrachloride, tetrachloroethylene, chlorobenzene, and ionic liquids may be used as the extraction solvent, while acetone, methanol, ethanol, acetonitrile, or tetrahydrofuran may play the role of dispensers. One of the more interesting varieties of the DLLME technique involves the solidification of a floating organic drop (DLLME-SFO) [31]. In DLLME-SFO, dodecanol or 2-dodecanol are usually used as extraction solvents. After centrifugation, the floating organic phase is solidified quickly by cooling (e.g., in an ice bath). The solidified extraction solvent with the isolated analytes is separated, melted at room temperature, and then subjected to analysis.

2.2.3. Hollow-Fiber Liquid-Phase Microextraction

The idea of a supported liquid membrane (SLM) was for first time integrated with single-use extraction units for liquid–liquid–liquid microextraction (LLLME) by Pedersen-Bjergaard and his co-workers in 1999 [32]. The authors employed a polypropylene hollow fiber as the membrane for extraction of the model compound (methamphetamine) from body fluids. In SLM, an organic solvent (e.g., 1-octanol) is usually impregnated in the small pores of a hollow fiber, which protects the extracting solvent, thus permitting extraction only on the surface of the solvent immobilized in the membrane pores. Hollow-fiber liquid-phase microextraction (HF-LPME), as a mode of liquid-phase microextraction, was firstly published by Shen and Lee in 2002 [33]. To impregnate the pores of the fiber wall with an appropriate solvent, the needle tip inserted into the hollow fiber is immersed in an organic solvent for several minutes. In order to remove the excess organic solvent from the inside of the fiber, water is injected to flush it before being removed from the solvent. For the extraction, the prepared fiber is immersed in an aqueous sample while the organic solvent in the syringe is injected completely into the hollow fiber. During the extraction, the solution is agitated using a magnetic stirrer. After the extraction process, the solvent with the isolated analyte is withdrawn into the syringe and subsequently injected into a GC instrument. The disadvantages of HF-LPME include the relatively low repeatability of the extraction process due to the air bubbles forming on the hollow-fiber surface (reduction of transport speed), and the formation of a membrane barrier between the sample and the acceptor solvent (usually an organic solvent), reducing the rate of extraction [30]. Electromembrane extraction (EME) is an extended concept of HF-LPME, which was introduced in 2006 [34]. In the EME approach, a charged analyte is extracted from a sample solution through the SLM into an acceptor solution, where its transfer is facilitated by an external electrical field applied across the SLM [35]. A review article concerning the EME technique, covering its principles, new SLMs, support materials for the SLM, new sample additives, and novel technical configurations, as well as the applications of EME in pharmaceutical analysis, was published [35].

3. Applications

Data concerning exemplary applications of solid-phase extractions and liquid-phase extractions for clinical and forensic analysis are gathered in Tables 1 and 2, respectively.

Table 1. Methods of sample preparation by solid-phase extraction.

Name of Analytes	Material	Extraction Efficiency (%)	Analytical Method	Precision (RSD %)	LOD/LOQ	Reference
Solid-Phase Microextraction (SPME) and Derived Techniques						
Methadone	Plasma	95–97	GC-FID	5	0.035/0.10 µg/L	[35]
	Urine			5.3	0.035/0.10 µg/L	
Beta-blockers (acebutolol, atenolol, fenoterol, nadolol, pindolol, procaterol, sotalol, timolol)	Plasma	98–115	LC-MS/MS	1.0–17.4	0.018–0.29/0.09–1.15 ng/mL	[36]
	Urine	85–119		0.8–20.7	0.018–0.31/0.06–1.04 ng/mL	
Chemotherapeutics (amoxicillin, cefotaxime, ciprofloxacin, daptomycin, fluconazole, gentamicin, clindamycin, linezolid, metronidazole, moxifloxacin)	Whole blood	65–83	LC-MS/MS	0.11–3.19	0.126–0.198/0.392–0.652 ng/mL	[37]
Testosterone	Saliva	99.5–101.9	LC-MS/MS	4.9–9.5	[a]/0.01 ng/mL	[38]
cortisol		102.4–105.5			[a]/0.03 ng/mL	
dehydroepiandrosterone		94.0–99.9			[a]/0.28 ng/mL	
Estron	Urine	75–103	HPLC-FLD [b]	3–19	3.03/10 µg/L	[39]
17 β-estradiol		87–101		1–9	0.03/0.1 µg/L	
17 µ-ethinylestradiol		71–92		2–186	0.15/0.5 µg/L	
Estrol		80–91		2–9	0.1/1 µg/L	
Microextraction in Packed Syringe (MEPS)						
Immunosuppressive drugs (cyclosporine, everolimus, sirolimus, tacrolimus)	Whole blood	102–103	LC-MS/MS	4.2–5.1	0.9/3.0 ng/mL	[40]
		103–109		3.5–13.7	0.15/0.5 ng/mL	
		102–108		3.6–8.9	0.15/0.5 ng/mL	
		103–106		2.0–9.1	0.15/0.5 ng/mL	
Ecgonine methyl ester, Benzoylecgonine, Cocaine, Cocaethylene	Urine	113–116	TOF-MS/DART [c]	-[a]	22.9/75.0 ng/mL	[41]
		93–104			23.7/65.0 ng/mL	
		100–109			4.0/95.0 ng/mL	
		100–105			9.8/75.0 ng/mL	
Linezolid	Plasma	99.1–108.3	LC-MS/MS	1.66–6.83	0.1407/0.3814 ng/mL	[42]
Amoxicillin		96.2–10.9		0.24–3.26	0.1341–0.4249 ng/mL	
Disposable Pipette Extraction (DPX, Other Acronyms: TIPS, PT-SPE)						
Carbamazepine	Urine	98.2–99.4	HPLC-UV	2.5	0.04/-[a] µg/L	[43]
Metoprolol	Plasma	101–103	LC-MS/MS	8–15	-[a]/5 nM	[44]
Pindolol		94–114		9–11	-[a]/5 nM	
Busulfan	Whole blood	99–113	LC-MS/MS	13–16	-[a]/5 nM	[45]
Cyclophosamid		103–110		7–12	-[a]/10 nM	

Table 1. Cont.

Name of Analytes	Material	Extraction Efficiency (%)	Analytical Method	Precision (RSD %)	LOD/LOQ	Reference
Dispersive Micro-Solid Phase Extraction (D-μ-SPE) and Magnetic Dispersive Extraction (MSPE)						
7-Aminoflunitrazepam		72.7–106.8	GC-MS (PTV-LVI [d]-GC-MS)	5.1–17.0	-[a]/0.2 ng/L	
Amitriptyline		108.1–139.7		1.92–15.7	-[a]/0.2 ng/L	
Carbamazepine		74.6–111.3		3.4–5.7	-[a]/0.2 ng/L	
Carbaryl		100.6–118.4		4.2–15.7	-[a]/0.3 ng/L	
Carbofuran	Whole blood (post-mortem)	93.3–143.5		5.4–9.0	-[a]/0.2 ng/L	[46]
Cocaine		103.2–116.3		0.94–4.45	-[a]/0.2 ng/L	
Diazepam		85.1–91.0		2.4–3.7	-[a]/0.2 ng/L	
Haloperidol		68.4–115.8		6.1–9.1	-[a]/0.3 ng/L	
MDMA		86.7–104.6		5.3–7.2	-[a]/0.3 ng/L	
Methiocarb		99.2–116.4		4.0–12.35	-[a]/0.3 ng/L	
Pirimicarb		107.3–119.8		2.6–17.1	-[a]/0.3 ng/L	
Terbufos		87.5–108.1		8.4–11.6	-[a]/0.2 ng/L	
Citalopram	Plasma urine	93.4–99	HPLC-UV	4.8–8.4	0.2–1.0 μg/L	[47]
Sertraline		94–98.4		4.3–9.2	0.3–0.7 μg/L	
Levofloxacin	Serum	78.7–83.4	HPLC-UV	4.1–4.8	-[a]	[48]
Stir Bar Sorptive Extraction (SBSE)						
Fluoxetine	Plasma	101.1–115.9	HPLC-FLD [b]	4.1–14.8	9.8/32.7 ng/mL	[49]
Losartan	Plasma	98–107	HPLC-UV	3–5	7 ng/mL/-[a]	[50]
Valsartan		98–117		3–8	27 ng/mL/-[a]	
Amphetamine	Urine	99.3–99.6	HPLC-UV	4.7–6.5	11/39.7 ng/mL	[51]
Methamphetamine		99.5–99.8		4.3–5.9	10/35.2 ng/mL	

[a] Lack of data; [b] fluorescence detector; [c] time-of-flight mass spectrometer with direct analysis in real time; [d] large volume injection programmed temperature vaporization.

Table 2. Methods of sample preparation by liquid-phase extraction.

Name of Analytes	Material	Extraction Efficiency (%)	Analytical Method	Precision (RSD %)	LOD/LOQ	Reference
Single-Drop Microextraction (SDME)						
Atorvastatin	Serum	87.6–105.6	LC–MS	4.2–6.9	0.02 ng/L/-[a]	[52]
Fluvastatin		88.3–92.8		3.4–5.9	2.12 ng/L/-[a]	
Lovastatin		84.5–104.2		4.2–6.9	0.03 ng/L/-[a]	
Mevostatin		87.4–96.3		4.6–6.4	2.0 ng/L/-[a]	
Simvastatin		82.2–97.8		4.8–6.7	0.03 ng/L/-[a]	
Caffeine	Horse urine	83.5 ± 1.22	OT–CEC[b]	-[a]	9.07/10.73 ng/mL	[53]
Cocaine		103.1 ± 2.14			8.27/9.783 ng/mL	
Ephedrine		80.3 ± 1.25			7.71/9.123 ng/mL	
Morphine		108.3 ± 2.03			5.12/5.983 ng/mL	
Piroxicam		94.8 ± 1.46			17.6/19.03 ng/mL	
Strychnine		98.8 ± 1.54			0.94/1.193 ng/mL	
Theophylline		87.5 ± 1.3			1.32/2.023 ng/mL	
Berberine	Urine	105.5–107.7	MEKC[c]–UV	4.6–8.8	0.2/1.5 ng/mL	[54]
Palmatine		88.5–95.0		7.3–12.4	0.5/0.7 ng/mL	
Tetrahydropalmatine		93.1–115.5		6.4–9.0	1.5/4.8 ng/mL	
Dispersive Liquid–Liquid Microextraction (DLLME)						
Aloe-emodin	Urine	87.1–105.0 (OS-DLLME[d])	HPLC–UV	7.2–8.7	0.07 ng/mL/-[a]	[55]
Chryophoanol					0.4 ng/mL/-[a]	
Danthon					0.08 ng/mL/-[a]	
Emodin					0.08 ng/mL/-[a]	
Physicon					0.1 ng/mL/-[a]	
Rhenib		94.8–103.0 (IL-DLLME[e])		5.7–6.4	0.3 ng/mL/-[a]	
					0.01 ng/mL/-[a]	
					0.08 ng/mL/-[a]	
					0.12 ng/mL/-[a]	
					1.0 ng/mL/-[a]	
					0.5 ng/mL/-[a]	
					0.5 ng/mL/-[a]	
Flurbiprofen	Urine	84.4 ± 1.5	HPLC–UV	8.1	16.3 ng/mL/-[a]	[56]
Indomethacin		84.61 ± 1.4		8.6	8.3 ng/mL/-[a]	
Ketoprofen		73.7 ± 1.4		2.5	32.0 ng/mL/-[a]	
Naproxen		76.6 ± 1.3		3.6	9.2 ng/mL/-[a]	

Table 2. Cont.

Name of Analytes	Material	Extraction Efficiency (%)	Analytical Method	Precision (RSD %)	LOD/LOQ	Reference
Carvedilol	Plasma	93–99	HPLC–UV	3.1–10.0	6/19 ng/mL	[57]
Diltiazem		93–100		4.9–13.4	3/9 ng/mL	
Metoprolol		94–104		5.1–11.1	2/7 ng/mL	
Propranolol		94–104		7.6–10.4	5/18 ng/mL	
Verapamil		90–100		2.0–5.4	3/10 ng/mL	
Hollow-Fiber Liquid-Phase Microextraction (HF-LPME) and Electromembrane Extraction						
Hydrochlorothiazide	Urine	91.5–92.5	HPLC–UV	4.4–7.9	-[a]/0.5 µg/L	[58]
Triamterene		89.5–89.0		5.8–9.3	-[a]/0.5 µg/L	
Ketamine	Urine	85.2–101.0	GC–MS	2.9–10.1	0.25/0.50 ng/mL	[59]
Norketamine		86.9–94.3		3.6–9.2	0.10/0.50 ng/mL	
Dehydronorketamine		64.6–69.7		9.3–16.9	0.10/0.50 ng/mL	
Cyproterone acetate	Urine	47.1	HPLC–UV	5.8	1.0/ µg/L	[60]
		46.3		6	0.5/ µg/L	
Dydrogesterone	Plasma	20.6		6.1	2.0/ µg/L	
		21.4		6.3	1.2/ µg/L	

[a] Lack of data; [b] open tubular capillary electrochromatography; [c] micellar electrokinetic chromatography; [d] organic solvent dispersive liquid–liquid microextraction; [e] ionic liquid dispersive liquid–liquid microextraction.

3.1. Solid-Phase Extractions

The SPME process may be fully automated and effectively accelerated using the 96-well plate format of thin-film solid-phase microextraction (TFME). A PPy–CH$_2$–COOH (nanostructured α-carboxy polypyrrol) coating for SPME fiber, used in headspace mode (HS-SPME), was synthesized and applied to the extraction of methadone at its trace level from plasma and urine samples [36]. The carboxy end-capped polypyrrole film was electrochemically deposited on a platinum wire, and a nano-fibrous structure with a diameter of 120 nm was obtained. The nanostructure of the film provided a high surface area that allowed for the high extraction efficiency of methadone. The extraction efficiency was 95–97%. Advantageous features of the used nanostructured fiber include high mechanical stability, strong adhesion of the coating to the substrate, fast extraction equilibrium and desorption, and relatively low cost.

Eight beta-blockers and bronchodilators were isolated from body fluids (plasma and urine) employing the 96-well plate format of the TFME system and an extraction phase made of hydrophilic–lipophilic balance particles (HLB) [37]. Methanol–acetonitrile (80:20 v/v) with 0.1% formic acid was used as a desorption solvent, which was compatible with the used mobile phase of LC–MS/MS method. The developed extraction method described in the form of a protocol is fast (time of full preparation of one sample less than 2 min), automated, and efficient. The authors also emphasized that the protocol could be modified through compromising sensitivity; finally, the proposed extraction approach could be shortened to 10–15 min for all 96 samples.

The combination of LC–MS/MS with SPME, involving polymeric sorption coatings with molecular imprints (MIP), was used for the simultaneous determination of 10 antibiotic drugs in-whole blood samples [38]. Three conducting polymers, polypyrrole, polythiophene, and poly(3-methylthiophene), used as coatings in SPME, were selected for the extraction of the target compounds. Optimization of the extraction conditions included parameters such as the extraction time, kind of desorption solution, and pH of organic extraction solvent. The obtained results showed that the tested antibiotics could be divided into three groups due to sorption capacity and selectivity to SPME coatings used and, thus, the appropriate sorbent was chosen for each group of analytes. According to the authors, the proposed MIP–SPME–LC–MS/MS method is suitable for therapeutic drug monitoring (TDM) of antibiotics in clinical laboratories, as well as in forensic laboratories for assays at higher concentration levels in body fluids.

Two modes of solid-phase microextraction (in-tube and thin-film SPME) were applied to the estimation of hormone (or similar substances) levels in saliva [38] and urine [39]. Testosterone, cortisol, and dehydroepiandrosterone were assayed by online in-tube SPME in combination with LC–MS/MS [39], and estrogens(estrone, 17β-estradiol, 17α-ethinylestradiol, and estriol) were determined employing a combination of thin-film SPME (TF-SPME) with a 96-well plate system and HPLC–FLD [8]. In the second case [39], a biosorbent—bract—as a novel extraction phase for TF-SPME, was used. In both cases, the methods were fast, with a high throughput and low sample consumption (100 μL of saliva and 37 μL of urine), and only ultrafiltration of samples was required prior to analysis.

Microextraction in packed sorbent (MEPS) is a new sample preparation technology, used in combination with liquid chromatography–mass spectrometry [40] or mass spectrometry alone [41]. An automated sample work-up and quantification of four immunosuppressive drugs in whole blood, using combined MEPS with LC–MS/MS, was presented [40]. The proposed method required 50 μL of whole-blood sample. The obtained quantitative results were in good agreement with a reference LC–MS/MS method involving protein precipitation. The analytical parameters of the developed method revealed that it can be useful for TDM of the studied immunosuppressive drugs.

The MEPS technology was also used in a fast screening of cocaine and its metabolites in human urine samples examined by direct analysis in a real-time source coupled to time-of-flight mass spectrometry (DART–TOF) [41]. As the type of sorbet is one of the most important parameters in solid-phase extraction, four various adsorbent materials (C8, ENV$^+$, Oasis MCX, Clean Screen DAU) were studied. The last material (Clean Screen DAU) worked best, showing satisfactory extraction efficiency for all studied analytes. In the described sample preparation method, a few microliters of the sample were used, and the extraction time was less than 2 min. Finally, the authors stated that coupling

MEPS technology to an autosampler and connecting it with DART/TOF mass spectrometer would allow for its full automation, making it a very useful tool for screening drugs of abuse in biological matrices.

Developing a new approach for antibiotic drug (linezolid and amoxicillin) determination in human plasma by LC/UV–MS/MS, three sample preparation methods were compared [42]. The pretreatments of samples were performed using protein precipitation (PP), solid-phase extraction (SPE), and microextraction in a packed syringe (MEPS). Among three studied extraction methods, MEPS appeared to be superior to PP and SPE, in terms of accuracy and precision.

A bio-inspired sponge, an amino-functionalized metal–organic framework (Zr–MOF–NH$_2$), was successfully applied as a sorbent in PT-SPE to extract carbamazepine from urine samples [43]. This extraction method was combined with the HPLC–UV method. The best extraction conditions were achieved when the sample volume was 100 µL, with sample pH adjusted to 7.5, and 5 mg of the sorbent and 10 µL of methanol as eluent solvent were used. The total time of analysis, including the sample preparation step, was less than 12 min. The sorbent was used for at least eight extractions without significant loss of its capacity and extraction repeatability.

Two automated clean-up methodologies, based on monolithic packed 96-tip sets and combined with LC–MS/MS, were employed for determination of the beta-blockers (pindolol and metoprolol) in plasma [44] and anti-cancer drugs (cyclophosphamide and busulfan) in whole-blood samples [45]. In the case of the beta-blocker analysis, a 100-µL sample volume was handled, and, for the determination of the anti-cancer drugs, 1 mL of a sample was needed. In both cases, sample preparation was performed in only about 2 min for 96 samples. However, in the case of whole-blood analysis, the monolithic packed 96 tips can be re-used up to only five times, inferior to the 100 times achieved with MEPS methodology [45]. Comparing the determination of the beta-blockers in plasma, using monolithic packed 96 tips, with the protein precipitation method (using 0.1% formic acid in acetonitrile), the tips method resulted in a higher (2–3 times) S/N ratio. Generally, the tips methodology provided better results in terms of selectivity, accuracy, and precision.

A d-SPE protocol based on the modified QuEChERS procedure (using 50 mg of PSA and 150 mg of anhydrous MgSO$_4$), followed by large volume injection programmed temperature vaporization (LVI–PTV)–gas chromatography–mass spectrometry (GC-MS) analysis, was developed for the simultaneous determination of 16 drugs and pesticides in postmortem blood samples without derivatization [46]. The validated d-SPE/LVI–PTV/GC–MS method is suitable for routine analysis in a forensic laboratory. The usability of the proposed method was confirmed by the analysis of 10 postmortem blood samples. Six samples contained cocaine, two contained MDMA, and two contained carbamazepine. Other found analytes were carbofuran, the metabolite 7-aminoflunitrazepam, amitriptyline, and diazepam.

Magnetically modified nanomaterials based on conductive polymers [47] and molecularly imprinted polymers (MIPs) [48] can be used as examples of new sorbents widely employed in MSPE for preconcentration and determination of pharmaceuticals in a complex biological matrix. Polypyrrole (PPy) with magnetic nanoparticles (MNPs)—Fe$_3$O$_4$—doped by sodium perchlorate (NaClO$_4$), exhibited high extraction efficiency: 93.4–99% and 94–98.4% for citalopram (CIT) and sertraline (STR), respectively. The isolation of the medicines was performed in the optimized extraction conditions: sample pH—9.0, sorbent amount—10 mg, sorption time—7 min, elution solvent (0.06 mol/L HCl in methanol) volume—120 µL, elution time—2 min. The performance of the developed extraction method was estimated by spiking CIT and STR at trace levels in urine and plasma samples [47].

Molecularly imprinted magnetic carbon nanotubes (MCNTs@MIPs) were applied to the MSPE of levofloxacin in serum samples [48]. The prepared sorbent was characterized by high specific surface area and high selectivity to template molecules of levofloxacin. Investigations indicated that the sorbent was very selective (exhibited excellent recognition) toward levofloaxacin. Under optimal extraction conditions (sorbent amount—15–20 mg, adsorption and desorption time—60 min, eluent—methanol/acetic (6:4, v/v), eluent volume—2 mL), extraction recoveries ranged from 78.7 ± 4.8% to 83 ± 4.1%. The sorbent stability was assessed as a reduction in average recovery after five cycles and was estimated at less than 7.6%.

A validated high-performance liquid chromatography with fluorescence detection (HPLC–FLD) method combined with stir bar sorption extraction (SBSE) was developed for determination of fluoxetine in human plasma [49]. The extractions of this antidepressant drug from plasma samples were performed using laboratory-made polydimethylsiloxane (PDMS) stir bars. Several factors such as temperature and time of sorption, stirring speed, and two modes of desorption (ultrasonic and magnetic stirring) were considered in the optimization of extraction conditions. The method was successfully applied to the analysis of real plasma samples originating from depressed patients treated with fluoxetine. Taking into account the time of sample preparation (3 h), low cost of each coated bar, and possibility to reuse it 50 times, the authors recommended the proposed method as a reliable tool for routine analysis. Novel coatings for bars in the SBSE approach were introduced for body fluid preparation for the determination of antihypertensive drugs [50] and illicit drugs [51]. A monolithic vinylpyrroli–doneethylene glycol dimethacrylate (VPD-EDMA) polymer was applied to the extraction of losartan and valsartan from plasma samples [50]. Compared to commercially available PDMS (polydimethylsiloxane) and PA (polyacrylamide) stir bars, the proposed coated stir bar exhibited higher physical stability and, therefore, it was suitable for the use of an ultrasonic stirring mode, which was profitable in terms of requiring less time and less solvent for the performed extraction. A nano graphene oxide sol–gel composite (NGO/sol–gel), used as the coating of a capillary glass tube stir bar, was employed for the isolation of amphetamine and methamphetamine from urine samples [51]. In both cases, the coated bars were coupled with HPLC in combination with relatively low-sensitivity UV detection, but the obtained results indicated the usefulness of both mentioned methods for the analysis of real samples in clinical and forensic laboratories.

3.2. Liquid-Phase Extractions

Jahan and co-workers [52] proposed a coupling microextraction device combining a microsyringe with a very short capillary. The designed device enabled manually controlling the shape and size of the aqueous organic droplet during the extraction process, as well as ensuring droplet stability even under vigorous stirring conditions. The performance of the developed method was checked using human serum samples spiked with five statins (lovastatin, simvastatin, mevastatin, Fluvastatin, and atorvastatin). Using 1.2 µL of a toluene–aqueous (0.2:1) droplet, a 350–1712-fold enrichment of the statins was achieved within four minutes.

A combination of SDME and open tubular capillary electrochromatography (OT–CEC) for the simultaneous determination of seven illicit drugs (caffeine, cocaine, ephedrine, morphine, piroxicam, strychnine, and theophylline) in horse urine was presented [53]. The extraction procedure enabled good compound recovery with good precision and appropriate reduction of interferences. The obtained enrichment factors were between 38 and 102 depending on the studied compound. The proposed SDME–OT–CEC method showed potential for application in toxicology investigations, as well as being a cheaper and more environmentally friendly method compared to LC–MS methods.

SDME was also used on-line coupled with sweeping micellar electrokinetic chromatography (MECK) for the isolation and preconcentration of three alkaloids: berberine, palmatine, and tetrahydropalmatine, present in human urine at trace level [54]. In this method, analytes were firstly extracted from a basic aqueous sample solution (donor phase) into n-octanol, and then back-extracted into the acidified aqueous solution (acceptor phase), which formed a droplet at the tip of a capillary. The acceptor phase was introduced into the capillary using the hydrodynamic injection mode and analyzed by sweeping MEKC. The extraction method was optimized considering a number of experimental factors: extraction solvent type, time of drop formation, stirring rate, duration of pre- and back-extraction, sample temperature, addition of NaCl, and composition of donor and acceptor phase. Under the optimized conditions, the developed method gave a 1583–3556-fold improvement in sensitivity for the studied analytes within 20 min.

Organic solvent dispersive liquid–liquid microextraction (OS–DLLME) and ionic liquid dispersive liquid–liquid microextraction (IL–DLLME) were applied to determine emodin and its six metabolites in urine samples. These two extraction approaches were evaluated and compared [55]. The analytes were assayed using HPLC–UV and HPLC–MS methods. In order to optimize DLLME, several parameters were studied and optimized, including kind of extraction solvent, volumes of the extraction solvent and the disperser solvent, extraction and centrifugation time, sample pH, and concentration of added NaCl. At optimal extraction conditions, the enrichment factors for emodin and its metabolites ranged from 90–295 for the OS–DLLME method and from 63–192 for the IL–DLLME method. Comparing the abovementioned extraction methods, it was concluded that IL–DLLME was more rapid and simple, as well as more repeatable, than OS–DLLME. However, OS–DLLME exhibited higher enrichment factors, a wider linear range, and lower limits of detection for the target compounds.

A one-step in-syringe set-up for DLLME with the use of an ionic liquid was proposed, and pros and cons of the novel approach were discussed [56]. Among the advantages may be listed the use of a simple extraction unit (only a conventional plastic syringe), avoiding the centrifugation step, reducing the extraction time, and being more suitable for the automation of the whole extraction procedure than in the conventional approach. The novel approach broadens the used solvent range (not only solvents denser than water can be exploited) by simply changing the orientation of the syringe during the phase separation step [56]. Among the limitations (at the present state), low extraction recoveries and low precision of the procedure may be mentioned. The suitability of the proposed approach was evaluated via the determination of four non-steroidal anti-inflammatory drugs (ketoprofen, naproxen, flurbiprofen, and indomethacin) spiked with urine samples. The obtained wide concentration range (0.02–10 µg/mL) of the target drugs allowed assaying them at therapeutic and toxic concentration levels.

DLLME in combination with HPLC–UV was applied to the determination of five antiarrhythmic drugs, metoprolol, propranolol, diltiazem, carvedilol, and verapamil, present in human plasma [57]. The method required 660 µL of plasma sample, and the complete separation of all the analytes was achieved within 7 min. In the extraction process, acetonitrile, used as a disperser solvent (resulting from the protein precipitation), was mixed with the extracting medium (dichloromethane) and then quickly injected into an aqueous basic solution. After centrifugation, the sedimented phase with the concentrated drugs was taken and evaporated to dryness. The dried residue was dissolved in 50 µL of deionized acidified water and subjected to HPLC analysis. The developed method was successfully applied to the analysis of the selected antiarrhythmic drugs in real plasma samples at ng/mL. According to the authors and considering the determined analytical parameters, the proposed method seems to be appropriate for routine analysis in drug analysis laboratories for pharmacokinetic and pharmacodynamics studies, as well as for therapeutic drug monitoring.

In hollow-fiber liquid-phase microextraction (HF-LPME), a porous membrane of the polypropylene hollow fiber with the pore size of 0.2 µm prevents the entry of macromolecular compounds in the sample solution into the acceptor phase present inside the fiber [58]. The HF-LPME technique may be realized using two modes: two-phase and three-phase. In two-phase HF-LPME, the target compound is enriched into the organic solvent as the acceptor phase and then determined by an appropriate chromatographic method. In three-phase HF-LPME, the analyte is extracted from the aqueous donor phase to organic phase, followed by back-extraction into the aqueous donor phase which is directly subjected to analysis. The applications of two-phase and three-phase HF-LPME for the isolation and preconcentration of hydrochlorothiazide (HYD) and triamterene (TRM) in urine samples were presented, respectively [58]. Under optimal conditions, enrichment factors of 128 and 239 were achieved for HYD and TRM, respectively. Finally, the proposed method was successfully applied to the analysis of TRM and HYD in a urine sample obtained from a volunteer who received 500 mg of triameterene-H. The performed analysis showed that the concentrations of HYD and TRM were 18.2 and 7.3 µg/L, respectively.

Using HF-LPME in the three-phase mode combined with GC–MS, ketamine (KT) and its main metabolites, norketamine (NK) and dehydronorketamine (DHNK), were determined in urine samples [59]. A "green chemistry" approach to the sample extraction involved using eucalyptus essential oil as a supported liquid membrane in HF-LPME. After drying, the acceptor phase with the isolated analytes was derivatized with trifluoroacetic anhydride before analysis by GC–MS. The usefulness of the developed method was confirmed by the analysis of real urine samples originating from two patients who were suspected of taking KT. KT, NK, and DHNK were determined in the concentrations 0.0873, 5.805, and 8.760 µg per 1 mL of urine, respectively (in one case) and 7.3, 5.3, and 6.8 ng per 1 mL of urine, respectively (in the second case).

A novel generation of deep eutectic solvent (DES) as an acceptor phase in three-phase HF-LPME was introduced for the extraction of two steroidal hormones, dydrogesterone (DYD) and cyproterone acetate (CPA), from urine and plasma samples [60]. The following factors influencing extraction efficiency were studied: nature and composition of DES, composition of supported liquid membrane, salt addition, length of hollow fiber, stirring rate, and duration of extraction. Under the optimized conditions, preconcentration factors ranged from 187 to 428. The method was validated via the analysis of real urine samples collected from two patients. In the case where patient 1 consumed one tablet (containing 2 mg of cyproterone acetate +0.035 mg of ethylene estradiol), the analysis showed that the medicine concentration was below the detection limit of the method. However, positive results of the analysis were achieved in the case of patient 2, who was subjected to hormonal therapy for a long time. The usefulness of the developed method was also confirmed via analysis of plasma samples spiked with 5 and 100 ppb of DYD and 10 and 200 ppb of CPA.

4. Summary and Conclusions

Microextraction methods constitute modern approaches for sample preparation in terms of efficiency, selectivity, simplicity, speed, enrichment capability, requirement of small sample volume, and congruence for automation and miniaturization. Due to low or very low (usually on the order of several hundred microliters) or no consumption (SPME combined with thermal desorption of analytes in GC injection port) of organic solvents, these techniques may be also classified as "green" approaches. The eight microextraction techniques belonging to solid-phase (SPME, MEPS, DPX, D-µ-SPE/MSPE, SBSE) and liquid-phase (SDME, DLLME, HF-LPME) extractions were briefly described and characterized. The usefulness of these techniques in the preparation of biological samples was supported by selected examples of their applications for the determination of medicines, drugs of abuse, and other bioactive compounds in body fluids. In the most reported cases, extraction recoveries were over 80% with an acceptable RSD lower than 10–15%. To quantify the studied bioactive compounds, the proposed microextraction techniques were usually combined with an appropriate chromatographic or capillary electrophoretic method. Using such combinations, limits of detection and quantification were at concentration levels equal to ppb or ppm. Considering sample consumption, solid-phase extractions generally require lower volumes, especially MEPS and DPX techniques, where sample volume consumption range from 5 to 600 µL. It is also worth noting that some proposed microextraction methods, at the optimized conditions, may be effective enough (they have a high enough enrichment capacity) to achieve LODs at low ppm levels of the studied compounds, even in combination with a relatively low-sensitivity UV detection method.

Microextraction techniques began being introduced during the last decade of the last century, and, based on the growing number of publications associated with them, it seems that they represent the current direction of development of sample preparation methods. Considering their numerous advantages, including "green" characteristics, without loss of analytical performance, it seems that these approaches will be extensively developed and employed in many important analytical areas, such as clinical and forensic investigations, as well as in environmental and food analysis.

Funding: This research received no external funding.

Conflicts of Interest: The authors declare no conflict of interest.

References

1. Niu, Z.; Zhang, W.; Yu, C.; Zhang, J.; Wen, Y. Recent advances in biological sample preparation methods coupled with chromatography, spectrometry and electrochemistry analysis techniques. *TrAC Trends Anal. Chem.* **2018**, *102*, 123–146. [CrossRef]

2. Wen, Y.; Chen, L.; Li, J.; Liu, D.; Chen, L. Recent advances in solid-phase sorbents for sample preparation prior to chromatographic analysis. *TrAC Trends Anal. Chem.* **2014**, *59*, 26–41. [CrossRef]
3. Rocío-Bautista, P.; Taima-Mancera, I.; Pasán, J.; Pino, V. Metal-Organic Frameworks in Green Analytical Chemistry. *Separations* **2019**, *6*, 33. [CrossRef]
4. Kissoudi, M.; Samanidou, V. Recent Advances in Applications of Ionic Liquids in Miniaturized Microextraction Techniques. *Molecules* **2018**, *23*, 1437. [CrossRef] [PubMed]
5. Zilfidou, E.; Kabir, A.; Furton, K.G.; Samanidou, V. Fabric Phase Sorptive Extraction: Current State of the Art and Future Perspectives. *Separations* **2018**, *5*, 40. [CrossRef]
6. Zohdi, Z.; Hashemi, M.; Uheida, A.; Moein, M.M.; Abdel-Rehim, M. Graphene oxide tablets for sample preparation of drugs in biological fluids: Determination of omeprazole in human saliva for liquid chromatography tandem mass spectrometry. *Molecules* **2019**, *24*, 1191. [CrossRef] [PubMed]
7. Mafraa, G.; Spudei, D.; Brognoli, R.; Merib, J.; Carasek, E. Expanding the applicability of cork as extraction phase for disposable pipette extraction in multiresidue analysis of pharmaceuticals in urine samples. *J. Chromatogr. B* **2018**, *1102–1103*, 159–166. [CrossRef]
8. Carmoa, S.N.; Meribb, J.; Carasek, E. Bract as a novel extraction phase in thin-film SPME combined with 96-well plate system for the high-throughput determination of estrogens in human urine by liquid chromatography coupled to fluorescence detection. *J. Chromatogr. B* **2019**, *1118–1119*, 17–24. [CrossRef]
9. Ríos-Gómez, J.; Fresco-Cala, B.; García-Valverde, M.T.; Lucena, R.; Cárdenas, S. Carbon nanohorn suprastructures on a paper support as a sorptive phase. *Molecules* **2018**, *23*, 1252. [CrossRef]
10. Arthur, C.L.; Pawliszyn, J. Solid phase microextraction with thermal desorption using fused silica optical fibers. *Anal. Chem.* **1990**, *62*, 2145–2148. [CrossRef]
11. Lord, H.; Pawliszyn, J. Evolution of solid-phase microextraction technology. *J. Chromatogr. A* **2000**, *885*, 153–193. [CrossRef]
12. Gómez-Ríos, G.A.; Mirabelli, M.F. Solid Phase Microextraction-mass spectrometry: Metanoia. *TrAC Trends Anal. Chem.* **2019**, *112*, 201–211. [CrossRef]
13. Kataoka, H. Automated sample preparation using in-tube solid-phase microextraction and its application—A review. *Anal. Bioanal. Chem.* **2002**, *373*, 31–45. [CrossRef] [PubMed]
14. Li, D.; Zou, J.; Cai, P.S.; Xiong, C.M.; Ruan, J.L. Preparation of magnetic ODS-PAN thin films for microextraction of quetiapine and clozapine in plasma and urine samples followed by HPLC-UV detection. *J. Pharm. Biomed. Analysis* **2016**, *125*, 319–328. [CrossRef]
15. Jiang, R.; Pawliszyn, J. Thin-film microextraction offers another geometry for solid phase microextraction. *TrAC Trends Anal. Chem.* **2012**, *39*, 245–253. [CrossRef]
16. Abdel-Rehim, M. New trend in sample preparation: On-line microextraction in packed syringe for liquid and gas chromatography applications I. Determination of local anaesthetics in human plasma samples using gas chromatography–mass spectrometry. *J. Chromatogr. B* **2004**, *801*, 317–321. [CrossRef]
17. Brewer, W.E. Disposable Pipette Extraction. U.S. Patent 6,566,145 B2, 20 May 2003.
18. Laxman Kole, P.; Venkatesh, G.; Kotecha Ravi Sheshala, J. Recent advances in sample preparation techniques for effective bioanalytical methods. *Biomed. Chromatogr.* **2011**, *25*, 199–217. [CrossRef]
19. Seidi, S.; Tajik, M.; Baharfar, M.; Rezazadeh, M. Micro solid-phase extraction (pipette tip and spin column) and thin film solid-phase microextraction: Miniaturized concepts for chromatographic analysis. *TrAC Trends Anal. Chem.* **2019**, *118*, 810–827. [CrossRef]
20. Cristiane Mozaner Bordin, D.; Nogueira Rabelo Alves, M.; Geraldo de Campos, E.; Spinosa De Martinis, B. Disposable pipette tips extraction: Fundamentals, applications and state of the art. *J. Sep. Sci.* **2016**, *39*, 1168–1172. [CrossRef]
21. Khezeli, T.; Daneshfar, A. Development of dispersive micro-solid phase extraction based on micro and nano sorbents. *TRAC Trends Anal. Chem.* **2017**, *89*, 99–118. [CrossRef]
22. Chisvert, A.; Cárdenas, S.; Lucena, R. Dispersive micro-solid phase extraction. *TrAC Trends Anal. Chem.* **2019**, *112*, 226–233. [CrossRef]
23. Baltussen, E.; Sandra, P.; David, F.; Cramers, C. Stir bar sorptive extraction (SBSE), a novel extraction technique for aqueous samples: Theory and principles. *J. Microcolumn Sep.* **1999**, *11*, 737–747. [CrossRef]
24. Farajzadeh, M.A.; Djozan, D.; Nouri, N.; Bamorowat, M.; Shalamzari, M.S. Coupling stir bar sorptive extraction-dispersive liquid–liquid microextraction for preconcentration of triazole pesticides from aqueous samples followed by GC-FID and GC-MS determinations. *J. Sep. Sci.* **2010**, *33*, 1816–1828. [CrossRef] [PubMed]

25. Cárdenas, S.; Lucena, R. Recent Advances in Extraction and Stirring Integrated Techniques. *Separations* **2017**, *4*, 6. [CrossRef]
26. Liu, H.H.; Dasgupta, P.K. Analytical chemistry in a drop. Solvent extraction in a microdrop. *Anal. Chem.* **1996**, *68*, 1817–1821. [CrossRef] [PubMed]
27. Jeannot, M.A.; Cantwell, F.F. Solvent microextraction into a single drop. *Anal. Chem.* **1996**, *68*, 2236–2240. [CrossRef]
28. Wen, Y.; Li, J.; Ma, J.; Chen, L. Recent advances in enrichment techniques for trace analysis in capillary electrophoresis. *Electrophoresis* **2012**, *33*, 2933–2952. [CrossRef]
29. Tang, S.; Qi, T.; Ansah, P.D.; Fouemina, J.C.N.; Shen, W.; Basheer, C.; Lee, H.K. Single-drop microextraction. *TrAC Trends Anal. Chem.* **2018**, *108*, 306–313. [CrossRef]
30. Rezaee, M.; Assadi, Y.; Milani, H.; Aghaee, E.; Ahmadi, F.; Berijani, S. Determination of organic compounds in water using dispersive liquid-liquid microextraction. *J. Chromatogr. A* **2006**, *1116*, 1–9. [CrossRef]
31. Zuo, M.; Cheng, J.; Matsadiq, G.; Liu, L.; Li, M.-L.; Zhang, M. Application of dispersive liquid–liquid microextraction based on solidification of floating organic droplet multi-residue method for the simultaneous determination of polychlorinated biphenyls, organochlorine, and pyrethroid pesticides in aqueous sample. *CLEAN—Soil Air Water* **2012**, *40*, 1326–1333. [CrossRef]
32. Pedersen-Bjergaard, S.; Rasmussen, K.E. Liquid-liquid-liquid microextraction for sample preparation of biological fluids prior to capillary electrophoresis. *Anal. Chem.* **1999**, *71*, 2650–2656. [CrossRef] [PubMed]
33. Shen, G.; Lee, H.K. Hollow fiber-protected liquid-phase microextraction of triazine herbicides. *Anal. Chem.* **2002**, *74*, 648–654. [CrossRef] [PubMed]
34. Pedersen-Bjergaard, S.; Rasmussen, K.E. Electrokinetic migration across artificial liquid membranes—New concept for rapid sample preparation of biological fluids. *J. Chromatogr. A* **2006**, *1109*, 183–190. [CrossRef] [PubMed]
35. Pedersen-Bjergaard, S.; Huang, C.; Gjelstad, A. Electromembrane extraction–Recent trends and where to go. *J. Pharm. Anal.* **2017**, *7*, 141–147. [CrossRef] [PubMed]
36. Ebrahimzadeh, H.; Mehdinia, A.; Kamarei, F.; Moradi, E. A Sensitive Method for the Determination of Methadone in Biological Samples Using Nano-Structured α-Carboxy Polypyrrol as a Sorbent of SPME. *Chromatographia* **2012**, *75*, 149–155. [CrossRef]
37. Goryński, K.; Kiedrowicz, A.; Bojko, B. Development of SPME-LC-MS method for screening of eight beta-blockers and bronchodilators in plasma and urine samples. *J. Pharmac. Biomed. Anal.* **2016**, *127*, 147–155. [CrossRef]
38. Szultka, M.; Krzemiński, R.; Jackowski, M.; Buszewski, B. Simultaneous determination of selected chemotherapeutics in human whole blood by molecularly imprinted polymers coated solid phase microextraction fibers and liquid chromatography-tandem mass spectrometry. *J. Chromatogr. B* **2013**, *940*, 66–76. [CrossRef]
39. Kataoka, H.; Ehara, K.; Yasuhara, R.; Saito, K. Simultaneous determination of testosterone, cortisol, and dehydroepiandrosterone in saliva by stable isotope dilution on-line in-tube solid-phase microextraction coupled with liquid chromatography–tandem mass spectrometry. *Anal. Bioanal. Chem.* **2013**, *405*, 331–340. [CrossRef]
40. Said, R.; Pohanka, A.; Abdel-Rehim, M.; Beck, O. Determination of four immunosuppressive drugs in whole blood using MEPS and LC-MS/MS allowing automated sample work-up and analysis. *J. Chromatogr. B* **2012**, *897*, 42–49. [CrossRef]
41. Jagerdeo, E.; Abdel-Rehim, M. Screening of cocaine and its metabolites in human urine samples by direct analysis in real-time source coupled to time-of-flight mass spectrometry after online preconcentration utilizing microextraction by packed sorbent. *J. Am. Soc. Mass Spectrom.* **2009**, *20*, 891–899. [CrossRef]
42. Szultka, M.; Krzemiński, R.; Szeliga, J.; Jackowski, M.; Buszewski, B. A new approach for antibiotic drugs determination in human plasma by liquid chromatography-mass spectrometry. *J. Chromatogr. A* **2013**, *1272*, 41–49. [CrossRef] [PubMed]
43. Rezaei Kahkha, M.R.; Oveisi, A.R.; Kaykhaii, M.; Rezaei Kahkha, B. Determination of carbamazepine in urine and water samples using amino-functionalized metal-organic framework as sorbent. *Chem. Cent. J.* **2018**, *12*, 77. [CrossRef] [PubMed]
44. Abdel-Rehim, M.; Persson, C.; Zeki, A.; Blomberg, L. Evaluation of monolithic packed 96-tips and liquid chromatography-tandem mass spectrometry for extraction and quantification of pindolol and metoprolol in human plasma samples. *J. Chromatogr. A* **2008**, *1196–1197*, 23–27. [CrossRef] [PubMed]

45. Skoglund, C.; Bassyouni, F.; Abdel-Rehim, M. Monolithic packed 96-tips set for high-throughput sample preparation: Determination of cyclophosphamide and busulfan in whole blood samples by monolithic packed 96-tips and LC-MS. *Biomed. Chromatogr.* **2013**, *27*, 714–719. [CrossRef]
46. Ettore, F.J.; Eloisa, D.C. Simultaneous determination of drugs and pesticides in postmortem blood using dispersive solid-phase extraction and large volume injection-programmed temperature vaporization-gas chromatography-mass spectrometry. *Forensic Sci. Inter.* **2018**, *290*, 318–326. [CrossRef]
47. Asgharinezhad, A.A.; Karami, S.; Ebrahimzadeh, H.; Shekari, N.; Jalilian, N. Polypyrrole/magnetic nanoparticles composite as an efficient sorbent for dispersive micro-solid-phase extraction of antidepressant drugs from biological fluids. *Int. J. Pharm.* **2015**, *494*, 102–112. [CrossRef]
48. Xiao, D.; Wang, C.; Dai, H.; Peng, J.; He, J.; Zhang, K.; Kong, S.; Qiu, P.; He, H. Applications of magnetic surface imprinted materials for solid phase extraction of levofloxacin in serum samples. *J. Mol. Recognit.* **2015**, *28*, 277–284. [CrossRef]
49. Marques, L.A.; Nakahara, T.T.; Madeira, T.B.; Almeida, M.B.; Monteiro, A.M.; de Almeida Silva, M.; Carrilho, E. Optimization and validation of an SBSE–HPLC–FD method using laboratory-made stir bars for fluoxetine determination in human plasma. *Biomed. Chromatogr.* **2018**, *33*, e4398. [CrossRef]
50. Babarahimi, V.; Talebpour, Z.; Haghighi, F.; Adib, N.; Vahidi, H. Validated determination of losartan and valsartan in human plasma by stir bar sorptive extraction based on acrylate monolithic polymer, liquid chromatographic analysis and experimental design methodology. *J. Pharmac. Biomed. Anal.* **2018**, *153*, 204–213. [CrossRef]
51. Taghvimia, A.; Hamishehkarb, H. Developed nanocarbon-based coating for simultaneous extraction of potent central nervous system stimulants from urine media by stir bar sorptive extraction method coupled to high performance liquid chromatography. *J. Chromatogr. B* **2019**, *1125*, 121721. [CrossRef]
52. Jahan, S.; Xie, H.; Zhong, R.; Yan, J.; Xiao, H.; Fan, L.; Cao, C. A highly efficient three-phase single drop microextraction technique for sample preconcentration. *Analyst* **2015**, *140*, 3193–3200. [CrossRef] [PubMed]
53. Stege, P.W.; Lapierre, A.V.; Martinez, L.D.; Messina, G.A.; Sombra, L.L. A combination of single-drop microextraction and open tubular capillary electrochromatography with carbon nanotubes as stationary phase for the determination of low concentration of illicit drugs in horse urine. *Talanta* **2011**, *86*, 278–283. [CrossRef] [PubMed]
54. Gao, W.; Chen, G.; Chen, Y.; Li, N.; Chen, T.; Hu, Z. Selective extraction of alkaloids in human urine by on-line single drop microextraction coupled with sweeping micellar electrokinetic chromatography. *J. Chromatogr. A* **2011**, *1218*, 5712–5717. [CrossRef]
55. Tian, J.; Chen, X.; Bai, X. Comparison of dispersive liquid-liquid microextraction based on organic solvent and ionic liquid combined with high-performance liquid chromatography for the analysis of emodin and its metabolites in urine samples. *J. Sep. Sci.* **2012**, *35*, 145–152. [CrossRef] [PubMed]
56. Cruz-Vera, M.; Lucena, R.; Cárdenas, S.; Valcárel, M. One-step in syringe ionic liquid-based dispersive liquid-liquid microextraction. *J. Chromatogr. A* **2009**, *1216*, 6459–6465. [CrossRef]
57. Jouyban, A.; Sorouraddin, M.H.; Farajzadeh, M.A.; Somi, M.H.; Fazeli-Bakhtiyari, R. Determination of five antiarrhythmic drugs in human plasma by dispersive liquid-liquid microextraction and high-performance liquid chromatography. *Talanta* **2015**, *134*, 681–689. [CrossRef]
58. Panahi, H.A.; Ejlali, M.; Chabouk, M. Two-phase and three-phase liquid-liquid microextraction of hydrochlorothiazide and triamterene in urine samples. *Biomed. Chromatogr.* **2015**, *30*, 1022–1028. [CrossRef]
59. Valle de Bairros, A.; Lanaro, R.; Menck de Almeida, R.; Yonamine, M. Determination of ketamine, norketamine and dehydronorketamine in urine by hollow-fiber liquid-phase microextraction using an essential oil as supported liquid membrane. *Forensic Sci. Inter.* **2014**, *243*, 47–54. [CrossRef]
60. Khataei, M.M.; Yamini, Y.; Nazaripour, A.; Karimi, M. Novel generation of deep eutectic solvent as an acceptor phase in three-phase hollow fiber liquid phase microextraction for extraction and preconcentration of steroidal hormones from biological fluids. *Talanta* **2018**, *178*, 473–480. [CrossRef]

© 2019 by the authors. Licensee MDPI, Basel, Switzerland. This article is an open access article distributed under the terms and conditions of the Creative Commons Attribution (CC BY) license (http://creativecommons.org/licenses/by/4.0/).

Review

Evolution of Environmentally Friendly Strategies for Metal Extraction

Govind Sharma Shyam Sunder [1], Sandhya Adhikari [1], Ahmad Rohanifar [1], Abiral Poudel [1] and Jon R. Kirchhoff [1,2,3,*]

1. Department of Chemistry and Biochemistry, College of Natural Sciences and Mathematics, University of Toledo, Toledo, OH 43606, USA; Govind-Sharma.Shyam-Sunder@rockets.utoledo.edu (G.S.S.S.); Sandhya.Adhikari@rockets.utoledo.edu (S.A.); Ahmad.RohaniFar@rockets.utoledo.edu (A.R.); Abiral.Poudel@rockets.utoledo.edu (A.P.)
2. School of Green Chemistry and Engineering, University of Toledo, Toledo, OH 43606, USA
3. Dr. Nina McClelland Laboratory for Water Chemistry and Environmental Analysis, University of Toledo, Toledo, OH 43606, USA
* Correspondence: jon.kirchhoff@utoledo.edu

Received: 9 October 2019; Accepted: 3 December 2019; Published: 6 January 2020

Abstract: The demand for the recovery of valuable metals and the need to understand the impact of heavy metals in the environment on human and aquatic life has led to the development of new methods for the extraction, recovery, and analysis of metal ions. With special emphasis on environmentally friendly approaches, efforts have been made to consider strategies that minimize the use of organic solvents, apply micromethodology, limit waste, reduce costs, are safe, and utilize benign or reusable materials. This review discusses recent developments in liquid- and solid-phase extraction techniques. Liquid-based methods include advances in the application of aqueous two- and three-phase systems, liquid membranes, and cloud point extraction. Recent progress in exploiting new sorbent materials for solid-phase extraction (SPE), solid-phase microextraction (SPME), and bulk extractions will also be discussed.

Keywords: metal extraction; liquid–liquid extraction; solid-phase extraction; solid-phase microextraction; green extraction methods

1. Introduction

Metals are ubiquitous in nature serving as essential elements for human health and critical materials for modern industrialization and urbanization. While some metals such as iron are necessary for human health, many metals are toxic, and can cause physical problems such as diarrhea, nausea, asthma, kidney malfunction, different cancers, and even death [1]. Arsenic, cadmium, chromium, mercury, and lead are commonly known as heavy metals—or metalloids in the case of arsenic—and have the greatest toxicity. The maximum limits in drinking water for these metal ions according to the World Health Organization (WHO) are 10, 3, 50, 6, and 10 µg L^{-1}, respectively [2]. The harmful effect of arsenic can mostly affect skin, respiratory, and cardiovascular systems. Elevated risk of skin and lung cancers has been reported among people who were exposed to arsenic from working in mining and smelting areas where inorganic arsenic was inhaled [3]. Cadmium and lead are harmful for the nervous system. Mercury used in electrical devices, dental fillings, Hg vapor lamps, solders, and X-ray tubes has a strong attraction to biological tissues and is carcinogenic, mutagenic, and teratogenic [4]. The Flint water crisis in 2014 affected about 100,000 people when lead from aging pipes leached into the water supply and contaminated the drinking water. This poignant example illustrates the importance of careful monitoring of heavy metals (HMs) in water systems and investigating new technologies to extract and remove them [5,6].

Other metals such as cobalt, copper, iron, and zinc have higher threshold limits. The maximum limit for copper in drinking water is 2 mg L^{-1} according to the WHO [2]. No guideline values are provided for iron and zinc in drinking water, however, high concentrations of these elements may still cause adverse health effects or, at a minimum, an unacceptable taste for consumers [2]. The recovery, removal, and recycling of valuable metals, including gold, platinum, and rare earth elements, from natural and secondary sources such as industrial wastes is also important for their economic, strategic, and national security value. These critical elements have important applications in metallurgy and the biomedical and electronics industries [7–12].

Several methods have been used for extraction and removal of metals from different sources of water, including microfiltration [13], chemical precipitation [14], coagulation and flocculation [15], electrochemical removal [16], liquid–liquid extraction [17,18], osmosis [19], crystallization and distillation [20], photocatalysis [21], and adsorption. In this review, we focus on several techniques for extraction, determination, and removal of metals, including heavy and valuable metals, from water samples. In particular, extraction methods that aim to provide environmentally friendly, simpler and faster techniques are discussed. Approaches include recent advances in primarily liquid–liquid and solid-phase extraction. Comparison of their advantages and disadvantages will be made to illustrate efforts to develop more environmentally friendly methods.

2. Liquid-Based Extraction

Numerous liquid-based techniques like liquid–liquid extraction (LLE) [9,22–24], chemical precipitation [14,25], and cloud point extraction [26] have been utilized for extraction of metal ions from aqueous media. Among the listed methods, LLE is based on analyte partitioning between two immiscible phases. Conventional LLE is widely used for separations and preconcentration, including extraction and recovery of metal species from aqueous media by the addition of organic solvents [24]. This technique has significant advantages. These include rapid extraction kinetics, the ability to choose selective solvents, amenability to large-scale separation, and easy and flexible implementation. In spite of these advantages, traditional LLE has several drawbacks, including the extensive use of volatile and flammable organic solvents, which are potential health and environmental hazards [27]. Moreover, from the economic point of view, LLE is quite expensive because of the cost of organic extractants and their disposal [28]. These drawbacks can be circumvented by embracing new methods that provide simple, low-cost, fast, sensitive, and accurate analyses in a more environmentally friendly manner. Various advancements in liquid-based extraction for metal ions, such as aqueous biphasic and triphasic extraction, cloud point extraction, and liquid membrane extraction, are discussed herein.

2.1. Aqueous Biphasic Systems

Aqueous biphasic systems (ABS, Figure 1) forms when two immiscible aqueous-based solutions are mixed together at a certain temperature [29]. ABS have gained more attention for metal extraction since 1984 when Zvarova and co-workers successfully extracted copper, zinc, cobalt, iron, indium, and molybdenum using a polyethylene glycol (PEG) 2000–ammonium sulfate–water system in the presence of ammonium thiocyanate and sulfuric acid [30]. Because organic solvents are not required in ABS, it has several advantages over traditional solvent extraction. ABS are less toxic, more economical, biocompatible, and have a reduced environmental risk [31]. Furthermore, numerous inorganic anions can be used as water-soluble extractants resulting in metal ion partitioning between two immiscible aqueous phases, which reduces dehydration effects [32].

ABS can be formed by various mechanisms and thus are tunable to the desired extraction. Biphasic systems composed of polymer–polymer [33], polymer–salt [34], salt–salt [35], ionic liquid–salt [36], and surfactant-based systems [37] have been reported. In addition to these, other phase-forming elements are amino acids, alcohols, and carbohydrates. In ABS, factors governing the metal ion extraction include molecular weight and polymer type [38], Gibbs free energy of hydration [32], medium pH [39], presence and absence of an extracting agent [32,40], and temperature [41]. Examples of metal ion extractions using different ABS are given in Table 1.

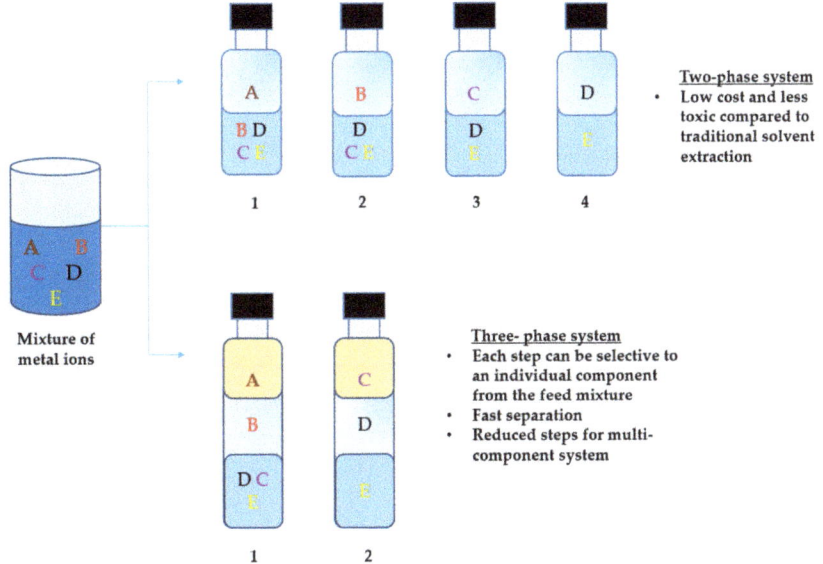

Figure 1. Schematic representations of a two- versus three-phase system for metal ion extraction.

Table 1. Aqueous biphasic systems for metal ion extraction.

Targeted Metal(s)	ABS Composition	Extraction Agent	Detection	Ref.
	Surfactant–Salt			
Zn^{2+}	Triton X-100 [a], $MgSO_4$	PAN [n]	UV–Vis	[42]
Mo^{6+}, W^{6+}	Triton X-100, $(NH_4)_2SO_4$	None	ICP–AES [r]	[43,44]
	Polymer–Salt			
Hg^{2+}, Zn^{2+}, Co^{2+}	PEG 6000 [b], Na_2CO_3	None	AAS [s]	[45]
Mn^{2+}, Fe^{3+}, Co^{2+}, Ni^{2+}, Cu^{2+}, Zn^{2+}, Cd^{2+}, Li^+	PEG 4000, Na_2SO_4	None	AAS	[46]
Fe^{3+}, Co^{2+}, Ni^{2+}, Cu^{2+}, Zn^{2+}, Cd^{2+}	L35 [c], Na_2SO_4	1N2N [o], SCN^-, I^-	AAS	[47]
Cd^{2+}, Ni^{2+}	L35, $LiSO_4$	KI, TTL [p]	AAS	[48]
Zn^{2+}, Cd^{2+}, Hg^{2+}, Pb^{2+}, Bi^{3+}	PEG 1550, Na_2SO_4, $NaNO_3$, $(NH_4)_2SO_4$	NaX, X = I^-, Cl^-, Br^-, SCN^-	FTIR [t]	[49]
Co^{2+}, Fe^{3+}, Ni^{2+}	PEO [d] 1500, $(NH_4)_2SO_4$, H_2O	KSCN	FAAS [u]	[50]
Hg^{2+}	PEG 5000, Na_2SO_4	NaX, X = I^-, Cl^-, Br^-	Packard Cobra II Auto-γ-Spectrometer	[51]
Co^{2+}, Ni^{2+}, Cd^{2+}	L64 [e], $Na_2C_4H_4O_6$	1N2N	FAAS	[52]
Ca^{2+}	L64, sodium tartrate	None	FAAS	[53]
As^{3+}	L64, $(NH_4)_2SO_4$, H_2O	APDC [q]	ICP–OES [v]	[54]

Table 1. Cont.

Targeted Metal(s)	ABS Composition	Extraction Agent	Detection	Ref.
	Salt–Salt			
Cd^{2+}	TBAB [f], $(NH_4)_2SO_4$	None	AAS	[55]
	Ionic Liquid–Salt			
Ni^{2+}, Co^{2+}	(P44414) [g](Cl), NaCl	None	NR	[56]
Co^{2+}, Fe^{3+}, Nd^{3+}, Sm^{3+}	Cyphos IL 101 [h], NaCl	None	ICP–OES, TXRF [w]	[57,58]
Sc^{3+}	$(P_{444}C_1COOH)Cl$ [i], NaCl	None	TXRF	[59]
Au^+	1-alkyl-3-methylimidazolium bromide, K_2HPO_4	None	AAS	[60]
Co^{2+}	$(HMIM)(BF_4)$ [j], NaCl	None	ICP–OES	[61]
Pr^{3+}	$(A336)(NO_3)$ [k], $NaNO_3$	None	UV–Vis	[62]
Nd^{3+}	(P4444) [l](NO_3), NaCl	None	ICP–MS	[63]
	Miscellaneous			
Au^{3+}	$(C_6mim)(C_{12}SO_3)$ [m], PEG 6000	None	UV–Vis	[64]

[a] octylphenolpolyethoxylene, [b] polyethylene glycol (average molecular mass 6000), [c] (ethylene oxide)$_{11}$ (propylene oxide)$_{16}$ (ethylene oxide)$_{11}$, [d] poly(ethylene oxide), [e] (ethylene oxide)$_{13}$-(propylene oxide)$_{30}$-(ethylene oxide)$_{13}$, [f] tetrabutylammonium bromide, [g] tributyl(tetradecyl)phosphonium, [h] tri(hexyl)tetradecylphosphonium chloride, [i] tri-n-butyl(carboxymethyl)phosphonium chloride, [j] 1-hexyl-3-methylimidazolium tetrafluoroborate, [k] tricaprylmethylammonium nitrate, [l] tetrabutylphosphonate, [m] 1-hexyl-3-methylimidazole dodecyl sulfonate, [n] 1-(2-pyridylazo)-2-naphthol, [o] 1-nitroso-2-naphthol, [p] tie-line length, [q] ammonium pyrrolidine dithiocarbamate, [r] inductively coupled plasma atomic emission spectrometry, [s] atomic absorption spectrophotometry, [t] fourier transform infrared spectrophotometry, [u] flame atomic absorption spectrophotometry, [v] inductively coupled plasma optical emission spectrometry, [w] total reflection X-ray fluorescence.

It is important to note that to extract a single target metal in each extraction step with a biphasic system, the extraction process for a specific metal from a mixture of metal ions must be highly selective, resulting in a potentially lengthy and costly method. A well-designed three-liquid-phase extraction system may overcome this disadvantage by selective separation and extraction of two or more targeted metals during a single extraction step.

2.2. Three-Liquid-Phase Extraction

Three-liquid-phase extraction (TLP, Figure 1) has been used for the isolation of organic macromolecules such as cellulose, enzymes, proteins, and metals [65,66]. This approach is based on the use of three immiscible liquid phases composed of different organic solvents, polymers, inorganic salts, water, or ionic liquids [67,68]. As the number of non-miscible phases is increased from two (biphasic) to three (triphasic), the steps required for separation decrease. Therefore, three metal cations can be separated simultaneously in a single step as shown in Figure 1. For example, in the case of a biphasic system, a mixture of five metals may require four steps for the separation, whereas for TLP, two steps may be sufficient. Different approaches have been considered to design a TLP system for metal extraction. These include one aqueous and two organic phases [69], one organic and two aqueous phases [70], and ionic liquid-based systems [71]. One recent study showed an improved extraction efficiency for Co^{2+} with a TLP system when directly compared to an ionic liquid ABS approach [61]. However, in TLP, the challenges associated with the use of organic phases are reintroduced. Examples of metal ion extraction using different TLP systems are tabulated in Table 2.

Table 2. TLP systems for metal ion extraction.

TLP Phases	TLP Component			Metal Extracted			Ref.
	Top	Middle	Bottom	Top	Middle	Bottom	
1 Organic 2 Aqueous							
	TRPO [a]	PEG-2000	$(NH_4)_2SO_4$, H_2O	Ti^{4+}	Fe^{3+}	Mg^{2+}	[72]
	S201 [b]	EOPO [j]	Na_2SO_4, H_2O	Pd^{2+}	Pt^{4+}	Rh^{3+}	[73,74]
	D2EHPA [c]	PEG	$(NH_4)_2SO_4$, H_2O	Cr^{3+}	Cr^{6+}	None	[75]
	Cyanex272 [d]	PEG	$(NH_4)_2SO_4$, H_2O	Yb^{3+}	Eu^{3+}	La^{3+}	[76]
	Cyanex272	PEG 2000	$(NH_4)_2SO_4$, H_2O	Yb^{3+}, Eu^{3+}	Fe^{3+}, Si^{4+}	La^{3+}, Al^{3+}	[77]
	PC-88A [e]	PEG 2000	$(NH_4)_2SO_4$, H_2O	Eu^{3+}	Al^{3+}, Si^{4+}, Fe^{3+}	La^{3+}, Yb^{3+}	[78]
	Xylene, (D2EHPA)	PEG	$(NH_4)_2SO_4$, H_2O	Mn^{2+}	Co^{2+}	Ni^{2+}	[79]
	N1923 [f]	PEG	$(NH_4)_2SO_4$, H_2O	V^{5+}	Cr^{6+}	Al^{3+}	[80]
2 Organic 1 Aqueous							
	S201	(Sugaring out) CH_3CN	glucose, H_2O	Pd^{2+}	Pt^{4+}	Rh^{3+}	[81]
	S201	(Salting out) CH_3CN	NaCl, H_2O	Pd^{2+}	Pt^{4+}	Rh^{3+}	[69]
TLP Systems with Ionic Liquids							
	H_2O	(HMIM) (BF_4) [k]	NaCl	None	Co^{2+}	None	[61]
	TOPO [g]	H_2O	(Bmim) (PF_6) [l]	Mn^{2+}, Zn^{2+}, Cd^{2+}, Pb^{2+}	None	Cu^{2+}, Ni^{2+}	[71]
	S201	H_2O	$(C_4mim)(PF_6)$ [l]	Pd^{2+}	Rh^{3+}	Pt^{4+}	[82]
	TBP [h], (P66614) (Tf_2N) [i]	H_2O	(Hbet) (Tf_2N) [m]	Sn^{2+}	Sc^{3+}	Y^{3+}	[66]

[a] trialkylphosphine oxide, [b] diisoamyl sulfide/nonane, [c] di(2-ethylhexyl)phosphoric acid, [d] bis(2,4,4-trimethylpentyl)phosphinic acid, [e] 2-ethylhexylphosphoric acid mono(2-ethylhexyl)ester, [f] primary amine, [g] tri-n-octylphosphine oxide, [h] tri-n-butyl phosphate, [i] trihexyl(tetradecyl)phosphonium bis(trifluoromethylsulfonyl)imide, [j] polyethylene oxide-polypropylene oxide, [k] 1-hexyl-3-methylimidazolium tetrafluoroborate, [l] 1-butyl-3-methylimidazolium hexafluorophosphate, [m] betainium bis(trifluoromethylsulfonyl) imide.

2.3. Cloud Point Extraction (CPE)

The cloud point is the point where a solution mixture turns cloudy due to diminished solubility of one component after changes to experimental conditions such as pressure, temperature, and inclusion of additives [83]. For example, this clouding process can result in the formation of two distinct phases of nonionic and zwitterionic surfactants in which one is a surfactant-rich phase and the other has a concentration close to the critical micelle concentration [84]. The surfactant-rich phase obtained at the cloud phase condition functions to extract and preconcentrate various inorganics [85]. This phase extracts metal cations and is dispersed in the aqueous phase formed after phase separation. Detection of the cloud point occurs by various techniques (e.g., light scattering or particle counting, turbidimetry, refractometry, thermo-optical methods, and viscometry) [86]. CPE shows great promise as a more environmentally friendly method for heavy metal extractions [87]. Kazi et al. have studied extraction of Al^{3+} by the cloud point technique where 8-hydroxyquinone was added to coordinate Al^{3+} while the surfactant octylphenoxypolyethoxyethanol (Triton X-114) was added to extract and entrap the complex [88]. Similarly, Zhao et al. studied the extraction of Cd^{2+}, Co^{2+}, Ni^{2+}, Pb^{2+}, Zn^{2+}, and Cu^{2+} using a dual-CPE technique [89]. The main advantage of CPE over other techniques is the use of water instead of organic solvents [90]. CPE is also easy to manipulate, is fast, requires minimal expense, and offers high analyte recovery [85,91].

2.4. Liquid Membrane Extraction

Membrane-based extraction is a non-equilibrium process that has been developed as an important green strategy for recovery of rare earth elements [92]. Different types of liquid membranes (LM) have been reported, such as bulk liquid membrane (BLM) [93], emulsion liquid membrane (ELM) [94], supported liquid membrane (SLM) [95], and hollow fiber-supported liquid membrane (HFSLM) [96]. Their advantages and disadvantages are summarized in Table 3. Various metal ions from common metals (copper, nickel, and cobalt) [97] and valuable metals (platinum, gold) [98,99] to radioactive species (uranium) [100] have been extracted using LM techniques. As noted in Table 3, there are several concerns regarding membrane stability when organic solvents are used.

Table 3. Liquid membrane systems for metal ion extraction.

Type of LM	Overview	Advantages	Disadvantages	Ref.
SLM	Hydrophobic membrane impregnated with an organic solvent is squeezed between an aqueous feed and stripping solution	Simplicity of operation Low operating cost	Emulsion formation of liquid membrane phase in water Instability	[95]
HFSLM	Hollow fiber is used as microporous hydrophobic membrane and impregnated with LM phase	High interfacial area-to-volume ratio	Lower transport rate than SLM	[96]
ELM	Water/organic/water (W/O/W) or organic/water/organic (O/W/O) with a thin middle LM phase	High transfer rates	Continuous operation is difficult to achieve as settling stage is performed after extraction. Long contact of emulsions with water in feed stream results in swelling and rupture due to the difference in osmotic pressure, shear forces, and static pressure between the feed and stripping phase	[101]
BLM	An aqueous feed and stripping phase separated by bulk organic LM phase	High transfer rate	Less interfacial area-to-volume ratio results in low fluxes	[93]

2.5. Summary

In summary, LLE methods often require several extractions for complete recovery of targeted metals. Thus, LLE is often replaced by solid-phase extraction (SPE) methods to achieve higher efficiency and recovery. SPE is advantageous because consumption of organic solvent can be minimized [102]. Additionally, errors from inaccurately measured extraction volumes, especially when multiple extraction steps are required with LLE, are minimized with SPE as it does not require phase separation [103].

3. Solid-Phase Extraction

Solid-phase extraction (SPE, Figure 2) is one of the most popular sample pretreatment and separation techniques because of its simplicity, low cost, high preconcentration factors, selectivity, and versatility. Furthermore, the availability of a wide variety of sorbent materials and the ability to use only minimal amounts, or in some cases, no organic solvents, makes SPE a very environmentally friendly technique [102,104,105]. Most of the benefits of SPE methods are governed by the physical and chemical nature of the sorbent [104,106]. Recent development and applications of a number of new sorbent materials for metal extraction, such as nanosorbent materials, polymers, metal oxides, magnetic materials, metal organic frameworks (MOFs), and bioadsorbents, are discussed herein.

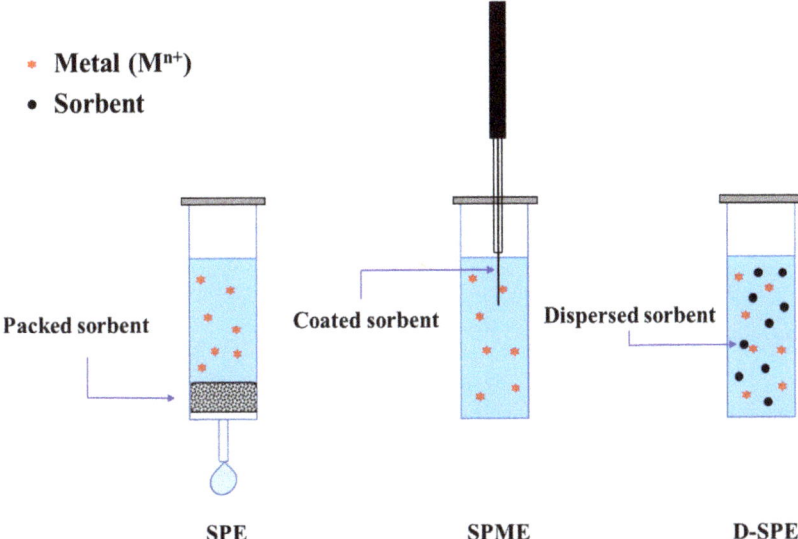

Figure 2. Schematic representation of solid-phase extraction (SPE), solid-phase microextraction (SPME, direct immersion only), and dispersive solid-phase extraction (D-SPE).

3.1. Nanosorbent Materials

Nanosorbent materials such as carbon nanotubes (CNTs) [107], graphene oxide (GO), silica [108], chitosan [109], and activated carbon [110,111] are particularly useful due to their large surface areas compared to their particle volume. Thus, they are excellent candidates as sorbent materials for metals since the high surface area provides a greater number of active sites leading to enhanced extraction efficiency. Recently, Gouda et al. developed a sorbent material based on multiwalled carbon nanotubes impregnated with 2-(2-benzothiazolylazo)orcinal (BTAO) for preconcentration of cadmium, copper, nickel, lead, and zinc from food and water samples prior to determination by flame atomic absorption [112]. Similarly, carbon nanotubes impregnated with tartrazine [113], polyaniline [114], and di-(2-ethyl hexyl phosphoric acid) [115] have been utilized as sorbent materials for preconcentration, separation, and determination of metals. Moreover, Awual et al. synthesized ligand-impregnated conjugate nanomaterials for the extraction of mercury from aqueous solution [116]. Metal oxides such as Al_2O_3 [117], TiO_2 [118], and SiO_2 [119] have been used for metal extraction due to their physical stability, cost-effectiveness, and high surface area [118]. Other examples are shown in Table 4. The utilization of nanosorbent materials is attributed to their high surface area, ease of modification, and nonspecific adsorption with metals [120,121]. However, limitations include low selectivity and, in some cases, low stability and limited reusability of the material.

Table 4. Nanomaterial-based solid sorbents.

Sorbent	Extraction Method	Target Metal(s)	Reusability	SC ^{f} (mg g^{-1})	Ref.
Tyre-based activated carbon	SPE-FAAS ^{d}	As^{5+}, Cd^{2+}, Cr^{3+}, Cu^{2+}, Fe^{3+}, Mn^{2+}, Ni^{2+}, Pb^{2+}, Zn^{2+}	NR	NR	[122]
Dowex 50W-x8 & Chelex-100	SPE	Cd^{2+}, Co^{2+}, Cr^{3+}, Cu^{2+}, Fe^{3+}, Ni^{2+}, Pb^{2+}, Zn^{2+}	Stable up to 150 elution cycles	NR	[123]
ZnFe$_2$O$_4$ nanotubes (ZFONTs)	DMSPE ^{e}	Co^{2+}, Ni^{2+}, Mn^{2+}, Cd^{2+}	NR	Co^{2+}-30.09 Ni^{2+}-28.4 Mn^{2+}-35.4 Cd^{2+}-27.9	[124]
Agarose-g-PMMA ^{a}	DMSPE	Cd^{2+}, Ni^{2+}, Cu^{2+}, Zn^{2+}	NR	Cd^{2+}-31.8 Ni^{2+}-42.5 Cu^{2+}-48.3 Zn^{2+}-34.3	[125]
Activated carbon	DSPE	Cu^{2+}	Stable up to 6 cycles	1.6	[126]
MWCNTs ^{b}	DMSPE	Cr^{6+}	NR	NR	[127]
GO-MWCNTs-DETA ^{c}	SPE	Cr^{3+}, Fe^{3+}, Pb^{2+}, Mn^{2+}	NR	Cr^{3+}-5.4 Fe^{3+}-13.8 Pb^{2+}-6.6 Mn^{2+}-9.5	[128]

^{a} poly(methyl methacrylate) grafted agarose, ^{b} multiwalled carbon nanotubes, ^{c} diethylenetriamine, ^{d} flame atomic absorption spectrometry, ^{e} dispersive magnetic SPE, ^{f} sorption capacity, NR: not reported.

3.2. Polymer-Based Materials

Some of the limitations found with nanosorbent materials have been addressed by employing specially designed sorbent materials based on chelating resins [129–132], polymers with chelating units [133,134], ion imprinted polymers [135–138], and polymeric ionic liquids [135,139,140].

Polymeric chelating materials, unlike the inorganic nanosorbents, have the advantage of tunability in functionalization using unique chelating groups to obtain enhanced selectivity and extraction efficiencies for metals. Recently, Nunes et al. developed a greener SPE approach for the extraction of Zn and Ni by employing nylon-6 nanofibers modified with di-(2-ethylhexyl) phosphoric acid [141]. The experimental results suggested that these polymeric nanofibers were cost-effective because of their reusability even after ten cycles of extraction in addition to being ecofriendly due to the absence of organic solvents. The same polymeric material was also used for SPE of indium from LCD screens [142]. Furthermore, polymeric materials based on ionic liquids also were utilized as effective sorbent materials for extraction of metals. For example, a polymeric ionic liquid containing 3-(1-ethyl imidazolium-3-yl)propyl-methacrylamido bromide and ethylene dimethacrylate was specifically developed by Zhang et al. for extraction of antimony employing a stir cake sorptive extraction method [143]. Table 5 summarizes several additional examples of polymer sorbents including ion-imprinted polymer (IIP) materials for SPE of metals.

Table 5. Polymer-based sorbent materials for metal ion extraction.

Sorbent Material	Extraction Method	Target Metal(s)	Flow Rate	Extraction Time	Ref.
Copolymer Strata™-X resin	On-line SPE	Cd^{2+}, Pb^{2+}, Cu^{2+}, Cr^{6+}	NR	1.5	[144]
mGO/SiO$_2$@coPPy-Th [a]	MSPE [b]	Cd^{2+}, Pb^{2+}, Cu^{2+}, Cr^{3+}, Zn^{2+}	NR	6.5 min	[145]
Thallium ion-imprinted polymer	SPE [c]	Tl^{3+}	NR	30 min	[146]
Copolymer of 4-Vinylpyridine and Ni-Dithizone	SPE	Ni^{2+}	0.2 mL min^{-1}	NR	[147]
(EGDMA-MAH/Ni) [d] imprinted polymer	SPE	Ni^{2+}	0.5 mL min^{-1}	NR	[148]
Double imprinted chitosan-succinate polymer	SPE	Cu^{2+}	NR	NR	[149]
Dual imprinted polymers of Cd	SPE	Cd^{2+}	3.0 mL min^{-1}	20 min	[150]
Poly(GMA [e]-co-EDMA [f])-IDA [g]	SPE	Cu^{2+}, Pb^{2+}, Cd^{2+}	10 µL s^{-1}	NR	[151]
Nylon 6-DEHPA [h]	SPE	Zn^{2+}, Ni^{2+}	NR	7.5 min	[141]

[a] SiO$_2$-coated magnetic graphene oxide modified with polypyrrole–polythiophene, [b] magnetic solid-phase extraction, [c] solid-phase extraction, [d] ethyleneglycoldimethacrylate-methacryloylhistidinedihydrate nickel(II), [e] glycidyl methacrylate, [f] ethylene dimethacrylate, [g] iminodiacetate, [h] di-(2-ethyl)phosphoric acid.

3.3. Metal–Organic Frameworks

Metal–organic frameworks (MOFs) consist of metal ions and organic linkers that are strongly bonded together. These materials have been used as effective sorbents in various applications due to their highly porous structure and the ability to be synthesized in various shapes and sizes [152,153]. Recently, Tadjarodi et al. designed a magnetic nanocomposite sorbent from HKUST-1 MOF combined with Fe$_3$O$_4$@4-(5)-imidazoledithiocarboxylic acid (Fe$_3$O$_4$@DTIM) for SPE of Hg^{2+} in canned tuna and fish samples [154]. The sorbent selectivity towards Hg^{2+} was due to the presence of sulfur atoms in DTIM. Also, the magnetic Fe$_3$O$_4$ nanoparticles facilitated separation from samples by simply applying an external magnetic field while the MOF prevented aggregation of Fe$_3$O$_4$ nanoparticles by acting as spacers and a support matrix with the MOF cavities providing increased surface area to enhance sorption capacity. Similarly, Esmaeilzadeh developed a MOF with iron-based magnetic nanoparticles decorated with tetraethyl orthosilicate to create a silica layer on the surface [155]. The nanoparticles were subsequently functionalized with morin (2-(2,4-dihydroxyphenyl)-3,5,7-trihydroxychromen-4-one) as a chelating agent to develop a MIL-101(Fe)/Fe$_3$O$_4$@morin nanocomposite for the selective extraction and speciation of V^{4+} and V^{5+}. In this case, the silica layer provided stability for the Fe$_3$O$_4$ nanoparticles in acidic conditions as well as allowed for further functionalization. MIL-101(Fe) also prevented aggregation of the nanoparticles by acting as a spacer and support. In addition, Nasir et al. developed a two dimensional leaf shaped zeolite imidazolate frame work (2D ZIF-L) for arsenite adsorption [156]. Table 6 shows recently reported MOFs as effective sorbents for the SPE of metals.

Table 6. MOF sorbent materials for metal extraction.

Sorbent Material	Extraction Method	Target Metal(s)	Reusability	SC [f] (mg g^{-1})	Ref.
UiO-66 [a] -NH$_2$	SPE	Cd^{2+}, Cr^{3+}, Pb^{2+}, Hg^{2+}	NR	Cd^{2+}-49 Cr^{3+}-117 Pb^{2+}-232 Hg^{2+}-769	[157]
KNiFC [b] Fe$_3$O$_4$/KNiFC	MSPE	Cs$^+$	5	153 and 109	[158]
Fe$_3$O$_4$@ZIF-8 [c]	SPE	As^{5+}	NR	0.035–0.036	[159]
ZIF-8@cellulose	SPE	Cr^{6+}	NR	NR	[160]
FJI-H12 [d]	SPE	Hg^{2+}	NR	440	[161]
Fe$_3$O$_4$/IRMOF-3 [e]	MSPE	Cu^{2+}	10	2.4	[162]
UiO-66-OH	SPE	Th^{4+}	25	47.5	[163]

[a] zirconium-based, [b] potassium nickel hexacyanoferrate, [c] zeolitic imidazolate framework-8, [d] Co(II) and 2,4,6-tri(1-imidazolyl)-1,3,5-triazine, [e] iso-reticular MOFs, [f] sorption capacity.

3.4. Magnetic-Based Materials

In the process of developing more environmentally friendly methods, incorporation of magnetic materials such as iron oxide nanoparticles into sorbent composites has increased in recent years. Magnetic materials are utilized to readily extract target metal ions from complex matrices followed by sorbent separation from samples by an external magnetic field. Following desorption of the metals, the sorbent can be recovered and effectively recycled. Magnetic nanoparticles have been combined with carbon-based [164], ionic liquid [165,166], MOF [167], and polymer [168] materials for magnetic SPE of metals. Several such examples are given in Tables 4–6, while other unique recent studies using magnetic-based materials are described below and in Table 7.

Shirani et al. developed a magnetic sorbent based on an ionic liquid linked to magnetic multiwalled carbon nanotubes for simultaneous separation and determination of cadmium and arsenic in food samples using electrothermal atomic absorption spectrometry [169]. Habila et al. synthesized a sorbent material based on Fe$_3$O$_4$@SiO$_2$@TiO$_2$, which shows unique magnetic, photocatalytic and acid resistant properties, and was used for the preconcentration of copper, zinc, cadmium, and lead prior to ICP–MS analysis [170]. The advantage of this sorbent material was it not only allowed extraction of toxic heavy metals from complex matrices, but also assisted the simultaneous degradation of the organic matrix to aid preconcentration. Additionally, Molaei et al. utilized a copolymer based on polypyrrole and polythiophene (PPy–PTh) layered on the surface of SiO$_2$-coated magnetic graphene oxide for the extraction of trace amounts of copper, lead, chromium, zinc, and cadmium from water and agricultural samples [145].

Table 7. Magnetic-based sorbent materials for metal ion extraction.

Sorbent Material	Extraction Method	Target Metal(s)	WS [f] pH	SC [g] (mg g^{-1})	Ref.
CEMNPs [a]	MSPE	Cu^{2+}, Co^{2+}, Cd^{2+}	9.0	Cu^{2+}-3.21 Co^{2+}-1.23 Cd^{2+}-1.77	[171]
Co-IDA [b]	MSPE	Cu^{2+}	7.5	NR	[172]
M-PhCP [c]	MSPE	Cd^{2+}, Pb^{2+}	6.0	NR	[173]
Fe_3O_4@MOF-235(Fe)-OSO_3H	MSPE	Cd^{2+}	3.0	NR	[167]
(Fe_3O_4-ethylenediamine)/MIL-101(Fe)	MSPE	Cd^{2+}, Pb^{2+}, Zn^{2+}, Cr^{3+}	6.1	Cd^{2+}-155 Pb^{2+}-198 Zn^{2+}-164 Cr^{3+}-173	[174]
Fe_3O_4@TAR [d]	MSPE	Cd^{2+}, Pb^{2+}, Ni^{2+}	6.2	185–210	[175]
MOF Fe_3O_4-Pyridine	MSPE	Cd^{2+}, Pb^{2+}	6.3	186–198	[176]
SH-Fe_3O_4/$Cu_3(BTC)_2$ [e]	MSPE	Pb^{2+}	6.0	198	[177]

[a] carbon-encapsulated magnetic nanoparticles, [b] magnetic cobalt nanoparticles functionalized with iminodiacetic acid, [c] magnetic phosphorous-containing polymer, [d] thiazolylazo resorcinol, [e] mercapto groups modified with benzene tricarboxylic acid, [f] working solution pH, [g] sorption capacity.

3.5. Ion Exchange

Ion exchange is another technique that can be used for the removal of metals, though it depends on the solution composition [178]. Moreover, other factors like the capacity and selectivity of sorbent material, pH, temperature, and solution salinity also play important roles in the ion exchange process [179]. Recently Murray et al. studied the removal of Pb^{2+}, Cu^{2+}, Zn^{2+}, and Ni^{2+} from natural water with polymeric submicron ion exchange resins [180]. Similarly, Vergili et al. found good extraction properties with a weak acid cation resin for the sorption of Pb^{2+} from industrial wastewater [181].

3.6. Ligand Binding

Simple coordination chemistry, where a ligand with affinity for a metal binds and forms a complex, is a useful method to selectively isolate a metal from aqueous solution. There are numerous organic chelating agents for heavy and precious metal extraction. The overall challenge is achieving selectivity for a single metal or class of metals. Depending on the strength of binding, recovery of the isolated metal ion can also be difficult. Recent studies show dithiocarbamate ligands as one of the most useful materials to coordinate and extract transition metals from aqueous solution [182]. Because of the presence of various hybridized states of nitrogen and sulfur and the tendency to share electrons between the nitrogen and sulfur with metal ions, the removal of heavy metals by these ligands has been demonstrated [183–186]. They also are known to form colored metal complexes, which makes detection and analysis relatively easy [187]. Table 8 provides examples of ion exchange and ligand binding techniques for metal extraction.

Table 8. Ion exchange and ligand binding techniques for metal ion extraction.

Techniques	Substance Used	SC [h]	Target Metal(s)	Ref.
Ion Exchange Membrane	Cellulose nanofiber modified with PAA [a] and PGMA [b]	160 mg g^{-1}	Cd^{2+}	[188]
	Chitisan/PVA [c]/Zeolite nanofiber	NR	Cr^{6+}, Fe^{3+}, Ni^{2+}	[189]
	PAN [d]/GO [e]/Fe_3O_4 nanofiber	799.4 mg g^{-1} of Pb^{2+}, 911.9 mg g^{-1} of Cr^{6+}	Pb^{2+}, Cr^{6+}	[190]
Ligand Binding	PMHS [f]-g-PyPz [g] PMHS-g-PyPz(OEt)$_2$	0.24 mmol (Co^{2+}) and 1.48 mmol (Cu^{2+}) g^{-1} of polymer	Cu^{2+}, Cd^{2+}, Cr^{3+}, Ni^{2+}, Co^{2+}	[191]
	N,N'-dialkyl-N,N'-diaryl-1,10-phenanthroline-2,9-dicarboxamides	NR	Lanthanides	[192]
	N,N'-dimethyl-1,4-piperazines	NR	Zn^{2+}, Cu^{2+}, Mn^{2+}, Li^+, Ni^{2+}, Mg^{2+}	[193]

[a] poly(acrylic acid), [b] poly(glycidylmethacrylate), [c] polyvinyl alcohol, [d] polyacrylonitrile, [e] graphene oxide, [f] poly(methylhydrosiloxane), [g] pyridine–pyrazole, [h] sorption capacity.

3.7. Solid-Phase Microextraction

Although SPE has advantages over LLE, more progress is needed in the development of more ecofriendly and cost-effective approaches to further reduce the amount of organic solvents and sorbent material, as well as to minimize cost, analysis time, and disposal of waste chemicals. Such considerations have led to the development of greener alternatives such as SPME (Figure 2), which was developed and introduced by Pawliszyn in 1990 [194]. SPME is a fiber-based version of SPE that has benefits over other extraction techniques because the sample and solvent amounts are reduced, liquid, solid and gas samples can be analyzed with higher sensitivity and cost-effectiveness, and the use of organic solvents is minimized. Briefly, a fiber-based material is used as the sorbent to extract molecules by direct immersion of the fiber into the sample solution (Figure 2) or into the headspace above the solution. Once analytes partition into the sorbent, the fiber is removed for desorption and analysis. Direct coupling to analytical instrumentation is then possible to achieve simultaneous preconcentration and determination of target species, thus reducing the analysis time [195–199].

There are limited reports on the use of SPME for metal ion detection and analysis using HPLC and GC [200,201]. For SPME–HPLC, determination of metal ions is limited to commercial adsorbents [200]. Derivatization is required to obtain a hydrophobic organometallic compound to achieve adsorption onto the fibers and desorption after injection into a SPME–HPLC chamber. Difficulties with slow analyte diffusion in HPLC complicate the analysis of metals. One notable example of SPME–HPLC was reported by Kaur et al., in which a complex of thiophenaldehyde-3-thiosemicarbazone with cobalt, nickel, copper, and palladium was followed by UV detection [202]. SPME coupled to GC is limited to volatile species, which also often requires derivatization prior to detection [203]. Apart from the need for derivatization, there are other challenges including fiber-to-fiber variation, carry over problems, relatively high cost, reusability and recycling of the coating material, instrumental compatibility and, most importantly, delicate fibers or fragile coatings [152,199,204].

A recent goal is the desire to use SPME for the direct extraction and analysis of metal ions without the need for derivatization or complicated procedures. Rahmi et al. developed a novel SPME approach for trace metal analysis by modifying the inner wall of a syringe filter tip with a monolithic chelating moiety [205]. Twenty-two elements, including titanium, iron, cobalt, nickel, copper, gallium, cadmium, tin, and rare earth elements, were extracted prior to ICP–MS analysis with extraction efficiencies higher than 80%. Rohanifar et al. developed a versatile, easily tunable, cost-effective, greener approach for SPME of heavy metals from natural waters [133]. In this study, pencil lead was used as a substrate as an alternative to a commercially available SPME fiber or a metal wire, which significantly reduced the cost. The pencil lead was coated by electropolymerization with a sorbent composite containing polypyrrole, carbon nanotubes, and different metal chelating ligands. The resultant fiber was then used for direct immersion SPME of heavy metals followed by determination by ICP–MS (Figure 3). The chelating ligand was trapped inside the polymer matrix, which effectively captured the metal from the solution. Metals were therefore preconcentrated onto the fiber and then released in an analysis solution by treatment with acid. A composite containing polypyrrole/carbon nanotubes/1,10-phenanthroline demonstrated exceptional extraction efficiencies for silver, cadmium, cobalt, iron, nickel, lead, and zinc in several sample matrices. The accuracy of the method was validated by the analysis of a certified reference standard. Analyses were accomplished in a minimum amount of aqueous solution and were thus very environmentally friendly.

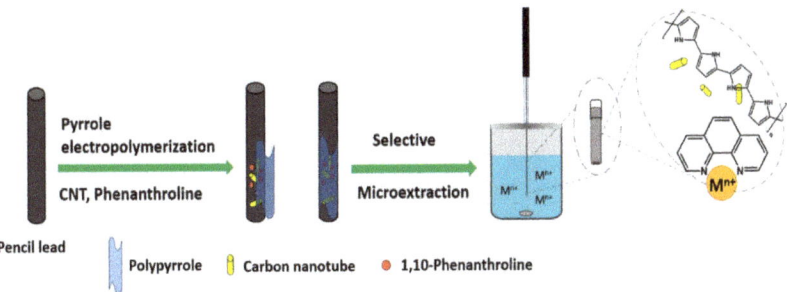

Figure 3. Schematic representation of the creation of an SPME fiber by electropolymerization and its application for metal extraction. Reprinted with permission from [133].

3.8. Dispersive Solid-Phase Extraction

Dispersive solid-phase extraction (D-SPE, Figure 2) is another variation of solid-phase extraction where a micron-sized sorbent is dispersed in the sample solution. This approach eliminates the need to optimize the flow rate and potential backpressure issues with a packed SPE cartridge, especially with newer nano-based materials. Enhanced contact between the analytes and sorbent results in very efficient extractions [206]. New sorbents for D-SPE for metals are beginning to be reported that utilize materials that effectively and selectively capture metal ions by chelation. Sitko et al. described the synthesis of a graphene oxide sorbent modified with (3-mercaptopropyl)-trimethoxysilane for determination of Co^{2+}, Ni^{2+}, Cu^{2+} As^{3+}, Cd^{2+}, and Pb^{2+} by total reflection X-ray fluorescence [207]. Preconcentration and metal capture is quite straightforward, while the analysis step is solvent free. Similarly, dithiocarbamate functionalized $Al(OH)_3$–polyacrylamide was prepared and characterized for extraction of Cu^{2+} and Pb^{2+} [208]. As with SPME, the goal for D-SPE applications is to enhance selectivity for metal analysis with new selective sorbent materials. Recently, pyrrole was derivatized with carbon disulfide and chemically polymerized to obtain an air stable, water-insoluble, chelating polymer for extraction of soft metal ions [209]. Application of this new sorbent for D-SPE of Co^{2+}, Ni^{2+}, Cu^{2+}, Zn^{2+}, Cd^{2+}, and Pb^{2+} demonstrated excellent removal and recovery of these ions. The chelating polymer is reversible, releasing the captured metals after acid treatment for preconcentration prior to analysis by ICP–MS. D-SPE is also amenable to magnetic sorbent particles as demonstrated by the references in Table 4. Therefore, D-SPE shows tremendous promise for developing simple environmentally friendly methods to extract metals.

4. Bulk Sorbent Methods

4.1. Chemical Precipitation

Wastewater is a common medium that regularly is contaminated with heavy metal ions. To ensure safe re-entry into the environment, treated water must contain metal concentrations below an accepted level called the maximum contaminant level (MCL) for each metal ion [210,211]. Chemical precipitation is a useful approach to remove large amounts of heavy metals from inorganic waste materials and prevent contamination of the environment [211]. This technique removes ionic metal components after adding counter-ions to reduce their solubility in aqueous solution [212]. Dissolved metals are turned into insoluble components by a precipitating agent under favorable pH conditions [212]. Much research on chemical precipitation for metal extraction has been conducted because of the low cost and ease of implementation for large volumes of wastewater. However, disadvantages such as the inability to maintain pH for optimum precipitation, high volume of sludge production [213], and low selectivity of metal extraction [214] limits widespread use. The treatment method should not produce toxic chemical sludge such that disposal remains ecofriendly and cost-effective [215]. Several examples on the use of precipitating agents to extract various metals have been reported [216–219].

4.2. Biosorbent Extraction

Biosorbent extraction is particularly important for the removal of heavy metals from industrial effluents as this process utilizes readily available and inexpensive dead biomass compared to conventional sorbents [220]. Aquatic organisms like yeast, algae, and bacteria adsorb dissolved heavy metals and even radioactive elements found in their surroundings [221]. Dead fungal material, for example, does not result in increased toxicity with the extracted metal or adverse operating conditions. Furthermore, no nutrients are needed for dead mass and relatively simple non-destructive treatments are used for the recovery of bound metals, which are often in their anionic forms [220,222]. Natural biosorbents can be valuable low-cost alternatives for metal removal and cleanup, especially for developing countries with limited financial resources. In addition, recent review articles have discussed progress related to the development of ecofriendly phytoremediation and phytoextraction approaches for the removal of metals from contaminated environmental sites [223–225].

Kratochvil et al. studied the removal of molybdate (MoO_4^{2-}) with chitosan beads for up to 700 mg g^{-1} of molybdate [220]. Similarly, removal of Cr^{6+} by peat moss [226] and corncobs [227] was achieved with excellent results. Marine green algae, due to presence of different proteins, lipids, or polysaccharides on the cell wall surface, show good metal binding strength [228]. Hence, for effective removal of heavy metals even at low levels, biosorbents are considered as an emerging technology [229]. However, despite the availability of large quantities of biomass, selection of the most suitable type of biomass is still a challenge. Slight variations in biomass properties can result in considerably different affinities for various metals, which also offers an opportunity to alter biomass properties to design new biosorbent materials. For example, Mallakpour et al. developed a new hydrogel nanocomposite biosorbent by embedding calcium carbonate nanoparticles into tragacanth gum for the removal of Pb^{2+} ions from water samples [230]. Similarly, pine (*Pinus sylvestris*) sawdust was modified with thiourea groups and utilized for the extraction of precious metals from industrial solutions [231]. Table 9 shows additional examples of recently reported natural biosorbent materials for extraction of metals.

Table 9. Biosorbent materials for metal ion extraction.

Biosorbent	SC [a]	Target Metal(s)	Ref.
Rice husk, palm leaf, water hyacinth	NR	Cu^{2+}, Co^{2+}, Fe^{3+}	[232]
Rhizopus arrhizus	180 mg g^{-1}	U^{6+}, Th^{4+}	[233]
Ascophyllum and *Sargassum*	30% of dry weight of biomass	Pb^{2+}, Cd^{2+}	[234]
Tobacco dust	39.6, 36.0, 29.6, 25.1, and 24.5 mg g^{-1}	Pb^{2+}, Cu^{2+}, Cd^{2+}, Zn^{2+}, Ni^{2+}	[235]
Sargassum filipendula	NR	Ag^{+}, Cd^{2+}, Cr^{3+}, Ni^{2+}, Zn^{2+}	[236]
Chlorella vulgaris	161.41 mg g^{-1} of Cr^{4+} and 169 mg g^{-1} of Pb^{2+}	Cr^{6+}, Pb^{2+}	[237,238]
Saccharomyces cerevisiae and *Rhizopus arrhizus*	Ranges from 31 to 180 mg g^{-1} for different metals	Cu^{2+}, Zn^{2+}, Cd^{2+}, U^{6+}	[239]
Alcaligenes sp.	66.7 mg g^{-1}	Pb^{2+}	[240]
Olive mill	Varies with pH and other conditions	Hg^{2+}, Pb^{2+}, Cu^{2+}, Zn^{2+}, Cd^{2+}	[241]
Parachlorella	NR	Y^{3+}, La^{3+}, Sm^{3+}, Dy^{3+}, Pr^{3+}, Nd^{3+}, Gd^{3+}	[242]

[a] sorption capacity.

5. Conclusions

Recovery of metals often requires extraction from complicated matrices in large quantities, while metal analysis is routinely sought at the trace level. In either case, strategies that are considered greener and minimize their impact on the environment drive development of emerging methods for metal extraction and analysis, many of which are described in this review. Much of the evolution of metal extraction and sample preparation has benefitted from the development and use of new materials. Aqueous two- and three-phase systems reduce the amount of organic solvents needed in LLE and include the use of ionic liquids, which offer the advantageous properties of low flammability and volatility, excellent solvating ability, and high thermal stability. Solid-phase extraction further reduces the need for organic solvents and utilizes novel materials based on adsorption, biosorption, ligand binding, and ion exchange. Extension of SPE into the micro-regime shows exciting promise for effective and selective SPME of metals. Initially, limited by the derivatization of metal ions to generate volatile or hydrophobic organometallic species for gas and liquid chromatographic analysis, new SPME coatings and materials take advantage of classical coordination chemistry to permit direct analysis of metal ions. Development of unique coordination type polymers, magnetic materials, and thin-film coatings for SPE and SPME shows great promise for highly selective and ecofriendly extraction methods for the recovery of valuable metals and for efficient sample preparation and preconcentration of a range of metals from complex matrices.

Author Contributions: Conceptualization, G.S.S.S., S.A., A.R., A.P. and J.R.K.; writing—original draft preparation, G.S.S.S., S.A., A.R., A.P. and J.R.K.; writing—review and editing, G.S.S.S., S.A., A.R., A.P. and J.R.K.; project administration, J.R.K. All authors have read and agreed to the published version of the manuscript.

Funding: This manuscript was prepared with no external funding.

Acknowledgments: The University of Toledo is acknowledged for providing support for the student co-authors of this review.

Conflicts of Interest: The authors declare no conflict of interest.

References

1. Ma, Y.; Egodawatta, P.; McGree, J.; Liu, A.; Goonetilleke, A. Human health risk assessment of heavy metals in urban stormwater. *Sci. Total Environ.* **2016**, *557–558*, 764–772. [CrossRef]
2. *Guidelines for Drinking-Water Quality*, 4th ed.; WHO: Geneva, Switzerland, 2011; pp. 315–442.
3. *IARC Monographs on the Evaluation of Carcinogenic Risks to Humans; Some Drinking-Water Disinfectants and Contaminants, Including Arsenic*; IARC (International Agency for Research on Cancer): Lyon, France, 2004.
4. Mohan, D.; Gupta, V.K.; Srivastava, S.K.; Chander, S. Kinetics of mercury adsorption from wastewater using activated carbon derived from fertilizer waste. *Colloids Surf. A* **2001**, *177*, 169–181. [CrossRef]
5. Pieper, K.J.; Tang, M.; Edwards, M.A. Flint water crisis caused by interrupted corrosion control: Investigating "Ground Zero" home. *Environ. Sci. Technol.* **2017**, *51*, 2007–2014. [CrossRef]
6. Baum, R.; Bartram, J.; Hrudey, S. The flint water crisis confirms that U.S. drinking water needs improved risk management. *Environ. Sci. Technol.* **2016**, *50*, 5436–5437. [CrossRef]
7. Boudesocque, S.; Mohamadou, A.; Conreux, A.; Marin, B.; Dupont, L. The recovery and selective extraction of gold and platinum by novel ionic liquids. *Sep. Purif. Technol.* **2019**, *210*, 824–834. [CrossRef]
8. Matsumiya, M.; Song, Y.; Tsuchida, Y.; Ota, H.; Tsunashima, K. Recovery of platinum by solvent extraction and direct electrodeposition using ionic liquid. *Sep. Purif. Technol.* **2019**, *214*, 162–167. [CrossRef]
9. Hidayah, N.N.; Abidin, S.Z. The evolution of mineral processing in extraction of rare earth elements using solid-liquid extraction over liquid-liquid extraction: A review. *Miner. Eng.* **2017**, *112*, 103–113. [CrossRef]
10. Jowitt, S.M.; Werner, T.T.; Weng, Z.; Mudd, G.M. Recycling of the rare earth elements. *Curr. Opin. Green. Sustain. Chem* **2018**, *13*, 1–7. [CrossRef]
11. Das, P.; Fatehbasharzad, P.; Colombo, M.; Fiandra, L.; Prosperi, D. Multifunctional magnetic gold nanomaterials for cancer. *Trends Biotechnol.* **2019**, *37*, 995–1010. [CrossRef]
12. Puja, P.; Kumar, P. A perspective on biogenic synthesis of platinum nanoparticles and their biomedical applications. *Spectrochim. Acta A* **2019**, *211*, 94–99. [CrossRef]
13. Du, R.; Gao, B.; Men, J. Microfiltration membrane possessing chelation function and its adsorption and rejection properties towards heavy metal ions. *J. Chem. Technol. Biotechnol.* **2019**, *94*, 1441–1450. [CrossRef]
14. Carolin, C.F.; Kumar, P.S.; Saravanan, A.; Joshiba, G.J.; Naushad, M. Efficient techniques for the removal of toxic heavy metals from aquatic environment: A review. *J. Environ. Chem. Eng.* **2017**, *5*, 2782–2799. [CrossRef]
15. Charerntanyarak, L. Heavy metals removal by chemical coagulation and precipitation. *Water Sci. Technol.* **1999**, *39*, 135–138. [CrossRef]
16. Duan, W.; Chen, G.; Chen, C.; Sanghvi, R.; Iddya, A.; Walker, S.; Liu, H.; Ronen, A.; Jassby, D. Electrochemical removal of hexavalent chromium using electrically conducting carbon nanotube/polymer composite ultrafiltration membranes. *J. Membr. Sci.* **2017**, *531*, 160–171. [CrossRef]
17. Sahraeian, T.; Sereshti, H.; Rohanifar, A. Simultaneous determination of bismuth, lead, and iron in water samples by optimization of USAEME and ICP–OES via experimental design. *J. Anal. Test.* **2018**, *2*, 98–105. [CrossRef]
18. Sereshti, H.; Far, A.R.; Samadi, S. Optimized ultrasound-assisted emulsification-microextraction followed by ICP-OES for simultaneous determination of lanthanum and cerium in urine and water samples. *Anal. Lett.* **2012**, *45*, 1426–1439. [CrossRef]
19. Vital, B.; Bartacek, J.; Ortega-Bravo, J.C.; Jeison, D. Treatment of acid mine drainage by forward osmosis: Heavy metal rejection and reverse flux of draw solution constituents. *Chem. Eng. J.* **2018**, *332*, 85–91. [CrossRef]
20. Lu, H.; Wang, J.; Wang, T.; Wang, N.; Bao, Y.; Hao, H. Crystallization techniques in wastewater treatment: An overview of applications. *Chemosphere* **2017**, *173*, 474–484. [CrossRef] [PubMed]
21. Wu, Q.P.; Zhao, J.; Qin, G.H.; Wang, C.Y.; Tong, X.L.; Xue, S. Photocatalytic reduction of Cr(VI) with TiO_2 film under visible light. *Appl. Catal. B- Environ.* **2013**, *142*, 142–148. [CrossRef]
22. Freiser, H. Extraction. *Anal. Chem.* **1968**, *40*, 522–553. [CrossRef]
23. Rydberg, J.; Cox, M.; Musikas, C.; Choppin, G.R. *Solvent Extraction Principles and Practices*, 2nd ed.; Marcel Dekker: New York, NY, USA, 2004.
24. Gunatilake, S. Methods of removing heavy metals from industrial wastewater. *J. Multidisciplin. Eng. Sci. Stud.* **2015**, *1*, 12–18.

25. Fu, F.; Wang, Q. Removal of heavy metal ions from wastewaters: A review. *J. Environ. Manage.* **2011**, *92*, 407–418. [CrossRef]
26. Bezerra, M.D.A.; Arruda, M.A.Z.; Ferreira, S.L.C. Cloud point extraction as a procedure of separation and pre-concentration for metal determination using spectroanalytical techniques: A review. *Appl. Spectrosc. Rev.* **2005**, *40*, 269–299. [CrossRef]
27. Bulgariu, L.; Bulgariu, D. Cd (II) extraction in PEG (1550)-$(NH_4)_2SO_4$ aqueous two-phase systems using halide extractants. *J. Serb. Chem. Soc.* **2008**, *73*, 341–350. [CrossRef]
28. Karmakar, R.; Sen, K. Aqueous biphasic extraction of metal ions: An alternative technology for metal regeneration. *J. Mol. Liq.* **2019**, *273*, 231–247. [CrossRef]
29. An, J.; Trujillo-Rodríguez, M.J.; Pino, V.; Anderson, J.L. Non-conventional solvents in liquid phase microextraction and aqueous biphasic systems. *J. Chromatogr. A* **2017**, *1500*, 1–23. [CrossRef] [PubMed]
30. Zvarova, T.I.; Shkinev, V.M.; Vorob'eva, G.A.; Spivakov, B.Y.; Zolotov, Y.A. Liquid-liquid extraction in the absence of usual organic solvents: Application of two-phase aqueous systems based on a water-soluble polymer. *Microchim. Acta* **1984**, *84*, 449–458. [CrossRef]
31. Chen, J.; Spear, S.K.; Huddleston, J.G.; Rogers, R.D. Polyethylene glycol and solutions of polyethylene glycol as green reaction media. *Green Chem.* **2005**, *7*, 64–82.
32. Rogers, R.D.; Bond, A.H.; Bauer, C.B.; Zhang, J.; Griffin, S.T. Metal ion separations in polyethylene glycol-based aqueous biphasic systems: Correlation of partitioning behavior with available thermodynamic hydration data. *J. Chromatogr. B* **1996**, *680*, 221–229. [CrossRef]
33. Sadeghi, R.; Maali, M. Toward an understanding of aqueous biphasic formation in polymer–polymer aqueous systems. *Polymer* **2016**, *83*, 1–11. [CrossRef]
34. Lahiri, S.; Roy, K. A green approach for sequential extraction of heavy metals from Li irradiated Au target. *J. Radioanal. Nucl. Chem.* **2009**, *281*, 531–534. [CrossRef]
35. Akama, Y.; Sali, A. Extraction mechanism of Cr(VI) on the aqueous two-phase system of tetrabutylammonium bromide and $(NH_4)_2SO_4$ mixture. *Talanta* **2002**, *57*, 681–686. [CrossRef]
36. Zafarani-Moattar, M.T.; Hamzehzadeh, S. Phase diagrams for the aqueous two-phase ternary system containing the ionic liquid 1-butyl-3-methylimidazolium bromide and tri-potassium citrate at T=(278.15, 298.15, and 318.15) K. *J. Chem. Eng. Data* **2008**, *54*, 833–841. [CrossRef]
37. Das, D.; Sen, K. Species dependent aqueous biphasic extraction of some heavy metals. *J. Ind. Eng. Chem.* **2012**, *18*, 855–859. [CrossRef]
38. Graber, T.A.; Andrews, B.A.; Asenjo, J.A. Model for the partition of metal ions in aqueous two-phase systems. *J. Chromatogr. B* **2000**, *743*, 57–64. [CrossRef]
39. Sun, P.; Huang, K.; Lin, J.; Liu, H. Role of hydrophobic interaction in driving the partitioning of metal ions in a PEG-based aqueous two-phase system. *Ind. Eng. Chem. Res.* **2018**, *57*, 11390–11398. [CrossRef]
40. de Lemos, L.R.; Santos, I.J.; Rodrigues, G.D.; da Silva, L.H.; da Silva, M.C. Copper recovery from ore by liquid-liquid extraction using aqueous two-phase system. *J. Hazard. Mater.* **2012**, *237–238*, 209–214. [CrossRef]
41. Zhang, T.; Li, W.; Zhou, W.; Gao, H.; Wu, J.; Xu, G.; Chen, J.; Liu, H.; Chen, J. Extraction and separation of gold (I) cyanide in polyethylene glycol-based aqueous biphasic systems. *Hydrometallurgy* **2001**, *62*, 41–46. [CrossRef]
42. Samaddar, P.; Sen, K. Species dependent sustainable preconcentration of zinc: Possible aspects of ABS and CPE. *J. Ind. Eng. Chem.* **2015**, *21*, 835–841. [CrossRef]
43. Zhang, Y.; Sun, T.; Lu, T.; Yan, C. Extraction and separation of tungsten(VI) from aqueous media with Triton X-100-ammonium sulfate-water aqueous two-phase system without any extractant. *J. Chromatogr. A* **2016**, *1474*, 40–46. [CrossRef]
44. Zhang, Y.Q.; Sun, T.C.; Hou, Q.X.; Guo, Q.; Lu, T.Q.; Guo, Y.C.; Yan, C.H. A green method for extracting molybdenum(VI) from aqueous solution with aqueous two-phase system without any extractant. *Sep. Purif. Technol.* **2016**, *169*, 151–157. [CrossRef]
45. Hamta, A.; Dehghani, M.R. Application of polyethylene glycol based aqueous two-phase systems for extraction of heavy metals. *J. Mol. Liq.* **2017**, *231*, 20–24. [CrossRef]
46. Shibukawa, M.; Nakayama, N.; Hayashi, T.; Shibuya, D.; Endo, Y.; Kawamura, S. Extraction behaviour of metal ions in aqueous polyethylene glycol–sodium sulphate two-phase systems in the presence of iodide and thiocyanate ions. *Anal. Chim. Acta* **2001**, *427*, 293–300. [CrossRef]

47. Rodrigues, G.D.; da Silva, M.D.C.H.; da Silva, L.H.M.; Paggioli, F.J.; Minim, L.A.; Reis Coimbra, J.S.D. Liquid–liquid extraction of metal ions without use of organic solvent. *Sep. Purif. Technol.* **2008**, *62*, 687–693. [CrossRef]
48. Lacerda, V.G.; Mageste, A.B.; Santos, I.J.B.; da Silva, L.H.M.; da Silva, M.D.C.H. Separation of Cd and Ni from Ni–Cd batteries by an environmentally safe methodology employing aqueous two-phase systems. *J. Power Sources* **2009**, *193*, 908–913. [CrossRef]
49. Bulgariu, L.; Bulgariu, D. Extraction of metal ions in aqueous polyethylene glycol–inorganic salt two-phase systems in the presence of inorganic extractants: Correlation between extraction behaviour and stability constants of extracted species. *J. Chromatogr. A* **2008**, *1196–1197*, 117–124. [CrossRef] [PubMed]
50. Patrício, P.D.R.; Mesquita, M.C.; da Silva, L.H.M.; da Silva, M.C.H. Application of aqueous two-phase systems for the development of a new method of cobalt(II), iron(III) and nickel(II) extraction: A green chemistry approach. *J. Hazard. Mater.* **2011**, *193*, 311–318. [CrossRef]
51. Rogers, R.D.; Griffin, S.T. Partitioning of mercury in aqueous biphasic systems and on ABEC™ resins. *J. Chromatogr. B* **1998**, *711*, 277–283. [CrossRef]
52. Rodrigues, G.D.; de Lemos, L.R.; da Silva, L.H.M.; da Silva, M.C.H. Application of hydrophobic extractant in aqueous two-phase systems for selective extraction of cobalt, nickel and cadmium. *J. Chromatogr. A* **2013**, *1279*, 13–19. [CrossRef]
53. Santos, L.H.; Carvalho, P.L.G.; Rodrigues, G.D.; Mansur, M.B. Selective removal of calcium from sulfate solutions containing magnesium and nickel using aqueous two phase systems (ATPS). *Hydrometallurgy* **2015**, *156*, 259–263. [CrossRef]
54. Assis, R.C.; de Araújo Faria, B.A.; Caldeira, C.L.; Mageste, A.B.; de Lemos, L.R.; Rodrigues, G.D. Extraction of arsenic(III) in aqueous two-phase systems: A new methodology for determination and speciation analysis of inorganic arsenic. *Microchem. J.* **2019**, *147*, 429–436. [CrossRef]
55. Akama, Y.; Ito, M.; Tanaka, S. Selective separation of cadmium from cobalt, copper, iron (III) and zinc by water-based two-phase system of tetrabutylammonium bromide. *Talanta* **2000**, *53*, 645–650. [CrossRef]
56. Onghena, B.; Opsomer, T.; Binnemans, K. Separation of cobalt and nickel using a thermomorphic ionic-liquid-based aqueous biphasic system. *Chem. Commun.* **2015**, *51*, 15932–15935. [CrossRef] [PubMed]
57. Wellens, S.; Thijs, B.; Binnemans, K. An environmentally friendlier approach to hydrometallurgy: Highly selective separation of cobalt from nickel by solvent extraction with undiluted phosphonium ionic liquids. *Green Chem.* **2012**, *14*, 1657–1665. [CrossRef]
58. Vander Hoogerstraete, T.; Wellens, S.; Verachtert, K.; Binnemans, K. Removal of transition metals from rare earths by solvent extraction with an undiluted phosphonium ionic liquid: Separations relevant to rare-earth magnet recycling. *Green Chem.* **2013**, *15*, 919–927. [CrossRef]
59. Depuydt, D.; Dehaen, W.; Binnemans, K. Solvent extraction of scandium(III) by an aqueous biphasic system with a nonfluorinated functionalized ionic liquid. *Ind. Eng. Chem. Res.* **2015**, *54*, 8988–8996. [CrossRef]
60. Yang, X.; Miao, C.; Sun, Y.; Lei, T.; Xie, Q.; Wang, S. Efficient extraction of gold(I) from alkaline aurocyanide solution using green ionic liquid-based aqueous biphasic systems. *J. Taiwan Inst. Chem. Eng.* **2018**, *91*, 176–185. [CrossRef]
61. Flieger, J.; Tatarczak-Michalewska, M.; Blicharska, E.; Madejska, A.; Flieger, W.; Adamczuk, A. Extraction of cobalt (II) using ionic liquid-based bi-phase and three-phase systems without adding any chelating agents with new recycling procedure. *Sep. Purif. Technol.* **2019**, *209*, 984–989. [CrossRef]
62. Sun, P.; Huang, K.; Liu, H. The nature of salt effect in enhancing the extraction of rare earths by non-functional ionic liquids: Synergism of salt anion complexation and Hofmeister bias. *J. Colloid Interf. Sci.* **2019**, *539*, 214–222. [CrossRef]
63. Chen, Y.; Wang, H.; Pei, Y.; Wang, J. A green separation strategy for neodymium (III) from cobalt (II) and nickel (II) using an ionic liquid-based aqueous two-phase system. *Talanta* **2018**, *182*, 450–455. [CrossRef]
64. Zheng, Y.; Tong, Y.; Wang, S.; Zhang, H.; Yang, Y. Mechanism of gold(III) extraction using a novel ionic liquid-based aqueous two phase system without additional extractants. *Sep. Purif. Technol.* **2015**, *154*, 123–127. [CrossRef]
65. Dutta, R.; Sarkar, U.; Mukherjee, A. Process optimization for the extraction of oil from Crotalaria juncea using three phase partitioning. *Ind. Crops Prod.* **2015**, *71*, 89–96. [CrossRef]

66. Vander Hoogerstraete, T.; Blockx, J.; De Coster, H.; Binnemans, K. Selective single-step separation of a mixture of three metal ions by a triphasic Iionic-liquid-water-ionic-liquid solvent extraction system. *Chem. Eur. J.* **2015**, *21*, 11757–11766. [CrossRef] [PubMed]
67. Shen, S.; Chang, Z.; Liu, J.; Sun, X.; Hu, X.; Liu, H. Separation of glycyrrhizic acid and liquiritin from Glycyrrhiza uralensis Fisch extract by three-liquid-phase extraction systems. *Sep. Purif. Technol.* **2007**, *53*, 216–223. [CrossRef]
68. Shen, S.; Chang, Z.; Liu, H. Three-liquid-phase extraction systems for separation of phenol and p-nitrophenol from wastewater. *Sep. Purif. Technol.* **2006**, *49*, 217–222. [CrossRef]
69. Zhang, C.; Huang, K.; Yu, P.; Liu, H. Salting-out induced three-liquid-phase separation of Pt (IV), Pd (II) and Rh (III) in system of S201– acetonitrile– NaCl– water. *Sep. Purif. Technol.* **2011**, *80*, 81–89. [CrossRef]
70. Grilo, A.L.; Raquel Aires-Barros, M.; Azevedo, A.M. Partitioning in aqueous two-phase systems: Fundamentals, applications and trends. *Sep. Purif. Rev.* **2016**, *45*, 68–80. [CrossRef]
71. Takata, T.; Hirayama, N. Organic-solvent/water/ionic-liquid triphasic system for the fractional extraction of divalent metal cations. *Anal. Sci.* **2009**, *25*, 1269–1270. [CrossRef]
72. Xie, K.; Huang, K.; Xu, L.; Yu, P.; Yang, L.; Liu, H. Three-liquid-phase extraction and separation of Ti(IV), Fe(III), and Mg(II). *Ind. Eng. Chem. Res.* **2011**, *50*, 6362–6368. [CrossRef]
73. Yu, P.; Huang, K.; Zhang, C.; Xie, K.; He, X.; Zhao, J.; Deng, F.; Liu, H. Block copolymer micellization induced microphase mass transfer: Partition of Pd(II), Pt(IV) and Rh(III) in three-liquid-phase systems of S201–EOPO–Na$_2$SO$_4$–H$_2$O. *J. Colloid Interf. Sci.* **2011**, *362*, 228–234. [CrossRef]
74. Yu, P.; Huang, K.; Liu, H.; Xie, K. Three-liquid-phase partition behaviors of Pt(IV), Pd(II) and Rh(III): Influences of phase-forming components. *Sep. Purif. Technol.* **2012**, *88*, 52–60. [CrossRef]
75. Xie, K.; Huang, K.; Yang, L.; Yu, P.; Liu, H. Three-liquid-phase extraction: A new approach for simultaneous enrichment and separation of Cr(III) and Cr(VI). *Ind. Eng. Chem. Res.* **2011**, *50*, 12767–12773. [CrossRef]
76. Sui, N.; Huang, K.; Zhang, C.; Wang, N.; Wang, F.; Liu, H. Light, middle, and heavy rare-earth group separation: A new approach via a liquid–liquid–liquid three-phase system. *Ind. Eng. Chem. Res.* **2013**, *52*, 5997–6008. [CrossRef]
77. Sui, N.; Huang, K.; Lin, J.; Li, X.; Wang, X.; Xiao, C.; Liu, H. Removal of Al, Fe and Si from complex rare-earth leach solution: A three-liquid-phase partitioning approach. *Sep. Purif. Technol.* **2014**, *127*, 97–106. [CrossRef]
78. Sui, N.; Huang, K.; Zheng, H.; Lin, J.; Wang, X.; Xiao, C.; Liu, H. Three-liquid-phase extraction and separation of rare earths and Fe, Al, and Si by a novel mixer–settler–mixer three-chamber integrated extractor. *Ind. Eng. Chem. Res.* **2014**, *53*, 16033–16043. [CrossRef]
79. Shirayama, S.; Uda, T. Simultaneous separation of manganese, cobalt, and nickel by the organic-aqueous-aqueous three-phase solvent extraction. *Metall. Mater. Trans. B* **2016**, *47*, 1325–1333. [CrossRef]
80. Sun, P.; Huang, K.; Wang, X.; Sui, N.; Lin, J.; Cao, W.; Liu, H. Three-liquid-phase extraction and separation of V(V) and Cr(VI) from acidic leach solutions of high-chromium vanadium–titanium magnetite. *Chin. J. Chem. Eng.* **2018**, *26*, 1451–1457. [CrossRef]
81. Dhamole, P.B.; Mahajan, P.; Feng, H. Phase separation conditions for sugaring-out in acetonitrile– water systems. *J. Chem. Eng. Data* **2010**, *55*, 3803–3806. [CrossRef]
82. Zhang, C.; Huang, K.; Yu, P.; Liu, H. Ionic liquid based three-liquid-phase partitioning and one-step separation of Pt(IV), Pd(II) and Rh(III). *Sep. Purif. Technol.* **2013**, *108*, 166–173. [CrossRef]
83. Nascentes, C.C.; Arruda, M.A.Z. Cloud point formation based on mixed micelles in the presence of electrolytes for cobalt extraction and preconcentration. *Talanta* **2003**, *61*, 759–768. [CrossRef]
84. Gullickson, N.; Scamehorn, J.; Harwell, J. *Surfactant-Based Separation Processes*; Marcel Dekker: New York, NY, USA, 1989.
85. Afkhami, A.; Madrakian, T.; Siampour, H. Flame atomic absorption spectrometric determination of trace quantities of cadmium in water samples after cloud point extraction in Triton X-114 without added chelating agents. *J. Hazard. Mater.* **2006**, *138*, 269–272. [CrossRef] [PubMed]
86. Samaddar, P.; Sen, K. Cloud point extraction: A sustainable method of elemental preconcentration and speciation. *J. Ind. Eng. Chem.* **2014**, *20*, 1209–1219. [CrossRef]
87. Paleologos, E.K.; Giokas, D.L.; Karayannis, M.I. Micelle-mediated separation and cloud-point extraction. *TrAC-Trends Anal. Chem.* **2005**, *24*, 426–436. [CrossRef]

88. Kazi, T.G.; Khan, S.; Baig, J.A.; Kolachi, N.F.; Afridi, H.I.; Kandhro, G.A.; Kumar, S.; Shah, A.Q. Separation and preconcentration of aluminum in parenteral solutions and bottled mineral water using different analytical techniques. *J. Hazard. Mater.* **2009**, *172*, 780–785. [CrossRef] [PubMed]
89. Zhao, L.; Zhong, S.; Fang, K.; Qian, Z.; Chen, J. Determination of cadmium(II), cobalt(II), nickel(II), lead(II), zinc(II), and copper(II) in water samples using dual-cloud point extraction and inductively coupled plasma emission spectrometry. *J. Hazard. Mater.* **2012**, *239–240*, 206–212. [CrossRef] [PubMed]
90. Citak, D.; Tuzen, M. A novel preconcentration procedure using cloud point extraction for determination of lead, cobalt and copper in water and food samples using flame atomic absorption spectrometry. *Food Chem. Toxicol.* **2010**, *48*, 1399–1404. [CrossRef] [PubMed]
91. Niazi, A.; Momeni-Isfahani, T.; Ahmari, Z. Spectrophotometric determination of mercury in water samples after cloud point extraction using nonionic surfactant Triton X-114. *J. Hazard. Mater.* **2009**, *165*, 1200–1203. [CrossRef]
92. Yang, X.; Fane, A.; Soldenhoff, K. Comparison of liquid membrane processes for metal separations: Permeability, stability, and selectivity. *Ind. Eng. Chem. Res.* **2003**, *42*, 392–403. [CrossRef]
93. Soniya, M.; Muthuraman, G. Comparative study between liquid–liquid extraction and bulk liquid membrane for the removal and recovery of methylene blue from wastewater. *J. Ind. Eng. Chem.* **2015**, *30*, 266–273. [CrossRef]
94. Benderrag, A.; Haddou, B.; Daaou, M.; Benkhedja, H.; Bounaceur, B.; Kameche, M. Experimental and modeling studies on Cd (II) ions extraction by emulsion liquid membrane using Triton X-100 as biodegradable surfactant. *J. Environ. Chem. Eng.* **2019**, *7*, 103166. [CrossRef]
95. Kocherginsky, N.; Yang, Q.; Seelam, L. Recent advances in supported liquid membrane technology. *Sep. Purif. Technol.* **2007**, *53*, 171–177. [CrossRef]
96. Parhi, P.K.; Behera, S.S.; Mohapatra, R.K.; Sahoo, T.R.; Das, D.; Misra, P.K. Separation and recovery of Sc(III) from Mg–Sc alloy scrap solution through hollow fiber supported liquid membrane (HFLM) process supported by Bi-functional ionic liquid as carrier. *Sep. Sci. Technol.* **2019**, *54*, 1478–1488. [CrossRef]
97. Duan, H.; Wang, Z.; Yuan, X.; Wang, S.; Guo, H.; Yang, X. A novel sandwich supported liquid membrane system for simultaneous separation of copper, nickel and cobalt in ammoniacal solution. *Sep. Purif. Technol.* **2017**, *173*, 323–329. [CrossRef]
98. Wongkaew, K.; Mohdee, V.; Pancharoen, U.; Arpornwichanop, A.; Lothongkum, A.W. Separation of platinum(IV) across hollow fiber supported liquid membrane using non-toxic diluents: Mass transfer and thermodynamics. *J. Ind. Eng. Chem.* **2017**, *54*, 278–289. [CrossRef]
99. Yang, X.; Zhang, Q.; Wang, Z.; Li, S.; Xie, Q.; Huang, Z.; Wang, S. Synergistic extraction of gold(I) from aurocyanide solution with the mixture of primary amine N1923 and bis (2-ethylhexyl) sulfoxide in supported liquid membrane. *J. Membr. Sci.* **2017**, *540*, 174–182. [CrossRef]
100. Panja, S.; Mohapatra, P.; Kandwal, P.; Tripathi, S. Uranium(VI) pertraction across a supported liquid membrane containing a branched diglycolamide carrier extractant: Part III: Mass transfer modeling. *Desalination* **2012**, *285*, 213–218. [CrossRef]
101. Hu, S.-Y.B.; Li, J.; Wiencek, J.M. Feasibility of surfactant-free supported emulsion liquid membrane extraction. *J. Colloid Interf. Sci.* **2003**, *266*, 430–437. [CrossRef]
102. Poole, C.F. New trends in solid-phase extraction. *TrAC-Trends Anal. Chem.* **2003**, *22*, 362–373. [CrossRef]
103. Risticevic, S.; Lord, H.; Gorecki, T.; Arthur, C.L.; Pawliszyn, J. Protocol for solid-phase microextraction method development. *Nat. Protoc.* **2010**, *5*, 122. [CrossRef]
104. Płotka-Wasylka, J.; Szczepańska, N.; de la Guardia, M.; Namieśnik, J. Modern trends in solid phase extraction: New sorbent media. *TrAC-Trends Anal. Chem.* **2016**, *77*, 23–43. [CrossRef]
105. Azzouz, A.; Kailasa, S.K.; Lee, S.S.; Rascón, A.J.; Ballesteros, E.; Zhang, M.; Kim, K.-H. Review of nanomaterials as sorbents in solid-phase extraction for environmental samples. *TrAC-Trends Anal. Chem.* **2018**, *108*, 347–369. [CrossRef]
106. Afkhami, A.; Bagheri, H. Preconcentration of trace amounts of formaldehyde from water, biological and food samples using an efficient nanosized solid phase, and its determination by a novel kinetic method. *Microchim. Acta* **2012**, *176*, 217–227. [CrossRef]
107. Ranjan, B.; Pillai, S.; Permaul, K.; Singh, S. Simultaneous removal of heavy metals and cyanate in a wastewater sample using immobilized cyanate hydratase on magnetic-multiwall carbon nanotubes. *J. Hazard. Mater.* **2019**, *363*, 73–80. [CrossRef] [PubMed]

108. Mahmoud, M.E.; Soliman, E.M. Silica-immobilized formylsalicylic acid as a selective phase for the extraction of iron(III). *Talanta* **1997**, *44*, 15–22. [CrossRef]
109. Wang, X.; Zheng, Y.; Wang, A. Fast removal of copper ions from aqueous solution by chitosan-g-poly(acrylic acid)/attapulgite composites. *J. Hazard. Mater.* **2009**, *168*, 970–977. [CrossRef] [PubMed]
110. Habila, M.; Yilmaz, E.; Alothman, Z.A.; Soylak, M. Flame atomic absorption spectrometric determination of Cd, Pb, and Cu in food samples after pre-concentration using 4-(2-thiazolylazo) resorcinol-modified activated carbon. *J. Ind. Eng. Chem.* **2014**, *20*, 3989–3993. [CrossRef]
111. Hashemi, B.; Rezania, S. Carbon-based sorbents and their nanocomposites for the enrichment of heavy metal ions: A review. *Microchim. Acta* **2019**, *186*, 578. [CrossRef] [PubMed]
112. Gouda, A.A.; Al Ghannam, S.M. Impregnated multiwalled carbon nanotubes as efficient sorbent for the solid phase extraction of trace amounts of heavy metal ions in food and water samples. *Food Chem.* **2016**, *202*, 409–416. [CrossRef] [PubMed]
113. Soylak, M.; Topalak, Z. Multiwalled carbon nanotube impregnated with tartrazine: Solid phase extractant for Cd(II) and Pb(II). *J. Ind. Eng. Chem.* **2014**, *20*, 581–585. [CrossRef]
114. Tajik, S.; Taher, M.A. A new sorbent of modified MWCNTs for column preconcentration of ultra trace amounts of zinc in biological and water samples. *Desalination* **2011**, *278*, 57–64. [CrossRef]
115. Vellaichamy, S.; Palanivelu, K. Preconcentration and separation of copper, nickel and zinc in aqueous samples by flame atomic absorption spectrometry after column solid-phase extraction onto MWCNTs impregnated with D2EHPA-TOPO mixture. *J. Hazard. Mater.* **2011**, *185*, 1131–1139. [CrossRef] [PubMed]
116. Awual, M.R.; Hasan, M.M.; Eldesoky, G.E.; Khaleque, M.A.; Rahman, M.M.; Naushad, M. Facile mercury detection and removal from aqueous media involving ligand impregnated conjugate nanomaterials. *Chem. Eng. J.* **2016**, *290*, 243–251. [CrossRef]
117. Zhang, H.; Gu, L.; Zhang, L.; Zheng, S.; Wan, H.; Sun, J.; Zhu, D.; Xu, Z. Removal of aqueous Pb(II) by adsorption on Al_2O_3-pillared layered MnO_2. *Appl. Surf. Sci.* **2017**, *406*, 330–338. [CrossRef]
118. Sharma, M.; Singh, J.; Hazra, S.; Basu, S. Adsorption of heavy metal ions by mesoporous ZnO and TiO_2@ZnO monoliths: Adsorption and kinetic studies. *Microchem. J.* **2019**, *145*, 105–112. [CrossRef]
119. Sobhanardakani, S.; Jafari, A.; Zandipak, R.; Meidanchi, A. Removal of heavy metal (Hg(II) and Cr(VI)) ions from aqueous solutions using Fe_2O_3@SiO_2 thin films as a novel adsorbent. *Process Saf. Environ.* **2018**, *120*, 348–357. [CrossRef]
120. Maya, F.; Palomino Cabello, C.; Frizzarin, R.M.; Estela, J.M.; Turnes Palomino, G.; Cerdà, V. Magnetic solid-phase extraction using metal-organic frameworks (MOFs) and their derived carbons. *TrAC-Trends Anal. Chem.* **2017**, *90*, 142–152. [CrossRef]
121. Herrero-Latorre, C.; Barciela-García, J.; García-Martín, S.; Peña-Crecente, R.M.; Otárola-Jiménez, J. Magnetic solid-phase extraction using carbon nanotubes as sorbents: A review. *Anal. Chim. Acta* **2015**, *892*, 10–26. [CrossRef]
122. Dimpe, K.M.; Ngila, J.C.; Nomngongo, P.N. Preparation and application of a tyre-based activated carbon solid phase extraction of heavy metals in wastewater samples. *Phys. Chem. Earth* **2018**, *105*, 161–169. [CrossRef]
123. Nomngongo, P.N.; Catherine Ngila, J.; Msagati, T.A.M.; Moodley, B. Preconcentration of trace multi-elements in water samples using Dowex 50W-x8 and Chelex-100 resins prior to their determination using inductively coupled plasma atomic emission spectrometry (ICP-OES). *Phys. Chem. Earth* **2013**, *66*, 83–88. [CrossRef]
124. Chen, S.; Yan, J.; Li, J.; Lu, D. Dispersive micro-solid phase extraction using magnetic $ZnFe_2O_4$ nanotubes as adsorbent for preconcentration of Co(II), Ni(II), Mn(II) and Cd(II) followed by ICP-MS determination. *Microchem. J.* **2019**, *147*, 232–238. [CrossRef]
125. Pourmand, N.; Sanagi, M.M.; Naim, A.A.; Ibrahim, W.A.W.; Baig, U. Dispersive micro-solid phase extraction method using newly prepared poly (methyl methacrylate) grafted agarose combined with ICP-MS for the simultaneous determination of Cd, Ni, Cu and Zn in vegetable and natural water samples. *Anal. Methods* **2015**, *7*, 3215–3223. [CrossRef]
126. Ebrahimi, B.; Mohammadiazar, S.; Ardalan, S. New modified carbon based solid phase extraction sorbent prepared from wild cherry stone as natural raw material for the pre-concentration and determination of trace amounts of copper in food samples. *Microchem. J.* **2019**, *147*, 666–673. [CrossRef]

127. Bahadir, Z.; Bulut, V.N.; Hidalgo, M.; Soylak, M.; Marguí, E. Determination of trace amounts of hexavalent chromium in drinking waters by dispersive microsolid-phase extraction using modified multiwalled carbon nanotubes combined with total reflection X-ray fluorescence spectrometry. *Spectrochim. Acta B* **2015**, *107*, 170–177. [CrossRef]
128. Zhu, X.; Cui, Y.; Chang, X.; Wang, H. Selective solid-phase extraction and analysis of trace-level Cr(III), Fe(III), Pb(II), and Mn(II) Ions in wastewater using diethylenetriamine-functionalized carbon nanotubes dispersed in graphene oxide colloids. *Talanta* **2016**, *146*, 358–363. [CrossRef] [PubMed]
129. Nomngongo, P.N.; Catherine Ngila, J.; Kamau, J.N.; Msagati, T.A.M.; Marjanovic, L.; Moodley, B. Pre-concentration of trace elements in short chain alcohols using different commercial cation exchange resins prior to inductively coupled plasma-optical emission spectrometric detection. *Anal. Chim. Acta* **2013**, *787*, 78–86. [CrossRef] [PubMed]
130. Pyrzyńska, K.; Wierzbicki, T. Pre-concentration and separation of vanadium on Amberlite IRA-904 resin functionalized with porphyrin ligands. *Anal. Chim. Acta* **2005**, *540*, 91–94. [CrossRef]
131. Ekinci, C.; Köklü, Ü. Determination of vanadium, manganese, silver and lead by graphite furnace atomic absorption spectrometry after preconcentration on silica-gel modified with 3-aminopropyltriethoxysilane. *Spectrochim. Acta B* **2000**, *55*, 1491–1495. [CrossRef]
132. AlSuhaimi, A.O.; AlRadaddi, S.M.; Al-Sheikh Ali, A.K.; Shraim, A.M.; AlRadaddi, T.S. Silica-based chelating resin bearing dual 8-Hydroxyquinoline moieties and its applications for solid phase extraction of trace metals from seawater prior to their analysis by ICP-MS. *Arab. J. Chem.* **2019**, *12*, 360–369. [CrossRef]
133. Rohanifar, A.; Rodriguez, L.B.; Devasurendra, A.M.; Alipourasiabi, N.; Anderson, J.L.; Kirchhoff, J.R. Solid-phase microextraction of heavy metals in natural water with a polypyrrole/carbon nanotube/1, 10–phenanthroline composite sorbent material. *Talanta* **2018**, *188*, 570–577. [CrossRef]
134. Wadhwa, S.K.; Tuzen, M.; Gul Kazi, T.; Soylak, M. Graphite furnace atomic absorption spectrometric detection of vanadium in water and food samples after solid phase extraction on multiwalled carbon nanotubes. *Talanta* **2013**, *116*, 205–209. [CrossRef]
135. Shakerian, F.; Kim, K.-H.; Kwon, E.; Szulejko, J.E.; Kumar, P.; Dadfarnia, S.; Haji Shabani, A.M. Advanced polymeric materials: Synthesis and analytical application of ion imprinted polymers as selective sorbents for solid phase extraction of metal ions. *TrAC-Trends Anal. Chem.* **2016**, *83*, 55–69. [CrossRef]
136. Rao, T.P.; Kala, R.; Daniel, S. Metal ion-imprinted polymers—Novel materials for selective recognition of inorganics. *Anal. Chim. Acta* **2006**, *578*, 105–116. [CrossRef] [PubMed]
137. Mafu, L.D.; Msagati, T.A.M.; Mamba, B.B. Ion-imprinted polymers for environmental monitoring of inorganic pollutants: Synthesis, characterization, and applications. *Environ. Sci. Pollut. Res.* **2013**, *20*, 790–802. [CrossRef] [PubMed]
138. Asgharinezhad, A.A.; Jalilian, N.; Ebrahimzadeh, H.; Panjali, Z. A simple and fast method based on new magnetic ion imprinted polymer nanoparticles for the selective extraction of Ni(ii) ions in different food samples. *RSC Adv.* **2015**, *5*, 45510–45519. [CrossRef]
139. Feist, B.; Sitko, R. Fast and sensitive determination of heavy metal ions as batophenanthroline chelates in food and water samples after dispersive micro-solid phase extraction using graphene oxide as sorbent. *Microchem. J.* **2019**, *147*, 30–36. [CrossRef]
140. Kim, B.-K.; Lee, E.J.; Kang, Y.; Lee, J.-J. Application of ionic liquids for metal dissolution and extraction. *J. Ind. Eng. Chem.* **2018**, *61*, 388–397. [CrossRef]
141. Nunes da Silva, F.; Bassoma, M.M.; Bertuol, D.A.; Tanabe, E.H. An eco-friendly approach for metals extraction using polymeric nanofibers modified with di-(2-ethylhexyl) phosphoric acid (DEHPA). *J. Cleaner Prod.* **2019**, *210*, 786–794. [CrossRef]
142. Cadore, J.S.; Bertuol, D.A.; Tanabe, E.H. Recovery of indium from LCD screens using solid-phase extraction onto nanofibers modified with Di-(2-ethylhexyl) phosphoric acid (DEHPA). *Process Saf. Environ.* **2019**, *127*, 141–150. [CrossRef]
143. Zhang, Y.; Mei, M.; Ouyang, T.; Huang, X. Preparation of a new polymeric ionic liquid-based sorbent for stir cake sorptive extraction of trace antimony in environmental water samples. *Talanta* **2016**, *161*, 377–383. [CrossRef]
144. Kazantzi, V.; Giakisikli, G.; Anthemidis, A. Reversed phase StrataTM-X resin as sorbent for automatic on-line solid phase extraction atomic absorption spectrometric determination of trace metals: Comparison of polymeric-based sorbent materials. *Int. J. Environ. Anal. Chem.* **2017**, *97*, 508–519. [CrossRef]

145. Molaei, K.; Bagheri, H.; Asgharinezhad, A.A.; Ebrahimzadeh, H.; Shamsipur, M. SiO$_2$-coated magnetic graphene oxide modified with polypyrrole–polythiophene: A novel and efficient nanocomposite for solid phase extraction of trace amounts of heavy metals. *Talanta* **2017**, *167*, 607–616. [CrossRef] [PubMed]
146. Arbab-Zavar, M.H.; Chamsaz, M.; Zohuri, G.; Darroudi, A. Synthesis and characterization of nano-pore thallium(III) ion-imprinted polymer as a new sorbent for separation and preconcentration of thallium. *J. Hazard. Mater.* **2011**, *185*, 38–43. [CrossRef] [PubMed]
147. Saraji, M.; Yousefi, H. Selective solid-phase extraction of Ni(II) by an ion-imprinted polymer from water samples. *J. Hazard. Mater.* **2009**, *167*, 1152–1157. [CrossRef] [PubMed]
148. Ersöz, A.; Say, R.; Denizli, A. Ni(II) ion-imprinted solid-phase extraction and preconcentration in aqueous solutions by packed-bed columns. *Anal. Chim. Acta* **2004**, *502*, 91–97. [CrossRef]
149. Birlik, E.; Ersöz, A.; Denizli, A.; Say, R. Preconcentration of copper using double-imprinted polymer via solid phase extraction. *Anal. Chim. Acta* **2006**, *565*, 145–151. [CrossRef]
150. Zhai, Y.; Liu, Y.; Chang, X.; Chen, S.; Huang, X. Selective solid-phase extraction of trace cadmium(II) with an ionic imprinted polymer prepared from a dual-ligand monomer. *Anal. Chim. Acta* **2007**, *593*, 123–128. [CrossRef]
151. Ribeiro, L.F.; Masini, J.C. Complexing porous polymer monoliths for online solid-phase extraction of metals in sequential injection analysis with electrochemical detection. *Talanta* **2018**, *185*, 387–395. [CrossRef]
152. Rocío-Bautista, P.; Pacheco-Fernández, I.; Pasán, J.; Pino, V. Are metal-organic frameworks able to provide a new generation of solid-phase microextraction coatings?—A review. *Anal. Chim. Acta* **2016**, *939*, 26–41. [CrossRef]
153. Rocío-Bautista, P.; Taima-Mancera, I.; Pasán, J.; Pino, V. Metal-organic frameworks in green analytical chemistry. *Separations* **2019**, *6*, 33. [CrossRef]
154. Tadjarodi, A.; Abbaszadeh, A. A magnetic nanocomposite prepared from chelator-modified magnetite (Fe$_3$O$_4$) and HKUST-1 (MOF-199) for separation and preconcentration of mercury(II). *Microchim. Acta* **2016**, *183*, 1391–1399. [CrossRef]
155. Esmaeilzadeh, M. A composite prepared from a metal-organic framework of type MIL-101(Fe) and morin-modified magnetite nanoparticles for extraction and speciation of vanadium(IV) and vanadium(V). *Microchim. Acta* **2019**, *186*, 14. [CrossRef] [PubMed]
156. Nasir, A.M.; Md Nordin, N.A.H.; Goh, P.S.; Ismail, A.F. Application of two-dimensional leaf-shaped zeolitic imidazolate framework (2D ZIF-L) as arsenite adsorbent: Kinetic, isotherm and mechanism. *J. Mol. Liq.* **2018**, *250*, 269–277. [CrossRef]
157. Saleem, H.; Rafique, U.; Davies, R.P. Investigations on post-synthetically modified UiO-66-NH$_2$ for the adsorptive removal of heavy metal ions from aqueous solution. *Microporous Mesoporous Mater.* **2016**, *221*, 238–244. [CrossRef]
158. Naeimi, S.; Faghihian, H. Performance of novel adsorbent prepared by magnetic metal-organic framework (MOF) modified by potassium nickel hexacyanoferrate for removal of Cs$^+$ from aqueous solution. *Sep. Purif. Technol.* **2017**, *175*, 255–265. [CrossRef]
159. Zou, Z.; Wang, S.; Jia, J.; Xu, F.; Long, Z.; Hou, X. Ultrasensitive determination of inorganic arsenic by hydride generation-atomic fluorescence spectrometry using Fe$_3$O$_4$@ZIF-8 nanoparticles for preconcentration. *Microchem. J.* **2016**, *124*, 578–583. [CrossRef]
160. Ma, S.; Zhang, M.; Nie, J.; Tan, J.; Song, S.; Luo, Y. Lightweight and porous cellulose-based foams with high loadings of zeolitic imidazolate frameworks-8 for adsorption applications. *Carbohydr. Polym.* **2019**, *208*, 328–335. [CrossRef]
161. Liang, L.; Chen, Q.; Jiang, F.; Yuan, D.; Qian, J.; Lv, G.; Xue, H.; Liu, L.; Jiang, H.-L.; Hong, M. In situ large-scale construction of sulfur-functionalized metal–organic framework and its efficient removal of Hg(ii) from water. *J. Mater. Chem.* **2016**, *4*, 15370–15374. [CrossRef]
162. Wang, Y.; Xie, J.; Wu, Y.; Hu, X. A magnetic metal-organic framework as a new sorbent for solid-phase extraction of copper(II), and its determination by electrothermal AAS. *Microchim. Acta* **2014**, *181*, 949–956. [CrossRef]
163. Moghaddam, Z.S.; Kaykhaii, M.; Khajeh, M.; Oveisi, A.R. Synthesis of UiO-66-OH zirconium metal-organic framework and its application for selective extraction and trace determination of thorium in water samples by spectrophotometry. *Spectrochim. Acta A* **2018**, *194*, 76–82. [CrossRef]

164. Wang, L.; Hang, X.; Chen, Y.; Wang, Y.; Feng, X. Determination of cadmium by magnetic multiwalled carbon nanotube flow injection preconcentration and graphite furnace atomic absorption spectrometry. *Anal. Lett.* **2016**, *49*, 818–830. [CrossRef]
165. Zhou, S.; Song, N.; Lv, X.; Jia, Q. Magnetic dual task-specific polymeric ionic liquid nanoparticles for preconcentration and determination of gold, palladium and platinum prior to their quantitation by graphite furnace AAS. *Microchim. Acta* **2017**, *184*, 3497–3504. [CrossRef]
166. Llaver, M.; Casado-Carmona, F.A.; Lucena, R.; Cárdenas, S.; Wuilloud, R.G. Ultra-trace tellurium preconcentration and speciation analysis in environmental samples with a novel magnetic polymeric ionic liquid nanocomposite and magnetic dispersive micro-solid phase extraction with flow-injection hydride generation atomic fluorescence spectrometry detection. *Spectrochim. Acta B* **2019**, *162*, 105705.
167. Moradi, S.E.; Haji Shabani, A.M.; Dadfarnia, S.; Emami, S. Sulfonated metal organic framework loaded on iron oxide nanoparticles as a new sorbent for the magnetic solid phase extraction of cadmium from environmental water samples. *Anal. Methods* **2016**, *8*, 6337–6346. [CrossRef]
168. Hassanpour, S.; Taghizadeh, M.; Yamini, Y. Magnetic Cr(VI) ion imprinted polymer for the fast selective adsorption of Cr(VI) from aqueous solution. *J. Polym. Environ.* **2018**, *26*, 101–115. [CrossRef]
169. Shirani, M.; Semnani, A.; Habibollahi, S.; Haddadi, H. Ultrasound-assisted, ionic liquid-linked, dual-magnetic multiwall carbon nanotube microextraction combined with electrothermal atomic absorption spectrometry for simultaneous determination of cadmium and arsenic in food samples. *J. Anal. At. Spectrom.* **2015**, *30*, 1057–1063. [CrossRef]
170. Habila, M.A.; Alothman, Z.A.; El-Toni, A.M.; Labis, J.P.; Soylak, M. Synthesis and application of Fe_3O_4@SiO_2@TiO_2 for photocatalytic decomposition of organic matrix simultaneously with magnetic solid phase extraction of heavy metals prior to ICP-MS analysis. *Talanta* **2016**, *154*, 539–547. [CrossRef]
171. Bystrzejewski, M.; Pyrzyńska, K.; Huczko, A.; Lange, H. Carbon-encapsulated magnetic nanoparticles as separable and mobile sorbents of heavy metal ions from aqueous solutions. *Carbon* **2009**, *47*, 1201–1204. [CrossRef]
172. Wei, Z.; Sandron, S.; Townsend, A.T.; Nesterenko, P.N.; Paull, B. Determination of trace labile copper in environmental waters by magnetic nanoparticle solid phase extraction and high-performance chelation ion chromatography. *Talanta* **2015**, *135*, 155–162. [CrossRef]
173. Yilmaz, E.; Alosmanov, R.M.; Soylak, M. Magnetic solid phase extraction of lead(ii) and cadmium(ii) on a magnetic phosphorus-containing polymer (M-PhCP) for their microsampling flame atomic absorption spectrometric determinations. *RSC Adv.* **2015**, *5*, 33801–33808. [CrossRef]
174. Babazadeh, M.; Hosseinzadeh-Khanmiri, R.; Abolhasani, J.; Ghorbani-Kalhor, E.; Hassanpour, A. Solid phase extraction of heavy metal ions from agricultural samples with the aid of a novel functionalized magnetic metal–organic framework. *RSC Adv.* **2015**, *5*, 19884–19892. [CrossRef]
175. Ghorbani-Kalhor, E. A metal-organic framework nanocomposite made from functionalized magnetite nanoparticles and HKUST-1 (MOF-199) for preconcentration of Cd(II), Pb(II), and Ni(II). *Microchim. Acta* **2016**, *183*, 2639–2647. [CrossRef]
176. Sohrabi, M.R.; Matbouie, Z.; Asgharinezhad, A.A.; Dehghani, A. Solid phase extraction of Cd(II) and Pb(II) using a magnetic metal-organic framework, and their determination by FAAS. *Microchim. Acta* **2013**, *180*, 589–597. [CrossRef]
177. Wang, Y.; Chen, H.; Tang, J.; Ye, G.; Ge, H.; Hu, X. Preparation of magnetic metal organic frameworks adsorbent modified with mercapto groups for the extraction and analysis of lead in food samples by flame atomic absorption spectrometry. *Food Chem.* **2015**, *181*, 191–197. [CrossRef]
178. Dabrowski, A.; Hubicki, Z.; Podkościelny, P.; Robens, E. Selective removal of the heavy metal ions from waters and industrial wastewaters by ion-exchange method. *Chemosphere* **2004**, *56*, 91–106. [CrossRef] [PubMed]
179. Figueiredo, B.R.; Cardoso, S.P.; Portugal, I.; Rocha, J.; Silva, C.M. Inorganic ion exchangers for cesium removal from radioactive wastewater. *Sep. Purif. Rev.* **2018**, *47*, 306–336. [CrossRef]
180. Murray, A.; Örmeci, B. Use of polymeric sub-micron ion-exchange resins for removal of lead, copper, zinc, and nickel from natural waters. *J. Environ. Sci.* **2019**, *75*, 247–254. [CrossRef]
181. Vergili, I.; Gönder, Z.B.; Kaya, Y.; Gürdağ, G.; Çavuş, S. Sorption of Pb (II) from battery industry wastewater using a weak acid cation exchange resin. *Process Saf. Environ.* **2017**, *107*, 498–507. [CrossRef]

182. Abu-El-Halawa, R.; Zabin, S.A. Removal efficiency of Pb, Cd, Cu and Zn from polluted water using dithiocarbamate ligands. *J. Taibah Univ. Sci.* **2017**, *11*, 57–65. [CrossRef]
183. Bai, L.; Hu, H.; Fu, W.; Wan, J.; Cheng, X.; Zhuge, L.; Xiong, L.; Chen, Q. Synthesis of a novel silica-supported dithiocarbamate adsorbent and its properties for the removal of heavy metal ions. *J. Hazard. Mater.* **2011**, *195*, 261–275. [CrossRef]
184. Gaur, J.; Jain, S.; Bhatia, R.; Lal, A.; Kaushik, N.K. Synthesis and characterization of a novel copolymer of glyoxal dihydrazone and glyoxal dihydrazone bis (dithiocarbamate) and application in heavy metal ion removal from water. *J. Therm. Anal. Calorim.* **2013**, *112*, 1137–1143. [CrossRef]
185. Fu, H.; Lv, X.; Yang, Y.; Xu, X. Removal of micro complex copper in aqueous solution with a dithiocarbamate compound. *Desalinat. Water Treat.* **2012**, *39*, 103–111. [CrossRef]
186. Li, Z. Synthesis of a carbamide-based dithiocarbamate chelator for the removal of heavy metal ions from aqueous solutions. *J. Ind. Eng. Chem.* **2014**, *20*, 586–590. [CrossRef]
187. Kanchi, S.; Singh, P.; Bisetty, K. Dithiocarbamates as hazardous remediation agent: A critical review on progress in environmental chemistry for inorganic species studies of 20th century. *Arab. J. Chem.* **2014**, *7*, 11–25. [CrossRef]
188. Chitpong, N.; Husson, S.M. Polyacid functionalized cellulose nanofiber membranes for removal of heavy metals from impaired waters. *J. Membr. Sci.* **2017**, *523*, 418–429. [CrossRef]
189. Habiba, U.; Afifi, A.M.; Salleh, A.; Ang, B.C. Chitosan/(polyvinyl alcohol)/zeolite electrospun composite nanofibrous membrane for adsorption of Cr^{6+}, Fe^{3+} and Ni^{2+}. *J. Hazard. Mater.* **2017**, *322*, 182–194. [CrossRef] [PubMed]
190. Koushkbaghi, S.; Jafari, P.; Rabiei, J.; Irani, M.; Aliabadi, M. Fabrication of PET/PAN/GO/Fe_3O_4 nanofibrous membrane for the removal of Pb(II) and Cr(VI) ions. *Chem. Eng. J.* **2016**, *301*, 42–50. [CrossRef]
191. Cegłowski, M.; Schroeder, G. Removal of heavy metal ions with the use of chelating polymers obtained by grafting pyridine–pyrazole ligands onto polymethylhydrosiloxane. *Chem. Eng. J.* **2015**, *259*, 885–893. [CrossRef]
192. Ustynyuk, Y.A.; Borisova, N.; Babain, V.; Gloriozov, I.; Manuilov, A.; Kalmykov, S.; Alyapyshev, M.Y.; Tkachenko, L.; Kenf, E.; Ustynyuk, N. N, N′-Dialkyl-N, N′-diaryl-1, 10-phenanthroline-2, 9-dicarboxamides as donor ligands for separation of rare earth elements with a high and unusual selectivity. DFT computational and experimental studies. *Chem. Commun.* **2015**, *51*, 7466–7469. [CrossRef]
193. Bérubé, C.; Cardinal, S.; Boudreault, P.-L.; Barbeau, X.; Delcey, N.; Giguère, M.; Gleeton, D.; Voyer, N. Novel chiral N,N′-dimethyl-1,4-piperazines with metal binding abilities. *Tetrahedron* **2015**, *71*, 8077–8084. [CrossRef]
194. Arthur, C.L.; Pawliszyn, J. Solid phase microextraction with thermal desorption using fused silica optical fibers. *Anal. Chem.* **1990**, *62*, 2145–2148. [CrossRef]
195. Souza Silva, E.A.; Risticevic, S.; Pawliszyn, J. Recent trends in SPME concerning sorbent materials, configurations and in vivo applications. *TrAC-Trends Anal. Chem.* **2013**, *43*, 24–36. [CrossRef]
196. Godage, N.H.; Gionfriddo, E. A critical outlook on recent developments and applications of matrix compatible coatings for solid phase microextraction. *TrAC-Trends Anal. Chem.* **2019**, *111*, 220–228. [CrossRef]
197. Spietelun, A.; Marcinkowski, Ł.; de la Guardia, M.; Namieśnik, J. Recent developments and future trends in solid phase microextraction techniques towards green analytical chemistry. *J. Chromatogr. A* **2013**, *1321*, 1–13. [CrossRef] [PubMed]
198. Armenta, S.; Garrigues, S.; de la Guardia, M. Green analytical chemistry. *TrAC-Trends Anal. Chem.* **2008**, *27*, 497–511. [CrossRef]
199. Płotka-Wasylka, J.; Szczepańska, N.; de la Guardia, M.; Namieśnik, J. Miniaturized solid-phase extraction techniques. *TrAC-Trends Anal. Chem.* **2015**, *73*, 19–38. [CrossRef]
200. Malik, A.K.; Kaur, V.; Verma, N. A review on solid phase microextraction—High performance liquid chromatography as a novel tool for the analysis of toxic metal ions. *Talanta* **2006**, *68*, 842–849. [CrossRef]
201. Mester, Z.; Sturgeon, R. Trace element speciation using solid phase microextraction. *Spectrochim. Acta B* **2005**, *60*, 1243–1269. [CrossRef]
202. Kaur, V.; Aulakh, J.S.; Malik, A.K. A new approach for simultaneous determination of Co(II), Ni(II), Cu(II) and Pd(II) using 2-thiophenaldehyde-3-thiosemicarbazone as reagent by solid phase microextraction–high performance liquid chromatography. *Anal. Chim. Acta* **2007**, *603*, 44–50. [CrossRef]

203. Mester, Z.; Pawliszyn, J. Electrospray mass spectrometry of trimethyllead and triethyllead with in-tube solid phase microextraction sample introduction. *Rapid Commun. Mass Spectrom.* **1999**, *13*, 1999–2003. [CrossRef]
204. Ridgway, K.; Lalljie, S.P.D.; Smith, R.M. Sample preparation techniques for the determination of trace residues and contaminants in foods. *J. Chromatogr. A* **2007**, *1153*, 36–53. [CrossRef]
205. Rahmi, D.; Takasaki, Y.; Zhu, Y.; Kobayashi, H.; Konagaya, S.; Haraguchi, H.; Umemura, T. Preparation of monolithic chelating adsorbent inside a syringe filter tip for solid phase microextraction of trace elements in natural water prior to their determination by ICP-MS. *Talanta* **2010**, *81*, 1438–1445. [CrossRef] [PubMed]
206. Chisvert, A.; Cárdenas, S.; Lucena, R. Dispersive micro-solid phase extraction. *TrAC-Trends Anal. Chem.* **2019**, *112*, 226–233. [CrossRef]
207. Sitko, R.; Janik, P.; Zawisza, B.; Talik, E.; Margui, E.; Queralt, I. Green approach for ultratrace determination of divalent metal ions and arsenic species using total-reflection X-ray fluorescence spectrometry and mercapto-modified graphene oxide nanosheets as a novel adsorbent. *Anal. Chem.* **2015**, *87*, 3535–3542. [CrossRef] [PubMed]
208. Liu, Y.; Qian, P.; Yu, Y.; Xiao, L.; Wang, Y.; Ye, S.; Chen, Y. Dithiocarbamate functionalized Al(OH)$_3$-polyacrylamide adsorbent for rapid and efficient removal of Cu(II) and Pb(II). *J. Appl. Polym. Sci.* **2017**, *134*, 45431. [CrossRef]
209. Rohanifar, A. Conductive Polymers for Electrochemical Analysis and Extraction. Ph.D. Dissertation, University of Toledo, Toledo, OH, USA, 2018.
210. Babel, S.; Kurniawan, T. Various treatment technologies to remove arsenic and mercury from contaminated groundwater: An overview. In Proceedings of the First International Symposium on Southeast Asian Water Environment, Bangkok, Thailand, 24–25 October 2003; pp. 433–440.
211. Kurniawan, T.A.; Chan, G.Y.S.; Lo, W.-H.; Babel, S. Physico–chemical treatment techniques for wastewater laden with heavy metals. *Chem. Eng. J.* **2006**, *118*, 83–98. [CrossRef]
212. Wang, L.K.; Vaccari, D.A.; Li, Y.; Shammas, N.K. *Physicochemical Treatment Processes*; Humana Press: Totowa, NJ, USA, 2005; pp. 141–197.
213. Lu, H.; Wang, Y.; Wang, J. Recovery of Ni^{2+} and pure water from electroplating rinse wastewater by an integrated two-stage electrodeionization process. *J. Cleaner Prod.* **2015**, *92*, 257–266. [CrossRef]
214. Sobianowska-Turek, A.; Szczepaniak, W.; Maciejewski, P.; Gawlik-Kobylińska, M. Recovery of zinc and manganese, and other metals (Fe, Cu, Ni, Co, Cd, Cr, Na, K) from Zn-MnO$_2$ and Zn-C waste batteries: Hydroxyl and carbonate co-precipitation from solution after reducing acidic leaching with use of oxalic acid. *J. Power Sources* **2016**, *325*, 220–228. [CrossRef]
215. Ahluwalia, S.S.; Goyal, D. Microbial and plant derived biomass for removal of heavy metals from wastewater. *Bioresour. Technol.* **2007**, *98*, 2243–2257. [CrossRef]
216. Tanong, K.; Tran, L.-H.; Mercier, G.; Blais, J.-F. Recovery of Zn(II), Mn(II), Cd(II) and Ni(II) from the unsorted spent batteries using solvent extraction, electrodeposition and precipitation methods. *J. Cleaner Prod.* **2017**, *148*, 233–244. [CrossRef]
217. Huang, Y.; Han, G.; Liu, J.; Chai, W.; Wang, W.; Yang, S.; Su, S. A stepwise recovery of metals from hybrid cathodes of spent Li-ion batteries with leaching-flotation-precipitation process. *J. Power Sources* **2016**, *325*, 555–564. [CrossRef]
218. Ghosh, P.; Samanta, A.N.; Ray, S. Reduction of COD and removal of Zn^{2+} from rayon industry wastewater by combined electro-Fenton treatment and chemical precipitation. *Desalination* **2011**, *266*, 213–217. [CrossRef]
219. Özverdi, A.; Erdem, M. Cu^{2+}, Cd^{2+} and Pb^{2+} adsorption from aqueous solutions by pyrite and synthetic iron sulphide. *J. Hazard. Mater.* **2006**, *137*, 626–632. [CrossRef] [PubMed]
220. Kratochvil, D.; Volesky, B. Advances in the biosorption of heavy metals. *Trends Biotechnol.* **1998**, *16*, 291–300. [CrossRef]
221. Nourbakhsh, M.; Sag, Y.; Özer, D.; Aksu, Z.; Kutsal, T.; Çaglar, A. A comparative study of various biosorbents for removal of chromium(VI) ions from industrial waste waters. *Process Biochem.* **1994**, *29*, 1–5. [CrossRef]
222. de Rome, L.; Gadd, G.M. Use of pelleted and immobilized yeast and fungal biomass for heavy metal and radionuclide recovery. *J. Ind. Microbiol.* **1991**, *7*, 97–104. [CrossRef]
223. Ashraf, S.; Ali, Q.; Zahir, Z.A.; Ashraf, S.; Asghar, H.N. Phytoremediation: Environmentally sustainable way for reclamation of heavy metal polluted soils. *Ecotoxicol. Environ. Saf.* **2019**, *174*, 714–727. [CrossRef] [PubMed]

224. Asgari Lajayer, B.; Khadem Moghadam, N.; Maghsoodi, M.R.; Ghorbanpour, M.; Kariman, K. Phytoextraction of heavy metals from contaminated soil, water and atmosphere using ornamental plants: Mechanisms and efficiency improvement strategies. *Environ. Sci. Pollut. Res.* **2019**, *26*, 8468–8484. [CrossRef]
225. Prabakaran, K.; Li, J.; Anandkumar, A.; Leng, Z.; Zou, C.B.; Du, D. Managing environmental contamination through phytoremediation by invasive plants: A review. *Ecol. Eng.* **2019**, *138*, 28–37. [CrossRef]
226. Sharma, D.; Forster, C. Removal of hexavalent chromium using sphagnum moss peat. *Water Res.* **1993**, *27*, 1201–1208. [CrossRef]
227. Bosinco, S.; Roussy, J.; Guibal, E.; Cloirec, P. Interaction mechanisms between hexavalent chromium and corncob. *Environ. Technol.* **1996**, *17*, 55–62. [CrossRef]
228. Pavasant, P.; Apiratikul, R.; Sungkhum, V.; Suthiparinyanont, P.; Wattanachira, S.; Marhaba, T.F. Biosorption of Cu^{2+}, Cd^{2+}, Pb^{2+}, and Zn^{2+} using dried marine green macroalga Caulerpa lentillifera. *Bioresour. Technol.* **2006**, *97*, 2321–2329. [CrossRef] [PubMed]
229. Gavrilescu, M. Removal of heavy metals from the environment by biosorption. *Eng. Life Sci.* **2004**, *4*, 219–232. [CrossRef]
230. Mallakpour, S.; Abdolmaleki, A.; Tabesh, F. Ultrasonic-assisted manufacturing of new hydrogel nanocomposite biosorbent containing calcium carbonate nanoparticles and tragacanth gum for removal of heavy metal. *Ultrason. Sonochem.* **2018**, *41*, 572–581. [CrossRef] [PubMed]
231. Losev, V.N.; Elsufiev, E.V.; Buyko, O.V.; Trofimchuk, A.K.; Horda, R.V.; Legenchuk, O.V. Extraction of precious metals from industrial solutions by the pine (Pinus sylvestris) sawdust-based biosorbent modified with thiourea groups. *Hydrometallurgy* **2018**, *176*, 118–128. [CrossRef]
232. Sadeek, S.A.; Negm, N.A.; Hefni, H.H.H.; Wahab, M.M.A. Metal adsorption by agricultural biosorbents: Adsorption isotherm, kinetic and biosorbents chemical structures. *Int. J. Biol. Macromol.* **2015**, *81*, 400–409. [CrossRef] [PubMed]
233. Tsezos, M.; Volesky, B. Biosorption of uranium and thorium. *Biotechnol. Bioeng.* **1981**, *23*, 583–604. [CrossRef]
234. Volesky, B.; Holan, Z. Biosorption of heavy metals. *Biotechnol. Prog.* **1995**, *11*, 235–250. [CrossRef]
235. Qi, B.; Aldrich, C. Biosorption of heavy metals from aqueous solutions with tobacco dust. *Bioresour. Technol.* **2008**, *99*, 5595–5601. [CrossRef]
236. Cardoso, S.L.; Costa, C.S.D.; Nishikawa, E.; da Silva, M.G.C.; Vieira, M.G.A. Biosorption of toxic metals using the alginate extraction residue from the brown algae Sargassum filipendula as a natural ion-exchanger. *J. Cleaner Prod.* **2017**, *165*, 491–499. [CrossRef]
237. Sibi, G. Biosorption of chromium from electroplating and galvanizing industrial effluents under extreme conditions using Chlorella vulgaris. *Green Ener. Environ.* **2016**, *1*, 172–177. [CrossRef]
238. El-Naas, M.; Al-Rub, F.A.; Ashour, I.; Al Marzouqi, M. Effect of competitive interference on the biosorption of lead (II) by Chlorella vulgaris. *Chem. Eng. Process.* **2007**, *46*, 1391–1399. [CrossRef]
239. Kuyucak, N.; Volesky, B. Biosorbents for recovery of metals from industrial solutions. *Biotechnol. Lett.* **1988**, *10*, 137–142. [CrossRef]
240. Jin, Y.; Yu, S.; Teng, C.; Song, T.; Dong, L.; Liang, J.; Bai, X.; Xu, X.; Qu, J. Biosorption characteristic of Alcaligenes sp. BAPb.1 for removal of lead(II) from aqueous solution. *3 Biotech* **2017**, *7*, 123. [CrossRef] [PubMed]
241. Pagnanelli, F.; Toro, L.; Vegliò, F. Olive mill solid residues as heavy metal sorbent material: A preliminary study. *Waste Manage.* **2002**, *22*, 901–907. [CrossRef]
242. Ponou, J.; Wang, L.P.; Dodbiba, G.; Okaya, K.; Fujita, T.; Mitsuhashi, K.; Atarashi, T.; Satoh, G.; Noda, M. Recovery of rare earth elements from aqueous solution obtained from Vietnamese clay minerals using dried and carbonized parachlorella. *J. Environ. Chem. Eng.* **2014**, *2*, 1070–1081. [CrossRef]

© 2020 by the authors. Licensee MDPI, Basel, Switzerland. This article is an open access article distributed under the terms and conditions of the Creative Commons Attribution (CC BY) license (http://creativecommons.org/licenses/by/4.0/).

Article

A Bottom-Up Approach for Data Mining in Bioaromatization of Beers Using Flow-Modulated Comprehensive Two-Dimensional Gas Chromatography/Mass Spectrometry

Andre Cunha Paiva [1], Daniel Simões Oliveira [2] and Leandro Wang Hantao [1,*]

[1] Institute of Chemistry, University of Campinas, Campinas, SP 13083-970, Brazil; cunhapaiva@gmail.com
[2] Confra da Mantiqueira, Córrego do Bom Jesus, MG 37605-000, Brazil; profdsimoes@gmail.com
* Correspondence: wang@unicamp.br; Tel.: +55-019-3521-3083

Received: 10 May 2019; Accepted: 19 August 2019; Published: 25 September 2019

Abstract: In this study, we report the combination of comprehensive two-dimensional gas chromatography (GC×GC) with multivariate pattern recognition through template matching for the assignment of the contribution of *Brazilian Ale 02* yeast strain to the aroma profile of beer compared with the traditional *Nottingham* yeast. Volatile organic compounds (VOC) from two beer samples, which were fermented with these yeast strains were sampled using headspace solid-phase microextraction (HS-SPME). The aroma profiles from both beer samples were obtained using GC×GC coupled to a fast scanning quadrupole mass spectrometer. Data processing performed through multiway principal components analysis succeeded in separating both beer samples based on yeast strain. The execution of a simple and reliable procedure succeeded and identified 46 compounds as relevant for sample classification. Furthermore, the bottom-up approach spotted compounds found exclusively in the beer sample fermented with the Brazilian yeast, highlighting the bioaromatization properties introduced to the aroma profile by this yeast strain.

Keywords: foodomics; Brazilian yeast; craft beer; sensomics

---

1. Introduction

The Brazilian beer market has been in the spotlight due to steady increase in beer consumption over the past few decades [1]. In this context, the search for better quality beers and new varieties of the raw materials (malt, water, hops and yeast) is fundamental. Additionally, new ingredients have been tested during brewing to produce unique beers with innovative flavor compounds [2].

Flavor compounds are characterized by their interaction with human olfactory system and it is through such chemistry that specific odor sensations are induced. Volatile organic compounds (VOC) are directly involved in the flavor sensation of beers. The chemical nature of VOC that are associated with flavor and aroma in beer is varied, including esters, aliphatic and aromatic alcohols, carbonyl-containing compounds and terpenoids [3,4]. Hence, beer is considered a complex matrix, even though the majority (92–95%) of the beverage is water [5]. Several qualitative studies have tried to profile the composition of beers, whether to characterize the VOC profile or to search for specific markers in aged beer [3,6–9].

The art of brewing involves careful control of several stages, and a paramount one is the fermentation, i.e., bioaromatization. Fermentation involves the addition of yeasts to the recipe to convert the sugars in ethanol and introduce aroma-related compounds derived from biological channels from the yeast's metabolism. Accordingly, important factors that can influence this metabolic balance include temperature, pH and yeast strain [5].

Research on non-conventional yeasts has taken place to effectively develop novel aroma profiles through bioaromatization [10]. Recently, a local startup (Yeastlab—Franca, Brazil) isolated a unique yeast strain from *Plinia cauliflora* (i.e., jabuticaba, a "Brazilian berry") peel that could be used for beer production, which was named *Brazilian Ale 02* yeast. Furthermore, to the best of our knowledge there are no reports evaluating the outcome of the bioaromatization properties of *Brazilian Ale 02* yeast strain to the aroma profile of a beer.

The bioaromatization properties study of *Brazilian Ale 02* yeast should be driven using appropriate sample preparation, which mitigates production of artifacts, to guarantee reliable results. Solid-phase microextraction (SPME) [11] fits these requirements since it eliminates the use of solvents and combines analyte isolation and pre-concentration into one simple step. The technique has been popular to analyze food products, including wine, cheese and beer [12]. Reports involving beer analysis usually use headspace solid-phase microextraction (HS-SPME) for targeted analysis. Headspace analysis is very beneficial to extent SPME fiber lifetime since non-volatile compounds such as sugars, proteins and polyphenols present in the beer are avoided in the extraction, which could damage the sorbent phase during direct immersion (DI) [13,14]. A wide variety of coatings can be used as extracting phases in SPME. Some reports applied a mixed fiber for beer analysis, i.e., divinylbenzene/carboxen/polydimethylsiloxane (DVB/CAR/PDMS), in which analyte adsorption dominates analyte sorption. These reports covered characterization of the VOC profile, off-flavors research or quantitation [9,12,15], demonstrating high reproducibility either in manual or automated methods [12].

Standard practices for VOC profiling in beers used conventional gas chromatography (1D-GC) [8,9,15–20]. However, peak overlap severely dampens qualitative and quantitative analyses in such complex samples [21]. Flow-modulated comprehensive two-dimensional gas chromatography coupled to mass spectrometry (FM-GC×GC-MS) enables proper sample separation and characterization due to the enhanced peak capacity of the composite system. Such technique uses two consecutive stages of separation with distinct selectivity. FM has been increasingly attractive to expert and non-expert users because of its operational simplicity, and capability to modulate compounds from a broad volatility range. Griffith et al. [22] demonstrated the increased peak capacity of differential flow modulation GC×GC. The report using reverse fill/flush (RFF) modulation showed the excellent precision and capacity to handle significant overloading peaks that exhibited poor peak shape and performance in forward fill/flush (FFF) modulation. Moreover, RFF modulation enables cost-effective and robust analysis, bridging the gap between powerful instrumentation and routine analysis.

Martins et al. [23] made use of GC×GC to trace the terpenic composition of 18 lager beers, revealing the presence of 94 mono and sesquiterpenic compounds. The second dimension was important to resolve key compounds such as 1,1,3,5-tetra-methylcyclohexane and β-ocimene. Similarly, GC×GC [4] allowed the assignment of 32 VOC related to beer aroma from Portuguese samples.

Chromatograms generated by FM-GC×GC-MS are structurally complex (i.e., four-way data) and contain a lot of information; thus, it is important to work with an appropriate data processing technique capable of extracting all meaningful and context-oriented information. Multiway principal components analysis (MPCA) is a well-established multivariate statistical method for data processing, which enables pixel-based pattern recognition [24].

In this article, we applied a bottom-up approach involving FM-GC×GC-MS and MPCA for assignment of yeast related volatile organic compounds in two beer samples fermented with distinguished yeast strains. Data processing had the aim of highlighting compounds that could be used to differentiate both beer samples.

2. Materials and Methods

2.1. Samples

Fresh beer samples were kindly provided by a local company (Cervejaria Confra da Mantiqueira—Campinas, Brazil). Comparative analysis was enabled by standardizing the brewing process. Brewing started from the same wort, which presented 14% of maltose as its original extract (OE). Fermentation/bioaromatization step lasted 7 days at 20 °C followed by a cold maturation step at 2 °C for 7 days. Sample uniqueness was established in the yeast strain used; each beer was fermented with a distinguished yeast strain. The yeast strains were the traditional *Nottingham* yeast (Lallemand—Montreal, QC, Canada) and the unique *Brazilian Ale 02* (Yeastlab Biotecnologia—Franca, Brazil). For discussion purposes, the beer produced with the Brazilian yeast will be named "BR" and with the English yeast will be named "ENG". The samples were CO_2 capped in amber bottles and stored in the refrigerator (4 °C) for up to 14 days. Analysis was performed immediately after opening the sample.

2.2. Materials and Extraction Devices

The linear temperature programmed retention indices (LTPRI) of the analytes were determined using a standard mixture of C_8-C_{20} *n*-alkanes (04070-1ML) (Merck—St. Louis, MO, USA). A 2-cm SPME fiber coated with 50/30 μm DVB/CAR/PDMS (57299-U) was used for analyte extraction (Merck). Magnetic screw caps with PTFE-PDMS septa (SU860101) and 20 mL glass vials (SU860097) (Merck) were used in the SPME extractions.

2.3. Solid-Phase Microextraction Method

The DVB/CAR/PDMS fiber was conditioned as recommended by the manufacturer. Sample degassing was performed using ultrasound for 10 min at room temperature. After degassing, 5 mL aliquots of beer were added to 20 mL vials. The vials contained pre-weighted 1.75 g of sodium chloride [17]. The samples were immediately stored in the refrigerator (−2 °C). Prior to extraction, a pre-equilibrium step of 20 min at 45 °C was applied. To ensure optimum extraction of VOC, an extraction profile was evaluated between 5 to 45 min at 45 °C. Longer extraction times were not evaluated to improve sample throughput. The agitation was maintained continuously at 500 rpm to ensure fast mass transfer between sample and HS. Four replicates were obtained for each sample. After extraction, sample introduction to the GC was attained by thermal desorption at the inlet, which operated in splitless mode at 260 °C with a sampling time of 1 min.

2.4. Gas Chromatography

The GC×GC analyses were performed on a flow-modulated instrument that comprised a TRACE 1300 gas chromatograph coupled with a fast scanning ISQ single quadrupole mass spectrometer (QMS) (Thermo Fisher Scientific—Waltham, MA, USA). The GC×GC was fitted with a split/splitless injector, which operated in splitless mode at 260 °C. The INSIGHT interface (SepSolve Analytical—Peterborough, UK) was used to perform differential flow modulation using the reverse fill/flush configuration. ChromSpace (SepSolve Analytical) was used to synchronize and control the INSIGHT modulator. Instrument control and data acquisition was performed using Xcalibur (Thermo Scientific—Waltham, MA, USA) software.

Column set consisted of two wall-coated open tubular (WCOT) capillary columns. The primary column was a 30 m × 0.25 mm-id (0.25 μm; β of 250) MEGA-5HT (MEGA srl—Legnano, Italy). The secondary column was a 5 m × 0.25 mm-id (0.25 μm; β of 250) HP-50+ (Agilent Technologies—Santa Clara, CA, USA). Oven temperature ramp was programmed from 45 °C (3 min) to 270 °C at 3 °C min^{-1}. Modulation period was set to 6.0 s with a re-injection (flush) pulse of 200 ms. A sampling loop of 50 μL was used for GC×GC modulation. Helium was used as auxiliary and carrier gas at constant flow rates of 12.5 mL/min and 1.00 mL/min, respectively. The mass range was set from 50 to 350 *m/z* units at

42 scans s^{-1}. Transfer line was operated at 250 °C and ion source at 200 °C. Electron ionization was performed at 70 eV and 150 µA.

2.5. Identification

GC Image (Zoex—Houston, TX, USA) was used for tentative identification of analytes by combining mass spectrum similarity searches and LTPRI filtering. Blob detection was done by setting the following parameters in GC Image: minimum area of 20, minimum volume of 50 and minimum peak value of 50. Tentative identification was executed by adopting minimum similarity match of 80% and ± 25 LTPRI deviation from NIST. Qualitative analysis was performed using NIST14 MS library (National Institute of Standards—Gaithersburg, MD, USA). The Good Scents Company was used to obtain the odor and flavor descriptors.

2.6. Pixel-Based Chemometric Analysis

The '.raw' Xcalibur (Thermo Scientific) native files were converted to '.cdf' ANDI/netCDF format using the File Converter plug-in. MATLAB R2017b (MathWorks—Natick, MA, USA) environment was used to perform multivariate data analysis. The netCDF files were imported to MATLAB to generate three-way data tensors. For additional detail on three-way data tensors refer to [25,26]. PLS Toolbox 7.5 (Eigenvector Research Inc.—Wenatchee, WA, USA) was used to perform the MPCA analysis and to highlight the relevant chemical information for sample classification based on yeast strain.

3. Results and Discussions

3.1. SPME Optimization

To guarantee the higher extraction rate with shortest time, the extraction profile was determined and the extraction time during SPME optimized for a representative beer. As shown in Figure 1, five extraction times were evaluated and total volume of peaks were used as response parameter to ensure adequate transfer of the analytes from the sample matrix to the extracting phase.

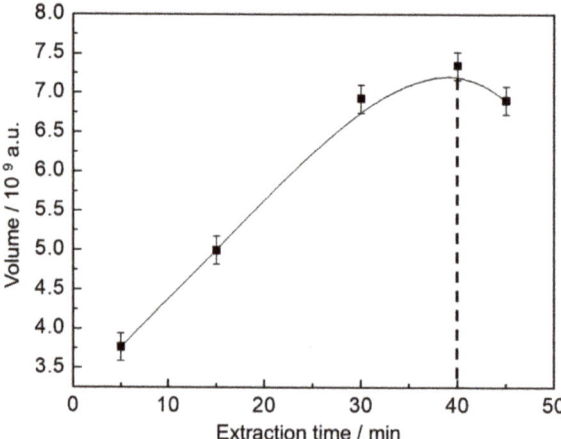

Figure 1. Extraction profile of volatile and semi-volatile organic compounds (VOC) from beer. The response is presented as the sum of the peak volumes from the GC×GC chromatogram.

The extraction profile reveals that the equilibration time for the analytes is reached within 40 min of extraction, after that the amount of analyte extracted is about the same. Therefore, to ensure sample throughput, extraction time was fixed in 40 min.

3.2. VOC Tracing Using GC×GC

Beer production involves a considerable number of important steps, which presents many variables that can affect the final composition of the beverage if they fluctuate from the expected values. A small fluctuation during beer production can be responsible for relevant changes between the ratio of its components or the presence of unique ones in the perceived flavor, altering the volatile organic compounds and organoleptic profile of the beer. The chromatograms obtained for BR and ENG beer samples, exhibited in Figure 2, shows the presence of approximately 210 compounds in the beer aroma. Main chemical classes of analytes, see Table 1, include alcohols, terpenoids, organic acids and esters, which are corroborated by previous findings [3,4].

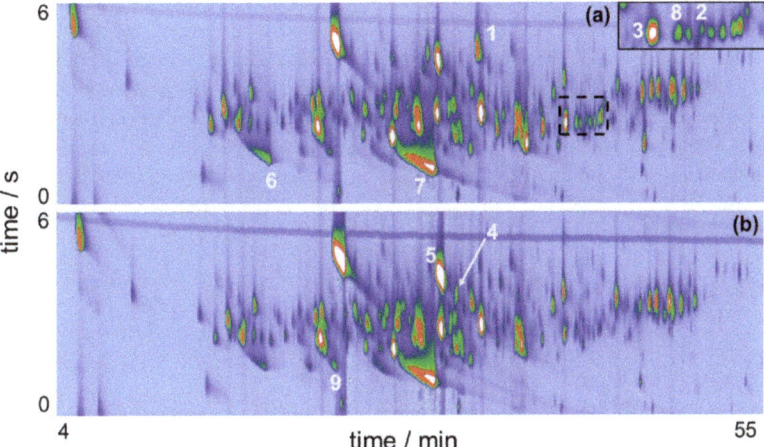

Figure 2. GC×GC-QMS chromatograms for two craft beer samples using HS-SPME and sampling with DVB/CAR/PDMS (a) BR sample; (b) ENG sample. BR corresponds to the beer brewed with the *Brazilian Ale 02* yeast and ENG corresponds to the beer brewed with the *Nottingham* yeast.

Esters are important contributors to flavor and in general are present in concentrations around their threshold, which means that minor changes in their concentration may dramatically affect beer flavor [27,28]. Usually they are compounds that impart a sweet-fruity aroma to beers and among them the ones with proven aroma activity can be synthesized intracellularly by fermenting yeast cells. The volatile portion of these compounds are produced in an enzyme-catalyzed condensation reaction between an activated fatty acid (acyl-Coa) and a higher alcohol [28]. Esters detected in the beer samples include 2-phenylethyl acetate and perillyl acetate, with first one contributing to honey attributes and the second one associated with woody and raspberry notes. Figure 2 also reveals a region, indicated by the dashed area between 40–46 min, with notable differences between the samples, with BR sample exhibiting compounds with higher intensities than ENG; among them are aromadendrene (analyte #2), which is associated with woody notes. Important terpene hydrocarbons also identified are γ-muurolene (analyte #8), responsible for herbal and spicy attributes, and α-humulene (analyte #3) a hop oil compound that impart woody notes in the aroma profile.

Another chemical class that plays an important role in the aromatic profile of beers are volatile phenols. 4-vinylguaiacol and 4-vinylphenol are the main phenols that impart flavor in beer. Usually their presence must be controlled, otherwise surpassing a specific concentration they are considered off-flavors that negatively affect beer quality [9,29]. 4-vinylguaiacol (analyte #1) exhibits a higher intensity in the BR sample, imparting a sweet smoky attribute to the organoleptic profile. Conversely, ENG sample exhibits higher intensities for methyl 2-nonynoate (analyte #4), a flavoring related substance that contributes with floral and tropical fruity notes, and 2-phenylethyl acetate (analyte #5) that imparts honey and floral rosy attributes.

These findings are corroborated by previous studies of the VOC profile of beers [16,23,30]. As a matter of fact, it is important to highlight that although approximately 210 peaks were detected in the chromatograms, only 79 (~38%) were successfully identified based on the selected qualitative parameters (80% of similarity and ± 25 LTPRI deviation from NIST Web Book). This highlights the need to couple GCxGC with high accuracy mass spectrometers to complement qualitative analyses.

It should be noted that hexanoic acid (analyte #6), heptanoic acid (analyte #9) and octanoic acid (analyte #7) were successfully identified, although their peaks exhibited deviations higher than 25 LTPRI units. Identification was confirmed by evaluating their peak shapes in the chromatograms and similarities between their experimental mass spectrums and NIST database. Figure 2 reveals the presence of pronounced peak tailing in the first dimension for these organic acids, two reasons are identified as responsible for it. First, slow mass transfer of the analytes from the SPME fiber to the GC column [31]. Second, peak tailing is related to non-linear chromatography of organic acids using a non-polar stationary phase used in the first dimension, a MEGA-5HT capillary column.

Figure 2 also exhibits the benefits that comprehensive two-dimensional gas chromatography brings to the analyses, mitigating possible co-elution between chromatographic peaks, which are resolved in the second dimension. For example, Figure 3 focuses on one of these regions of the chromatograms for BR sample.

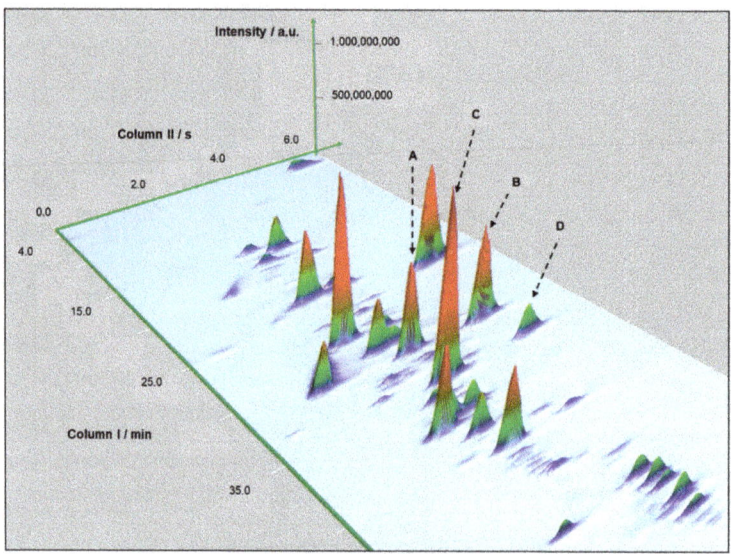

Figure 3. Example of resolved peaks in the second dimension of the chromatograms for BR. Second dimension resolved trans-geraniol (peak A) and 2-phenylethyl acetate (peak B). Additionally, (E)-methyl geranate (peak C) and 4-vinylguaiacol (peak D).

The selected region reveals the gain in resolution obtained due to increased peak capacity of the composite 2D-GC analysis. The second dimension mitigated the co-elution between trans-geraniol (peak A) and 2-phenylethyl acetate (peak B). In addition, (E)-methyl geranate (peak C) and 4-vinylguaiacol (peak D) were resolved in ^{2}D.

Although visual inspection of the chromatograms reveals variations between the chromatographic profile of the BR and ENG samples, to ensure that all the chemical information provided by the GCxGC-QMS will be properly analyzed, an appropriate data handling technique is necessary. Furthermore, as GCxGC-MS is an analytical technique that produces structurally complex and dense data, it is also important to guarantee that relevant information will not be lost or overlooked during data processing. Then, with the aim of an unbiased assessment during identification of compounds produced due to the use of the unique yeast strain, computational tools and a multivariate technique were required for proper pattern recognition.

3.3. Chemometrics

Development of an adequate protocol to evaluate BR and ENG beers was accomplished by MPCA. The main idea was to extract data information by using a multiway exploratory method, which benefits from all the available data dimensions and results in useful and easily understood plots. Principal components (PC) are generated based on the explained information extracted from the original data, with the first PC explaining most of the data variance and subsequent components explaining progressively less of the remaining variance [24].

In this pixel-based chemometric analysis four replicates from the same BR beer sample and four replicates from the same ENG beer sample form the sample group, and each variable refers to a pixel of the chromatogram's data obtained after raw data handling in MATLAB environment. MPCA basically decomposes the data into what is called scores and loadings. Through the scores information it is possible to identify the patterns within the samples, while the loadings can be used to pinpoint the chromatographic peaks responsible for such patterns.

The scores graph of total ion chromatogram (TIC) from PC1 and PC2, seen in Figure 4, reveals important information for pattern recognition. Its analysis reveals the presence of distinct traits between the VOC profile of BR and ENG beer samples. In this case, the information within PC1 is able to distinguish the samples groups (BR and ENG).

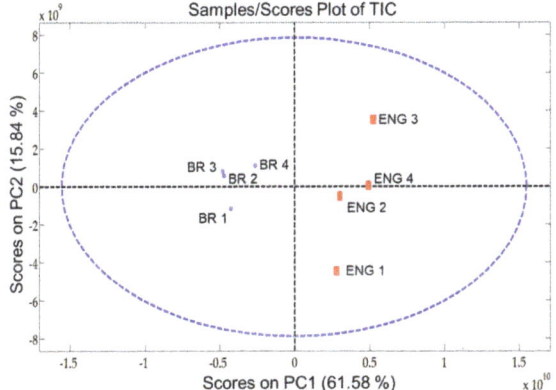

Figure 4. Scores plot of pixel-based two component Multiway principal components analysis (MPCA) from beer samples evaluated (BR and ENG) using the four-way data from GC×GC-QMS.

The loading's graph can be used to elucidate the variables/pixels that contributed to the pattern evidenced in Figure 4. For additional detail on template matching refer to Reference [32]. This information enables targeting the analytes that are responsible for beer differentiation and establishing the contributions the unique *Brazilian Ale 02* yeast strain brought to the VOC profile of the beer. As the BR samples are found in the negative part of PC1, loadings with negative values cover the VOC that are present in higher intensities in BR samples in comparison with ENG. By selecting the negative weights from the loadings of PC1, it was possible to plot a virtual chromatogram, which highlighted the peaks that differentiated BR and ENG samples. This virtual chromatogram was then used to create a template using GC Image software, Figure 5a. After creating the template, it was matched and applied onto the real sample chromatograms of BR samples. Hence, the analytes that varied the most between BR and ENG beer were identified, as shown in Figure 5b.

Table 1. Analytes identified in the volatile and semi-volatile fraction of the ENG and BR samples by HS-SPME and GC×GC-QMS. Linear temperature programmed retention index (LTPRI) were obtained from NIST. All analytes were identified by considering a minimum similarity match of 800 and a ± 25 LTPRI deviation from NIST Web Book. The detection of volatile organic compounds in BR and ENG samples is also indicated. Template matching refers to the assignment of class-specific peaks using a mixed data analysis approach, namely, multiway principal component analysis and template matching. The odor and flavor attribution were acquired from The Good Scents Company [33].

Compound	Formula	CAS	Match	LTPRI Exp	LTPRI NIST	Detection BR	Detection ENG	Template Matching BR	Template Matching ENG	Odor	Flavor
isoamyl acetate	$C_7H_{14}O_2$	123-92-2	885	841	876	x	x	x	x	sweet, fruity, banana solvent	sweet, fruity, banana-like with a green ripe nuance
2-methylpropyl isobutyrate	$C_8H_{16}O_2$	97-85-8	832	898	910	x	x	-	-	ethereal fruity, tropical fruit, pineapple, banana	fruity, pineapple, tropical fruit, ripe fruit
pentyl propionate	$C_8H_{16}O_2$	624-54-4	806	973	969	-	x	-	-	sweet, fruity, apricot, pineapple	-
β-pinene	$C_{10}H_{16}$	127-91-3	890	989	979	x	x	x	-	dry, woody, resinous pine and hay green	fresh, piney and woody, terpy and resinous with a slight minty, camphoraceous with a spicy nuance
β-myrcene	$C_{10}H_{16}$	123-35-3	849	990	991	-	x	-	-	peppery, terpene spicy, balsam	woody, citrus, fruity with a tropical mango and slight leafy minty nuances
ethyl hexanoate	$C_8H_{16}O_2$	123-66-0	859	1003	-	x	x	x	-	sweet, fruity, pineapple, waxy, green and banana	sweet, pineapple, fruity, waxy and banana with a green estry nuance
hexanoic acid	$C_6H_{12}O_2$	142-62-1	873	1045	990	x	x	x	-	sour, fatty, sweat and cheese	cheesy, fruity, phenolic fatty goaty
β-ocimene	$C_{10}H_{16}$	13877-91-3	828	1052	1037	-	x	-	-	citrus, tropical green terpene, woody green	green, tropical, woody with floral and vegetable nuances
ethyl 5-methylhexanoate	$C_9H_{18}O_2$	10236-10-9	857	1067	1072	x	x	-	-	-	-
linalool oxide	$C_{10}H_{18}O_2$	5989-33-3	843	1076	1074	x	x	-	-	earthy floral, sweet woody	-
hop ether	$C_{10}H_{16}O$	19901-95-2	869	1100	-	x	x	-	x	-	-
linalool	$C_{10}H_{18}O$	78-70-6	899	1107	1099	x	x	x	x	citrus, floral woody, green and blueberry	citrus, orange, lemon, floral, waxy, aldehydic and woody
heptanoic acid	$C_7H_{14}O_2$	111-14-8	834	1125	1078	x	x	-	-	rancid, sour, cheesy, sweat	waxy, cheesy, fruity, dirty and fatty

Table 1. *Cont.*

Compound	Formula	CAS	Match	LTPRI		Detection		Template Matching		Odor	Flavor
				Exp	NIST	BR	ENG	BR	ENG		
fenchol	$C_{10}H_{18}O$	1632-73-1	828	1127	-	x	-	-	-	camphor borneol pine, woody, dry, sweet, lemon	camphoreous cooling medicinal minty, earthy humus
trans-pinocarveol	$C_{10}H_{16}O$	547-61-5	864	1149	-	x	x	x	x	warm, woody and balsamic fennel	-
cis-isocarveol	$C_{10}H_{16}O$	22626-43-3	821	1166	-	x	x	-	-	-	-
2,6-dimethyl-1,5,7-octatrien-3-ol	$C_{10}H_{16}O$	29414-56-0	808	1168	-	-	x	x	x	camphoreous lime	-
borneol	$C_{10}H_{18}O$	507-70-0	811	1181	1167	x	-	-	-	pine, woody, camphor balsamic	-
terpinen-4-ol	$C_{10}H_{18}O$	562-74-3	825	1187	1177	x	x	-	-	pepper, woody, earth musty sweet	cooling, woody, earthy, clove spicy with a citrus undernote
ethyl octanoate	$C_{10}H_{20}O_2$	106-32-1	911	1201	1196	x	x	x	x	fruity, wine, waxy, sweet apricot, banana brandy and pear	sweet, waxy, fruity and pineapple with creamy, fatty, mushroom and cognac notes
terpineol	$C_{10}H_{18}O$	98-55-5	857	1206	1196	x	x	-	x	pine, terpene, lilac citrus, woody, floral	citrus, woody with a lemon and lime nuance. it has a slight soapy mouth feel
myrtenol	$C_{10}H_{16}O$	515-00-4	823	1207	1195	-	x	-	-	woody, pine, balsam sweet, mint	cooling, minty, camphoreous, green with a medicinal nuance
cis-geraniol	$C_{10}H_{18}O$	106-25-2	833	1236	-	x	x	-	x	sweet, neroli, citrus, magnolia	lemon, bitter, green and fruity with a terpy nuance
citronellol	$C_{10}H_{20}O$	1117-61-9	877	1238	-	x	x	x	x	citronella oil, rose leaf and oily petal	-
ethyl phenylacetate	$C_{10}H_{12}O_2$	101-97-3	815	1252	1246	x	-	x	-	sweet, floral, honey, rose and balsam cocoa	strong sweet rosy honey and balsamic cocoa-like with molasses and yeasty nuances
octanoic acid	$C_8H_{16}O_2$	124-07-2	911	1257	1180	x	x	x	x	fatty, waxy, rancid oily, vegetable cheesy	rancid soapy, cheesy, fatty brandy
trans-geraniol	$C_{10}H_{18}O$	106-24-1	930	1263	-	x	x	x	x	sweet, floral, fruity, rose, waxy and citrus	floral, rosy, waxy and perfume with a fruity, peach-like nuance

Table 1. Cont.

Compound	Formula	CAS	Match	LTPRI Exp	LTPRI NIST	Detection BR	Detection ENG	Template Matching BR	Template Matching ENG	Odor	Flavor
2-phenylethyl acetate	$C_{10}H_{12}O_2$	103-45-7	935	1266	1255	x	-	-	x	floral, rose, sweet, honey, fruity tropical	sweet, honey, floral, rosy with a slight green nectar fruity body and mouth feel
2-isopropenyl-5-methylhex-4-enal	$C_{10}H_{16}O$	-	832	1280	-	-	x	-	-	-	-
3-nonenoic acid, ethyl ester	$C_{11}H_{20}O_2$	91213-30-8	805	1286	-	x	x	-	-	-	-
trans-shisool	$C_{10}H_{18}O$	22451-48-5	885	1288	-	x	x	x	-	-	-
(Z)-3-decen-1-ol	$C_{10}H_{20}O$	10340-22-4	807	1290	-	-	x	-	-	-	-
1,10-decanediol	$C_{10}H_{22}O_2$	112-47-0	812	1290	-	x	-	-	x	-	-
methyl 2-nonynoate	$C_{10}H_{16}O_2$	111-80-8	800	1292	1311	-	x	-	x	floral, green, violet leaf, melon and cucumber	green melon, cucumber, violet tropical fruity
2-dodecanol	$C_{12}H_{26}O$	10203-28-8	839	1312	-	x	x	-	-	-	-
isocarveol	$C_{10}H_{16}O$	536-59-4	879	1313	-	x	x	-	x	green, linalool, terpineol and fatty	sweet, woody, aromatic spicy cardamom and green cumin like with dried orange peel and green waxy floral nuances
4-vinylguaiacol	$C_9H_{10}O_2$	7786-61-0	902	1323	-	x	-	x	-	dry, woody, fresh amber cedar and roasted peanut	smoky bacon
(E)-methyl geranate	$C_{11}H_{18}O_2$	1189-09-9	918	1329	-	x	x	x	x	waxy, green, fruity flower	-
myrtanyl acetate	$C_{12}H_{20}O_2$	29021-36-1	808	1351	-	x	-	-	-	-	-
citronellyl acetate	$C_{12}H_{22}O_2$	150-84-5	873	1353	-	x	x	x	x	floral green rose, fruity, citrus, woody and tropical fruit	floral, waxy, aldehydic, with green fruity nuances. fruity pear and apple like
α-ylangene	$C_{15}H_{24}$	14912-44-8	812	1371	1372	x	-	-	-	-	-
γ-nonanolactone	$C_9H_{16}O_2$	104-61-0	818	1374	1363	-	x	-	-	coconut, creamy, waxy sweet, buttery oily	coconut, creamy, waxy with fatty milky notes
α-copaene	$C_{15}H_{24}$	3856-25-5	861	1377	-	x	-	-	-	woody	-

Table 1. Cont.

Compound	Formula	CAS	Match	LTPRI		Detection		Template Matching		Odor	Flavor
				Exp	NIST	BR	ENG	BR	ENG		
ethyl trans-4-decenoate	$C_{12}H_{22}O_2$	76649-16-6	878	1381	-	x	x	x	-	green, fruity, waxy and cognac	fatty, waxy, green, pineapple and pear nuances
geranyl acetate	$C_{12}H_{20}O_2$	105-87-3	836	1381	1382	x	x	x	x	floral rose lavender, green and waxy	waxy, green, floral, oily and soapy with citrus and winey, rum nuances
ethyl 9-decenoate	$C_{12}H_{22}O_2$	67233-91-4	858	1389	1387	x	x	-	x	fruity, fatty	-
ethyl decanoate	$C_{12}H_{24}O_2$	110-38-3	910	1396	1396	x	-	x	-	sweet, waxy, fruity, apple grape and oily brandy	waxy, fruity, sweet apple
isopulegol acetate	$C_{12}H_{20}O_2$	57576-09-7	801	1420	-	-	x	-	-	woody, sweet peppermint, tropical	woody, berry green and camphoreous with a fruity nuance
bicyclo[4.1.0]heptane, 7-(1-methylethylidene)	$C_{10}H_{16}$	53282-47-6	831	1420	-	x	-	-	-	-	-
isocaryophyllene	$C_{15}H_{24}$	118-65-0	890	1423	1406	-	x	-	-	woody, spicy	-
caryophyllene	$C_{15}H_{24}$	87-44-5	939	1423	1419	x	x	x	-	sweet, woody, spice clove dry	spicy, clove, woody, nut skin and powdery peppery
isogermacrene D	$C_{15}H_{24}$	317819-80-0	872	1433	1448	x	x	x	-	-	-
p-mentha-1(7),8(10)-dien-9-ol	$C_{10}H_{16}O$	29548-13-8	845	1440	-	-	-	-	-	-	-
perillyl acetate	$C_{12}H_{18}O_2$	1511-96-3	863	1440	1436	x	x	-	x	fruity, woody and raspberry	ionone raspberry
isoamyl octanoate	$C_{13}H_{26}O_2$	2035-99-6	812	1448	1446	x	-	x	-	sweet, oily, fruity, green soapy, pineapple and coconut	sweet, fruity, waxy, pineapple, fruity and green with coconut and cognac nuances
α-humulene	$C_{15}H_{24}$	6753-98-6	915	1460	-	x	x	x	-	woody	-
γ-muurolene	$C_{15}H_{24}$	30021-74-0	887	1478	-	x	-	x	-	herbal, woody and spice	-
farnesyl butanoate	$C_{19}H_{32}O_2$	51532-27-5	811	1485	-	x	-	x	-	-	-
aromadendrene	$C_{15}H_{24}$	72747-25-2	861	1496	-	x	-	x	-	woody	-

Table 1. *Cont.*

Compound	Formula	CAS	Match	LTPRI Exp	LTPRI NIST	Detection BR	Detection ENG	Template Matching BR	Template Matching ENG	Odor	Flavor
guaia-1(10),11-diene	$C_{15}H_{24}$	-	875	1501	1509	x	-	-	-	-	-
neryl isobutanoate	$C_{14}H_{24}O_2$	2345-24-6	840	1509	1487	x	-	-	-	sweet, fresh, fruity, raspberry, strawberry, green	juicy and fruity, green, sweet, melon and waxy
neryl hexanoate	$C_{16}H_{28}O_2$	68310-59-8	865	1510	-	x	-	x	-	-	-
γ-amorphene	$C_{15}H_{24}$	6980-46-7	903	1517	1496	x	-	-	-	-	-
delta-amorphene	$C_{15}H_{24}$	16729-01-4	904	1522	-	x	-	x	-	-	-
cis-calamenene	$C_{15}H_{22}$	72937-55-4	869	1527	1531	x	-	x	-	-	-
cadinadiene-1,4	$C_{15}H_{24}$	16728-99-7	823	1537	1533	x	-	-	-	-	-
α-muurolene	$C_{15}H_{24}$	31983-22-9	836	1541	-	x	-	-	-	-	-
cis-7-hexadecene	$C_{16}H_{32}$	35507-09-6	884	1590	1566	x	-	-	-	-	-
caryophyllene oxide	$C_{15}H_{24}O$	1139-30-6	914	1590	1581	x	x	x	-	sweet, fresh, dry, woody and spicy	dry, woody, cedar old, carrot, ambrette amber
Ethyl dodecanoate	$C_{14}H_{28}O_2$	106-33-2	862	1594	1595	x	-	x	-	sweet, waxy, floral, soapy clean	waxy, soapy and floral with a creamy, dairy and fruity nuance
bisaboladien-4-ol	$C_{15}H_{26}O$	-	800	1608	-	x	x	-	x	-	-
humulol	$C_{15}H_{26}O$	28446-26-6	862	1613	-	-	x	-	-	-	-
humulene-1,2-epoxide	$C_{15}H_{24}O$	19888-34-7	914	1619	-	x	x	-	x	-	-
epicubenol	$C_{15}H_{26}O$	19912-67-5	870	1624	1627	x	x	-	-	-	-
cubenol	$C_{15}H_{26}O$	21284-22-0	878	1636	1642	x	x	-	-	spicy, herbal green tea	-
alloaromadendrene oxide	$C_{15}H_{24}O$	-	833	1645	-	x	x	-	x	-	-
τ-cadinol	$C_{15}H_{26}O$	-	902	1650	1640	x	x	-	-	-	-
intermedeol	$C_{15}H_{26}O$	6168-59-8	800	1667	1667	x	x	x	-	-	-
elemol acetate	$C_{17}H_{28}O_2$	-	800	1681	-	x	x	-	-	-	-
4(15),5,10(14)-germacratrien-1-ol	$C_{15}H_{24}O$	81968-62-9	800	1694	1695	x	x	-	x	-	-
10-heneicosene	$C_{10}H_{16}O$	95008-11-0	854	1792	-	-	x	-	-	-	-

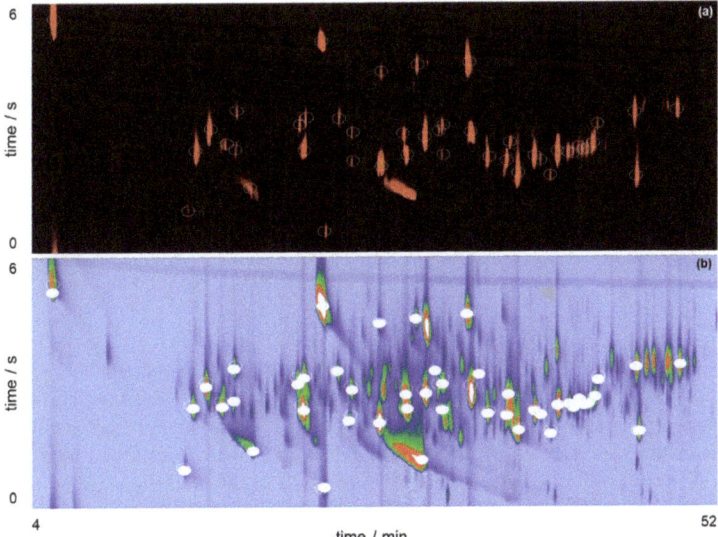

Figure 5. (a) Peak template creation using the loadings vector and the reconstructed chromatogram; (b) Process of template matching using GC Image highlighting potential unique compounds in the GC×GC-QMS total ion chromatogram of the BR beer sample.

Template matching revealed 46 compounds that were exclusively in one sample or in both, but with distinguished intensities. Table 1 summarizes all the compounds identified by the approach described herein.

According to the data of Table 1, ENG brewing enabled retention/formation of higher alcohols and monoterpene alcohols. Template matching identified compounds that were present in higher quantities in the ENG sample, including alcohols and esters. These chemical classes are known to enhance fruity notes to the sample; examples include cis-geraniol; trans-geraniol and 2-phenylethyl acetate.

Furthermore, the chemometric approach highlighted that the *Brazilian Ale 02* yeast supported the formation and/or retention of terpenic compounds in the beer. The terpenoids caryophyllene oxide, beta-pinene, and caryophyllene contribute to enhance spicy aromas and flavors in the beer, corroborating the information presented in the datasheet of the *Brazilian Ale 02* yeast [34]. Classification of terpenic compounds depends on the number of repetitions of the isoprene unit (C_5H_8), being monoterpenic compounds (two isoprene units), sequiterpenic compounds (three isoprene units) and so on [35]. Their presence in a beer comes mostly from the hops added during brewing, which act as preservative and impart the bitterness characteristic of beers [23]. A minor source to these chemical compounds may be the yeast's metabolism. King et al. [36] demonstrated that ale and lager yeasts are capable of biotransforming the monoterpene alcohols trans-geraniol and linalool into other monoterpene alcohols or esters.

Finally, template matching provided conditions to the identification of the bioaromatization properties introduced in the aroma profile of the beer by the unique Brazilian yeast. The approach enabled the identification of seven compounds that were found exclusively in the BR samples, chemical classes found include esters (ethyl phenylacetate; ethyl decanoate; isoamyl octanoate; ethyl dodecanoate), terpenes (γ-muurolene; aromadendrene) and a phenolic compound (4-vinylguaiacol). Unfortunately, due to synergistic and antagonistic effects it is difficult to know what will be the effect that different levels of higher alcohols and esters will have on the overall profile of the beer [27]. Therefore, the data analyzed provided general answers about the aroma that compounds can impart to the beer. These findings highlight the necessity of allying a sensory panel of specialists to taste beer samples and to elucidate the influence of such compounds to the aroma of beers.

Furthermore, there is a threshold in which a compound with a concentration lower than that will not be perceived, but also due to synergistic effects the presence of different esters can affect beer flavor even if they are well below their individual threshold concentrations [27]. This is valid for both desired and undesired molecules. Examples of undesired compounds (off-flavors) and their individual sensory threshold in beers include the esters isoamyl acetate (1400 ng mL^{-1}); ethyl acetate (15,000 ng mL^{-1}); ethyl hexanoate (200 ng mL^{-1}) and 2,3-butanedione (40 ng mL^{-1}) [37]. Ethyl hexanoate and isoamyl acetate were both detected in BR and ENG, which highlights how difficult it is to produce beers without off-flavors.

4. Conclusions

The steady increase in beer consumption over the past decades demonstrates the notorious importance that the beverage possesses in a global scenario. Therefore, the search for better quality beers is fundamental nowadays. In this study, we evaluated a bottom-up approach to reveal the novel contributions that the *Brazilian Ale 02* yeast could impart in the VOC profile of a beer. The protocol adopted succeeded in comparing both beer samples based on yeast strain, highlighting 46 compounds as relevant for sample classification. Furthermore, multiway principal components analysis discarded unrelated information from the GC×GC-MS chromatograms, which enabled simple and efficient data mining, even by non-expert users. It was possible to accurately identify seven key markers spotted in the VOC profile of the BR sample due to the use of the unique Brazilian yeast during brewing.

These outcomes highlighted the importance of combining high peak capacity techniques, as comprehensive two-dimensional gas chromatography, with appropriate data processing techniques such as MPCA. Finally, we expect to continue the research through the analyses of more samples and the alignment of a sensory panel of specialists to obtain comprehensive aroma-related information among beers.

Author Contributions: D.S.O. brewed the beer samples. A.C.P., D.S.O. and L.W.H. conceived and planned the experiments. A.C.P. performed the experiments, collected and processed all the data. A.C.P. and L.W.H. wrote the paper with input from all authors.

Funding: This research was funded by National Council for Scientific and Technological Development (CNPq 400182/2016-5), São Paulo Research Foundation (17/25490-1), and Unicamp (FAEPEX 519.292). Andre Cunha Paiva thanks the Coordination for the Improvement of Higher Education Personnel (CAPES 88882.329162/2019-01) for research fellowships.

Acknowledgments: We are indebted to Danilo Pierone and Felipe Lugão (Nova Analítica Imp. Exp.) for generously providing Thermo Scientific instruments.

Conflicts of Interest: The authors declare no conflicts of interest.

References

1. KIRIN HOLDINGS. Available online: https://www.kirinholdings.co.jp/english/news/2018/1220_01.html (accessed on 2 April 2019).
2. Vasconcelos, Y. Inovações cervejeiras. *Pesqui. Fapesp* **2017**, *251*, 18–25.
3. da Silva, G.A.; Augusto, F.; Poppi, R.J. Exploratory analysis of the volatile profile of beers by HS–SPME–GC. *Food Chem.* **2008**, *111*, 1057–1063.
4. Martins, C.; Brandão, T.; Almeida, A.; Rocha, S.M. Insights on beer volatile profile: Optimization of solid-phase microextraction procedure taking advantage of the comprehensive two-dimensional gas chromatography structured separation. *J. Sep. Sci.* **2015**, *38*, 2140–2148. [PubMed]
5. Venturini Filho, W.G. *Bebidas alcoólicas: Ciência e Tecnologia*, 2nd ed.; Edgard Blücher Ltda.: Sao Paulo, Brazil, 2016.
6. Vanderhaegen, B.; Neven, H.; Coghe, S.; Verstrepen, K.J.; Verachtert, H.; Derdelinckx, G. Evolution of Chemical and Sensory Properties during Aging of Top-Fermented Beer. *J. Agric. Food Chem.* **2003**, *51*, 6782–6790. [PubMed]

7. Saison, D.; De Schutter, D.P.; Delvaux, F.; Delvaux, F.R. Optimisation of a complete method for the analysis of volatiles involved in the flavour stability of beer by solid-phase microextraction in combination with gas chromatography and mass spectrometry. *J. Chromatogr. A* **2008**, *1190*, 342–349. [PubMed]
8. Rodrigues, J.A.; Barros, A.S.; Carvalho, B.; Brandão, T.; Gil, A.M.; Ferreira, A.C.S. Evaluation of beer deterioration by gas chromatography-mass spectrometry/multivariate analysis: A rapid tool for assessing beer composition. *J. Chromatogr. A* **2011**, *1218*, 990–996. [PubMed]
9. Pizarro, C.; Pérez-del-Notario, N.; González-Sáiz, J.M. Optimisation of a simple and reliable method based on headspace solid-phase microextraction for the determination of volatile phenols in beer. *J. Chromatogr. A* **2010**, *1217*, 6013–6021.
10. Basso, R.F.; Alcarde, A.R.; Portugal, C.B. Could non-Saccharomyces yeasts contribute on innovative brewing fermentations? *Food Res. Int.* **2016**, *86*, 112–120.
11. Arthur, C.; Pawliszyn, J. Solid Phase Microextraction with Thermal Desorption Using Fused Silica Optical Fibers. *Anal. Chem.* **1990**, *62*, 2145–2148.
12. Jeleń, H.H.; Majcher, M.; Dziadas, M. Microextraction techniques in the analysis of food flavor compounds: A review. *Anal. Chim. Acta* **2012**, *738*, 13–26.
13. Lord, H.; Pawliszyn, J. Evolution of solid-phase microextraction technology. *J. Chromatogr. A* **2000**, *885*, 153–193.
14. Kataoka, H.; Lord, H.L.; Pawliszyn, J. Applications of solid-phase microextraction in food analysis. *J. Chromatogr. A* **2000**, *880*, 35–62. [PubMed]
15. Sterckx, F.L.; Saison, D.; Delvaux, F.R. Determination of volatile monophenols in beer using acetylation and headspace solid-phase microextraction in combination with gas chromatography and mass spectrometry. *Anal. Chim. Acta* **2010**, *676*, 53–59. [PubMed]
16. Kobayashi, M.; Shimizu, H.; Shioya, S. Beer Volatile Compounds and Their Application to Low-Malt Beer Fermentation. *J. Biosci. Bioeng.* **2008**, *106*, 317–323. [PubMed]
17. Riu-Aumatell, M.; Miró, P.; Serra-Cayuela, A.; Buxaderas, S.; López-Tamames, E. Assessment of the aroma profiles of low-alcohol beers using HS-SPME–GC-MS. *Food Res. Int.* **2014**, *57*, 196–202.
18. Saison, D.; De Schutter, D.P.; Delvaux, F.; Delvaux, F.R. Determination of carbonyl compounds in beer by derivatisation and headspace solid-phase microextraction in combination with gas chromatography and mass spectrometry. *J. Chromatogr. A* **2009**, *1216*, 5061–5068. [PubMed]
19. Tian, J. Determination of several flavours in beer with headspace sampling-gas chromatography. *Food Chem.* **2010**, *123*, 1318–1321.
20. Leça, J.M.; Pereira, A.C.; Vieira, A.C.; Reis, M.S.; Marques, J.C. Optimal design of experiments applied to headspace solid phase microextraction for the quantification of vicinal diketones in beer through gas chromatography-mass spectrometric detection. *Anal. Chim. Acta* **2015**, *887*, 101–110.
21. Cajka, T.; Riddellova, K.; Tomaniova, M.; Hajslova, J. Recognition of beer brand based on multivariate analysis of volatile fingerprint. *J. Chromatogr. A* **2010**, *1217*, 4195–4203.
22. Griffith, J.F.; Winniford, W.L.; Sun, K.; Edam, R.; Luong, J.C. A reversed-flow differential flow modulator for comprehensive two-dimensional gas chromatography. *J. Chromatogr. A* **2012**, *1226*, 116–123.
23. Martins, C.; Brandão, T.; Almeida, A.; Rocha, S.M. Unveiling the lager beer volatile terpenic compounds. *Food Res. Int.* **2018**, *114*, 199–207. [PubMed]
24. Callao, M.P.; Ruisánchez, I. An overview of multivariate qualitative methods for food fraud detection. *Food Control* **2018**, *86*, 283–293.
25. Escandar, G.M.; Olivieri, A.C.; Faber, N.M.; Goicoechea, H.C.; Muñoz de la Peña, A.; Poppi, R.J. Second- and third-order multivariate calibration: Data, algorithms and applications. *TrAC Trends Anal. Chem.* **2007**, *26*, 752–765.
26. Olivieri, A.C. Analytical Advantages of Multivariate Data Processing. One, Two, Three, Infinity? *Anal. Chem.* **2008**, *80*, 5713–5720. [PubMed]
27. Meilgaard, M.C. Flavor chemistry in beer: Part I: Flavor interaction between principal volatiles. *Master Brew. Assoc. Am. Tech. Q.* **1975**, *12*, 107–117.
28. Verstrepen, K.J.; Derdelinckx, G.; Dufour, J.-P.; Winderickx, J.; Thevelein, J.M.; Pretorius, I.S.; Delvaux, F.R. Flavor-Active Esters: Adding Fruitiness to Beer. *J. Biosci. Bioeng.* **2003**, *96*, 110–118.

29. Zhu, M.; Cui, Y. Determination of 4-vinylgaiacol and 4-vinylphenol in top-fermented wheat beers by isocratic high performance liquid chromatography with ultraviolet detector. *Braz. Arch. Biol. Technol.* **2013**, *56*, 1018–1023.
30. Kishimoto, T.; Noba, S.; Yako, N.; Kobayashi, M.; Watanabe, T. Simulation of Pilsner-type beer aroma using 76 odor-active compounds. *J. Biosci. Bioeng.* **2018**, *126*, 330–338.
31. Snow, N.H.; Sinex, J.; Danser, M. Multiple Dimensions of Separations: SPME with GC×GC and GC×GC-TOF-MS. *Lc Gc Eur.* **2010**, *23*, 260–267.
32. Crucello, J.; Miron, L.F.O.; Ferreira, V.H.C.; Nan, H.; Marques, M.O.M.; Ritschel, P.S.; Zanus, M.C.; Anderson, J.L.; Poppi, R.J.; Hantao, L.W. Characterization of the aroma profile of novel Brazilian wines by solid-phase microextraction using polymeric ionic liquid sorbent coatings. *Anal. Bioanal. Chem.* **2018**, *410*, 4749–4762.
33. The Good Scents Company. Available online: http://www.thegoodscentscompany.com/ (accessed on 6 August 2019).
34. Yeastlab Biotecnologia. Available online: http://yeastlab.com.br/Files/YLB6001%20-%2001.pdf (accessed on 2 April 2019).
35. Kumari, S.; Priya, P.; Misra, G.; Yadav, G. Structural and biochemical perspectives in plant isoprenoid biosynthesis. *Phytochem. Rev.* **2013**, *12*, 255–291.
36. King, A.J.; Dickinson, J.R. Biotransformation of hop aroma terpenoids by ale and lager yeasts. *FEMS Yeast Res.* **2003**, *3*, 53–62. [PubMed]
37. Silva, G.C.; Silva, A.A.S.; Silva, L.S.N.; Godoy, R.L.D.O.; Nogueira, L.C.; Quitério, S.L.; Raices, R.S.L. Method development by GC–ECD and HS-SPME–GC–MS for beer volatile analysis. *Food Chem.* **2015**, *167*, 71–77. [PubMed]

© 2019 by the authors. Licensee MDPI, Basel, Switzerland. This article is an open access article distributed under the terms and conditions of the Creative Commons Attribution (CC BY) license (http://creativecommons.org/licenses/by/4.0/).

Article

Determination of Hydrophilic UV Filters in Real Matrices Using New-Generation Bar Adsorptive Microextraction Devices

Alessandra Honjo Ide and José Manuel Florêncio Nogueira *

Centro de Química e Bioquímica e Centro de Química Estrutural, Faculdade de Ciências, Universidade de Lisboa, 1749-016 Lisboa, Portugal; alessandrahide@gmail.com
* Correspondence: nogueira@fc.ul.pt; Tel.: +351-217500899

Received: 29 May 2019; Accepted: 19 August 2019; Published: 25 September 2019

Abstract: In the present contribution, new-generation bar adsorptive microextraction devices combined with microliquid desorption, followed by high-performance liquid chromatography–diode array detection (BAµE-µLD/HPLC–DAD) are proposed for the determination of two very polar ultraviolet (UV) filters (2-phenylbenzimidazole-5-sulfonic acid (PBS) and 5-benzoyl-4-hydroxy-2-methoxybenzenesulfonic acid (BZ4)) in aqueous media. Different sorbents were evaluated as BAµE coating phases, in which polystyrene–divinylbenzene polymer showed the best selectivity for the analysis of both UV filters, with average extraction efficiency of 61.8 ± 9.1% for PBS and 69.5 ± 4.8% for BZ4. The validated method showed great reproducibility for the analysis of PBS and BZ4 UV filters, providing suitable limits of detection (0.04 µg L^{-1} and 0.20 µg L^{-1}), as well as good linear dynamic ranges (0.16–16.0 and 0.8–80.0 µg L^{-1}), respectively. The proposed methodology was applied for monitoring the target analytes in several real matrices, including tap, sea, and estuarine waters, as well as wastewater samples. Despite some matrix effects being observed for some real samples, good selectivity and linearity were obtained. The present contribution showed an innovative analytical cycle that includes the use of disposable devices, which make BAµE much more user-friendly and suitable for the routine work, being a remarkable analytical alternative for trace analysis of priority compounds in real matrices.

Keywords: bar adsorptive microextraction; floating sampling technology; high-performance liquid chromatography; polar UV filters; real matrices

1. Introduction

In the last decades, many improvements have been introduced in the field of sample preparation, aiming at simplification, miniaturization, easy manipulation, as well as the reduction of the use of toxic organic solvents, in compliance with the green analytical chemistry (GAC) principles. In this context, the sorption-based techniques have played an important role, mainly for trace analysis of complex matrices. In this regard, solid phase microextraction (SPME) [1], stir bar sorptive extraction (SBSE) [2], and, more recently, bar adsorptive microextraction (BAµE) [3,4] are good examples of effective sample-enrichment approaches. In BAµE, a bar-shaped polypropylene device (7.5 mm in length and 3 mm in diameter) coated with a convenient sorbent phase is used, simultaneously with a conventional Teflon magnetic stirring bar at the bottom of the sampling flask. This technique operates under the floating sampling technology mode, which uses an analytical device lighter in weight in comparison with water density. The BAµE devices are easily lab-made by coating the polypropylene support bars with a convenient adhesive, where the sorbent materials (<5 mg) are fixed. BAµE is very cost-effective, and the great advantage over other sorption-based methodologies is that the analytical devices can be easily coated with the most selective sorbent phases, according to the characteristics of the target analytes. Furthermore, this technology is also in compliance with GAC principles, needing just around 100 µL of an appropriate organic solvent during the back-extraction stage, performed through liquid desorption (LD). In a previous work [5], novel improvements were introduced in this technique, by designing new-generation BAµE devices in order to reduce the number of steps involved during the back-extraction stage. It allowed the implementation of an innovative analytical cycle, with an effective microextraction stage together with a back-extraction stage performed in "only single LD step", for interface enhancement with the instrumental systems. From the analytical point of view, the new-generation BAµE approach is user-friendly, as well as much more competitive for the routine work over other well-established microextraction techniques. Furthermore, due to the simplicity, eco-friendliness and cost-effectiveness involved, the reutilization is not considered in this novel analytical approach. To show the versatility of the new-generation devices recently introduced [5], the present contribution aims to apply them to monitor trace levels of highly polar compounds, such as hydrophilic ultraviolet (UV) filters in real samples.

UV filters are a large group of chemical compounds used to block or absorb UV light and protect the skin from the harmful effects of sun radiation. They are extensively present in a variety of personal care products, such as sunscreen lotions, skin care, facial makeup, and lip care products. These compounds are considered persistent pollutants once they are continuously released into the aquatic ecosystems through recreational aquatic activities or industrial and urban wastewaters. As a result, UV filters have been found at trace levels in several environmental matrices, mainly in surface [6], sea [7], as well as tap [8] waters, and their effects and consequences are issues of increasing concern. Numerous reports can be found in the literature regarding the determination of lipophilic UV filters from environmental water samples using microextraction techniques [9–11]. Meanwhile, few studies aim at the analysis of very polar UV filters and, most of them, use non-green analytical techniques such as solid phase extraction [12,13]. Yet, only one paper can be found in the literature using a miniaturized technique for the extraction of these hydrophilic UV filters compounds, the stir bar sorptive-dispersive microextraction mediated by magnetic nanoparticles–nylon 6 composite [14]. Although it is efficient, this methodology requires the composite synthesis in three steps, including the synthesis of the nanoparticles, followed by coating with silica, and finally the synthesis of the polymeric network with the nanoparticles, which takes many hours and requires several chemicals.

In this work, the new-generation BAµE devices combined with high-performance liquid chromatography–diode array detection (BAµE-µLD/HPLC–DAD) is applied for the determination of two highly polar UV filters (2-phenyl-5-benzemidazolesulfonic acid (PBS) and 5-benzoyl-4-hydroxy-2-methoxy-benzenesulfonic acid, also known as benzophenone-4 (BZ4)) (Table 1) in real matrices. The optimization, validation, and application of the proposed analytical methodology to several types of matrices are fully discussed.

Table 1. Chemical structures, log K_{OW} and pK_a of the two ultraviolet (UV) filters under study. PBS: 2-phenyl-5-benzemidazolesulfonic acid. BZ4: 5-benzoyl-4-hydroxy-2-methoxy-benzenesulfonic acid.

UV Filter	Chemical Structure	Log K_{OW}	pK_a
PBS		−0.234	−0.87
BZ4		0.993	−0.70

2. Materials and Methods

2.1. Reagents, Standards, and Samples

Analytical standards of PBS (96%) and BZ4 (≥97.0%) were purchased from Sigma Aldrich (Steinheim, Germany). Methanol (MeOH, 99.8%) and acetic acid (99.5%) were obtained from Carlo Erba (Arese, Italy), sodium chloride (NaCl, 99.5%) was provided from Merck (Darmstadt, Germany), sodium hydroxide (NaOH, 98.0%) was purchased from BDH Chemicals (London, UK), hydrochloric acid (HCl, 37%) was provided from Panreac (Barcelona, Spain). Ultra-pure water was obtained from Mili-Q water Milipore purification systems (Billerica, MA, USA). Stock solutions of individual UV filters (1000 mg L^{-1}) were prepared in MeOH, and standard mixtures for instrumental calibration were prepared by appropriated dilution of the previous stock solutions. The polymeric phases used were N-vinylpyrrolidone–divinylbenzene (HLB; particle size 30, surface area 810 m^2 g^{-1} and pH stability 1–14, Waters, Milford, MA, USA) and polystyrene–divinylbenzene (PS-DVB; particle size 40–120 µm, surface area 1200 m^2 g^{-1} and pH stability: 1–13, Merck, Kenilworth, NJ, USA). The activated carbons, AC1 (surface area 1400 m^2 g^{-1}; pH_{PZC}: 2.2) and AC2 (surface area 1400 m^2 g^{-1}; pH_{PZC}: 6.4) were obtained from Salmon & Cia (Lisbon, Portugal). All water samples were obtained in the metropolitan area of Lisbon (Portugal). The tap water samples were obtained from our lab; the estuarine and sea water samples from the coast, and the wastewater samples from Alcantara wastewater treatment plant.

2.2. Experimental Set-Up

2.2.1. Optimization Assays

The new generation of BAµE devices were prepared "in house" by using nylon supports having bar-shaped geometry (7.5 mm in length and 1.0 mm in diameter) and an adhesive film coated with the powdered sorbents, previously cleaned with ultra-pure water. The detailed description of the devices manufacturing can be consulted in previous reports [3,5]. For the optimization assays, 25 mL of ultra-pure water spiked with the working standard solution (8.0 µg L^{-1} for PBS and 40.0 µg L^{-1} for BZ4), a BAµE device, and a conventional Teflon magnetic bar were introduced into glass sampling flasks. The microextraction stage was performed in a univariate optimization strategy using a multipoint agitation plate (Variomag H + P Labortechnik AG Multipoint 15, Germany) at room temperature (25 °C), under the floating sampling technology mode. Parameters such as stirring rate (750, 1000, and 1250 rpm), equilibrium time (1, 2, 4, 6, and 16 h), matrix pH (HCl/NaOH; 1.0, 2.0, 4.0, 6.0, and 8.0), organic modifier (MeOH; 0, 5, 10, and 15%, v/v), and ionic strength (NaCl; 0, 5, 10, 15, 20, and 30%, w/v) were systematically studied in triplicate. The back-extraction stage was performed using a single LD step, in which the devices were removed from the sampling flasks using clean tweezers and placed directly into inserts containing 100 µL of the stripping solvent (HPLC mobile phase) inside glass vials (2 mL), ensuring their total immersion. Then the vials were sealed using a handy crimper prior to ultrasonic treatment at room temperature. To evaluate the best LD conditions, different desorption times (10, 15, 30, 45, and 60 min) using ultrasonic treatment were systematically studied in triplicate. After the back-extraction assays, the vials became ready for analysis and were placed into the autosampler trays of the HPLC–DAD system. Blank assays were also performed in triplicate, using the procedure described above without spiking.

2.2.2. Validation Assays

For the method validation experiments, ultra-pure water samples were spiked with the working standard solutions of the UV filters at the desired concentrations (0.16–16.0 µg L^{-1} for PBS and 0.8–80.0 µg L^{-1} for BZ4), and the extraction and back-extraction assays were performed in triplicate, as described above, under experimental optimized conditions.

2.2.3. Real Samples Assays

The proposed methodology was applied for monitoring the two UV filters in real matrices using the standard addition method (SAM). For tap, surface, estuarine, sea water, and wastewater assays, 25 mL of sample, previously filtered with paper filters (125 mm of diameter, Cat. No. 1001 125, Whatman, UK), were spiked with the working standards solutions (0.4–8.0 µg L^{-1} for PBS and 2.0–40.0 µg L^{-1} for BZ4) and performed in triplicate using the optimized procedure, as described before. Blank assays were also analyzed without spiking.

2.3. Instrumental Set-Up

The analyses were performed using a HPLC–DAD Agilent 1100 Series system (Agilent Technologies, Waldbronn, Germany), constituted by vacuum degasser (G1322A), quaternary pump (G1311A), autosampler (G1313A), thermostated column compartment (G1316A), and a diode array detector (G1315B). The data acquisition and instrumental control were performed by the software LC3D Chemistation (versions Ver. A. 10.02, Agilent Technologies, Germany). The separation of the analytes was performed on a Kinetex C18 column (150 × 4.6 mm, 2.6 µm; Phenomenex, Torrance, CA, USA). The mobile phase consisted of an isocratic mixture of MeOH and 20 mM acetate buffer (pH 4.75 at a mixing ratio of 30:70 (v/v)) with a flow rate of 0.5 mL min^{-1}. The column temperature was maintained constant at 50 °C. The injection volume was 20 µL, with a draw speed of 200 µL min^{-1}. The detector was set at 300 nm. For identification purposed, the retention time and peak purity were compared with the UV/vis spectral reference data of each compound. For extraction efficiency calculation, peak areas obtained from each assay were compared with the peak areas of standard controls used for spiking. For quantification purposes on real matrices, calibration plots from the SAM were used.

3. Results and Discussion

3.1. HPLC–DAD Optimization

In the present contribution, two UV filters (PBS and BZ4, Table 1) were studied as target analytes. From the very beginning, the HPLC–DAD instrumental conditions were evaluated, such as UV/vis absorbance spectra, retention times, and resolution. A wavelength at 300 nm (λ_{max}) was selected as it maximized the DAD signal for both target compounds. Since the ionization equilibrium is temperature dependent and the two UV filters are very polar and ionizable compounds, temperature control can improve the HPLC selectivity [14]. Therefore, the column temperature was maintained constant at 50 °C, once better symmetrical peak shapes were achieved. Therefore, optimized HPLC–DAD conditions resulted in good resolution in a suitable analytical time (11.0 min). Instrumental sensitivity was evaluated through the limits of detection (LODs) and quantification (LOQs), obtained after the injection of diluted standard mixtures and calculated with a signal-to-noise ratio (S/N) of 3 and 10, respectively. From the data obtained, instrumental LODs and LOQs of 20 and 100 µg L^{-1}, as well as 60.0 and 300.0 µg L^{-1} were achieved, respectively. The instrumental calibration was assessed with standard mixtures ranging from 60.0 to 10,000 µg L^{-1} for PBS and 300.0 to 25,000 µg L^{-1} for BZ4, using six levels. The regression plots achieved showed good linearity responses (r^2 = 0.9982 for PBS and 0.9996 for BZ4).

3.2. Selection of the BAµE Coating Phase

One of the great advantages of the BAµE technique is that we can easily choose the most convenient sorbent phase, according to the best selectivity for the analytes under study. Therefore, two polymers (HLB and PS-DVB) and two activated carbons (AC1 and AC2) were evaluated, and the results are depicted in Figure 1. From the data obtained, PS-DVB presented the best extraction efficiency among all sorbents tested. The polymers used are reversed-phase-type, retaining the analytes according to the particle size, surface area, and mechanisms involved. Thus, the presence of aromatic rings both in their network and in the molecules involved seems to favour π-π and dipole–dipole-type interactions between them. Furthermore, the PS-DVB coating phase presented a better extraction efficiency, probably due to its higher surface area (1200 m^2 g^{-1}) when compared with the HLB polymer (810 m^2 g^{-1}). On the other hand, the activated carbons are porous solid materials that retain the solutes through electrostatic and/or dispersive interactions, according to the textural adsorption properties, as well as surface area and pore dimensions involved. Moreover, the interactions between the analytes and the sorbent are strongly influenced by the pH of the bulk solution. When the matrix pH is equal to the pH$_{PZC}$, the activated carbon surface becomes neutral, with equal number of positive and negative charges. If the value is lower than the pH$_{PZC}$, the activated carbon surface becomes positive and, if higher than the pH$_{PZC}$, turns into negative. At pH 2.0, the sulfonated groups of the analytes are deprotonated (pK_a < 1), wherein AC1 and AC2 are neutral and positively charged, respectively, in which electrostatic interactions may occur between both analytes and AC2. However, the attractive forces are probably, in some cases, so strong that both analytes are still retained in the activated carbon network, even after the back-extraction stage under ultrasonic treatment, resulting in lower extraction efficiency.

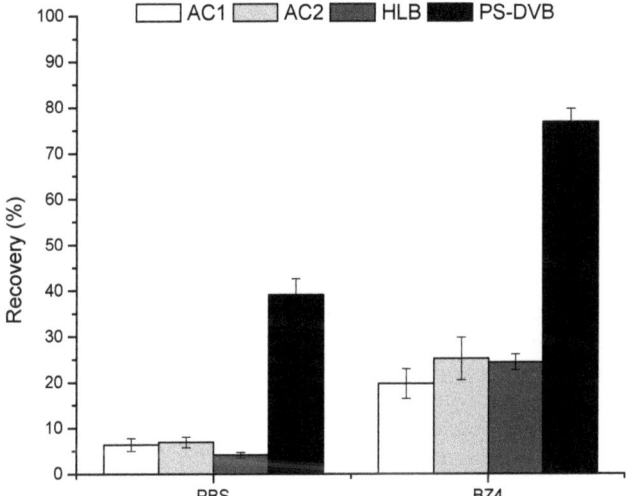

Figure 1. Average extraction efficiency obtained by BAµE-µLD/HPLC–DAD using different activated carbons and polymers sorbent phases for the microextraction of PBS and BZ4 in ultra-pure water, using standard experimental conditions. The error bars represent the standard deviation of 3 replicates.

3.3. Back-Extraction Stage

The back-extraction stage was always limitative in BAµE, since substantial manipulation, many times up to six steps (i.e., removing the device from the sampling flask, placing the device into a convenient organic solvent inside a vial, sonication treatment, removing the device from the vial, evaporating the organic microextract until dryness, solvent switch) were required. However, the implementation of an innovative analytical cycle allowed negligible manipulation during the back-extraction stage, which is now performed in "only single LD step", making BAµE a much more simple approach. In this technique, after the microextraction stage, the devices were placed into glass vial inserts, having 100 µL of mobile phase to desorb the target analytes, in which, after sealing and ultrasonic treatment, became ready for instrumental analysis. Therefore, desorption times of 10, 15, 30, and 60 min were assayed, in which the results (data not shown) proved that 10 min of ultrasonic treatment was enough to desorb both UV filters from the PS-DVB sorbent phase. Besides the results obtained, the back-extraction stage showed to be very user-friendly, once the several steps of cumbersome manipulation were avoided. Since the analytical device has small dimensions (7.5 × 1.0 mm), the desorption stage can be performed inside common glass vial inserts, having enough space for the needle of the conventional HPLC injection systems, with the possibility of automation for routine work. Moreover, due to the very easy preparation and low cost of the analytical devices, they are used only once, being discarded at the end of the sample analysis.

3.4. Microextraction Stage

Since BAµE is a static microextraction technique, the equilibrium between the matrix and the sorbent phase depends on the affinity of each analyte to the PS-DVB polymer. Therefore, the main experimental conditions that may affect the microextraction stage, including the kinetic (extraction time and agitation speed) and thermodynamic (pH, ionic strength, and polarity of the matrix) parameters, were evaluated to achieve the best extraction efficiency for the two target analytes.

Since the beginning, it was noticed that pH is a very important parameter, as the UV filters under study are extremely hydrophilic compounds. Thus, the effect of the matrix pH was assessed, ranging from 1.0 to 8.0. As the extraction efficiency decreased with the increase of the pH (Figure 2a), pH 2.0 was chosen for further studies. Due to the very low and even negative pK_a values involved (Table 1), the analytes were deprotonated even in low values of pH. However, maximum adsorption at more acidic pH values indicated that the low pH contributed to the increase of H^+ ions on the adsorbent surface, resulting in strong electrostatic attraction between the positively charged anionic PS-DVB surface and the deprotonated analyte ions. Subsequently, the equilibrium time and agitation speed were evaluated, where the former has a great effect on the microextraction kinetics, as it may limit the distribution of the target compounds between the matrix and the sorbent phase. On the other hand, the later influences the mass transfer (diffusion) of both UV filters towards the analytical device through the floating sampling technology approach. Therefore, equilibrium times between 1 and 16 h were studied, in which the results obtained revealed that 16 h was needed (Figure 2b), ensuring the best equilibrium conditions. Besides the slow kinetics demanding a substantial period of equilibrium time, this analytical approach can be performed overnight without any special requirements. Agitation speed was then assessed through assays performed at 750, 1000, and 1250 rpm, and, according to the data obtained (data not shown), an agitation speed of 750 rpm provided much higher stability on the analytical devices during the microextraction process and was chosen for the subsequent studies.

Finally, the matrix characteristics such as ionic strength and polarity were also evaluated, as they can significantly affect the performance of the sample enrichment. To check the effect of the ionic strength, assays were performed, with the addition of NaCl having concentrations ranging from 0 to 30%. From the data achieved (Figure 2c), the increment of the ionic strength up to 20% helped the microextraction process, decreasing the solubility of both organic compounds in aqueous media ("salting-out effect"), favoring their migration towards the sorbent phase. On the other hand, the matrix polarity was studied through the addition of MeOH up to 15%. The data obtained in Figure 2d showed that the average recoveries for both UV filters decreased significantly with the increment of the MeOH content, once it favored their solubility in the bulk solution.

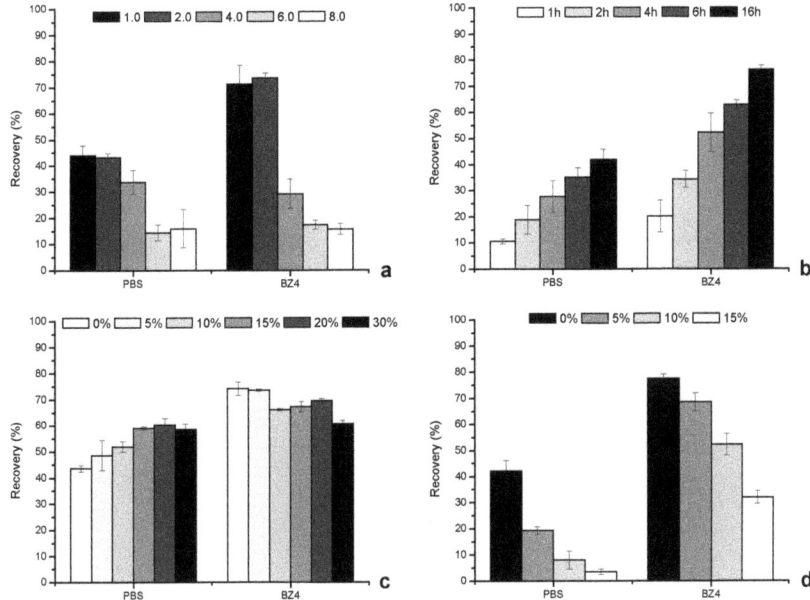

Figure 2. Effect of pH (**a**), equilibrium time (**b**), ionic strength (**c**), and matrix polarity (**d**) on the microextraction efficiency for both UV filters in ultra-pure water. The error bars represent the standard deviation of three replicates.

3.5. Method Validation

The optimized experimental conditions achieved for monitoring the UV filters in aqueous media were as follows; microextraction stage: PS-DVB phase, 16 h (750 rpm), pH 2.0, and 20% of NaCl; back-extraction stage: mobile phase (100 µL), 10 min under sonication. In order to validate the proposed methodology, assays were performed on 25 mL of ultra-pure water samples spiked with the target compounds at concentrations ranging from 0.16 to 16.0 µg L^{-1} for PBS and 0.80 to 80.0 µg L^{-1} for BZ4. The analytical limits were also evaluated and calculated, with S/N of 3 and 10, respectively. Table 2 summarizes the average extraction efficiency (61.8 ± 9.1% for PBS and 69.5 ± 4.8% for BZ4), LODs (0.04 µg L^{-1} for PBS and 0.20 µg L^{-1} for BZ4), LOQs (0.16 µg L^{-1} for PBS and 0.80 µg L^{-1} for BZ4), and determination coefficients (0.9985 for PBS and 0.9993 for BZ4) obtained for the proposed methodology to monitor the two UV filters, under optimized experimental conditions. Repeatability studies were then evaluated in terms of interday (three replicates a day in three consecutive days) and intraday (five replicates performed in the same day) assays, in which the data obtained showed RSD below 12%. From the validation data, the proposed methodology performed by BAµE(PS-DVB)-µLD/HPLC–DAD showed remarkable analytical performance for trace analysis of the two UV filters under study.

Table 2. Average recoveries, LODs (limits of detection), LOQs (limits of quantification), linear dynamic ranges, determination coefficients (r^2), and slopes obtained for the two UV filters by BAµE(PS-DVB)-µLD/HPLC–DAD, under optimized experimental conditions.

UV Filter	Extraction Efficiency (%)	LOD (µg L^{-1})	LOQ (µg L^{-1})	Dynamic Range	r^2	Slope
PBS	61.8 ± 9.1	0.04	0.16	0.16–16.0	0.9985	9.25
BZ4	69.5 ± 4.8	0.20	0.80	0.80–80.0	0.9993	2.47

By comparing the proposed methodology with other passive sampling approaches, such as stir bar sorptive-dispersive microextraction (SBSDME; recoveries yields of 58 ± 2 and 55 ± 5% and LODs of 2.8 µg L^{-1} and 1.6 µg L^{-1} for PBS and BZ4, respectively) [14], BAµE presented much higher extraction efficiency and lower LODs, being an effective analytical alternative for monitoring both UV filters in aqueous media.

The presence of the sulfonic groups in the chemical structure of both target analytes makes them highly soluble, hindering the corresponding microextraction. Consequently, few studies are reported in the literature regarding the analysis of very hydrophilic UV filters, and most of them use SPE as sample enrichment, which is more expensive, requires larger sample volumes, and is not eco-friendly. In short, the methodology proposed herein is a remarkable contribution as an alternative analytical methodology for monitoring trace levels of polar UV filters in aqueous media, besides being eco-friendly, user-friendly, very cost-effective and in compliance with the routine work.

3.6. Application to Real Matrices

In order to demonstrate the reliability of the proposed methodology for real matrices, BAµE(PS-DVB)-µLD/HPLC–DAD was applied to monitor both hydrophilic UV filters in tap, estuarine, and sea water, as well as wastewater samples. For such purpose and in order to suppress possible matrix interferences, SAM was used for quantitative purposes, by spiking the samples with four working standards, ranging from 0.4 to 8.0 µg L^{-1} for PBS and 2.0 to 40.0 µg L^{-1} for BZ4. Blank assays were also performed without spiking to ensure maximum control of the analytical methodology. As expected, in the water matrices studied, any of the analytes were detected (< LODs). As already reported in the literature [14], significant matrix effects were observed during the study of the real matrices, although good selectivity and linearity were achieved, with determination coefficients higher than 0.9940 (BZ4, wastewater sample), obtained from the regression plots (Table 3). Even so, the methodology proposed showed to be influenced by the substantial matrix effects observed, in particular, for the more complex samples, i.e., sea and estuarine water, as well as wastewater. As the UV filters involved are extremely hydrophilic, the higher is the complexity of the matrix, the lower is the average extraction efficiency. For instance, the sea and the estuarine samples studied presented efficiencies which were three-quarters of that obtained for ultra-pure water, and the wastewater presented about one-third. Figure 3 depicts chromatogram profiles obtained from assays performed on spiked (4.0 µg L^{-1} for PBS and 20.0 L^{-1} for BZ4) ultra-pure (a), tap (b), sea (c), and estuarine (d) waters, as well as wastewater (e) samples, in which very good selectivity is noticed.

Table 3. Determination coefficients (r^2), slopes, and contents obtained for the two UV filters by BAµE(PS-DVB)-µLD/HPLC–DAD in several real matrices, under optimized experimental conditions.

UV Filter	Tap Water	Sea Water	Estuarine Water	Wastewater
			r^2	
PBS	0.9999	0.9998	0.9975	0.9942
BZ4	0.9998	0.9998	0.9994	0.9940
			Slope	
PBS	15.06	10.21	9.25	5.25
BZ4	4.28	2.69	2.47	1.10
			Content (mg g^{-1})	
PBS	<LOD	<LOD	<LOD	<LOD
BZ4	<LOD	<LOD	<LOD	<LOD

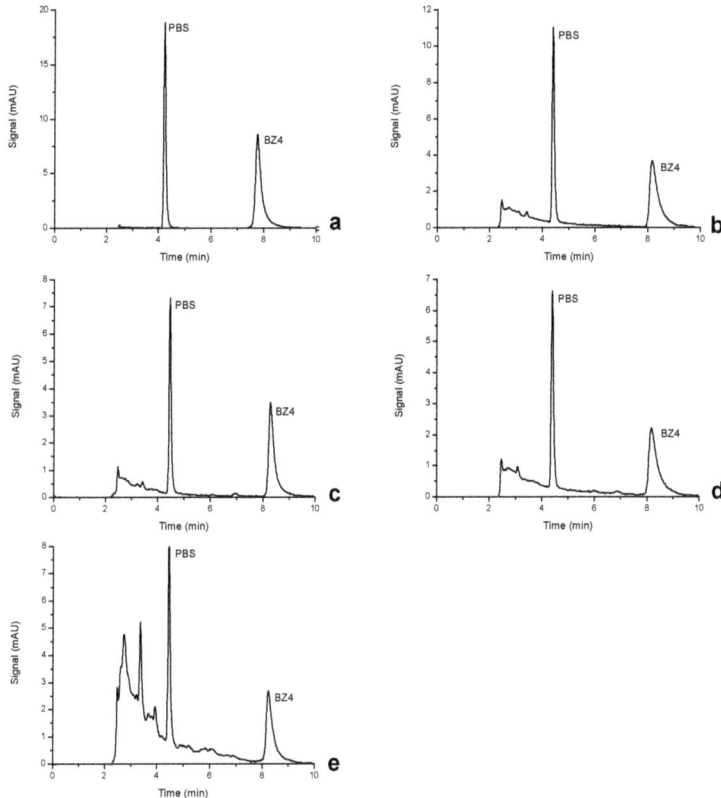

Figure 3. Chromatogram profiles obtained from assays performed in spiked ultra-pure (**a**), tap (**b**), sea (**c**) and estuarine water (**d**), as well as wastewater (**e**) samples by BAµE(PS-DVB)-µLD/HPLC–DAD, under optimized experimental conditions.

4. Conclusions

The proposed methodology uses the new-generation BAµE devices for trace analysis of two polar UV filters (2-phenyl-5-benzemidazolesulfonic acid and 5-benzoyl-4-hydroxy-2-methoxy-benzenesulfonic acid) in aqueous media. It was fully optimized, validated, and applied to real samples. An innovative analytical cycle that includes the use of disposable devices for an effective microextraction stage together with a back-extraction stage performed in "only single LD step", makes BAµE still more user-friendly, eco-friendly, as well as dedicated for routine analysis. Under optimized experimental conditions, good extraction efficiency, suitable detection limits and precision, as well as convenient linear dynamic ranges were obtained. From the data achieved, BAµE presents several advantages, such as the simple preparation of the analytical devices, the possibility to choose the best sorbent phase according to the characteristics of the target analytes, the low cost involved, the easy manipulation, as well as the ability to use negligible amounts of organic solvents, in compliance with GAC principles. In short, this approach is a remarkable analytical alternative for trace analysis of UV filters in aqueous media over other well-established microextraction techniques.

Author Contributions: Both authors contributed substantially to the work reported. A.H.I.: experimental, data analysis, writing and J.M.F.N.: supervision, writing and reviewing.

Funding: This research was funded by "Coordenação de Aperfeiçoamento de Pessoal de Nível Superior (CAPES BEX 0394-14-9) and "Fundação para a Ciência e a Tecnologia" (UID/MULTI/00612/2013).

Acknowledgments: The authors wish to thank "Coordenação de Aperfeiçoamento de Pessoal de Nível Superior" (Brazil) for the Ph.D. grant (CAPES BEX 0394-14-9) and "Fundação para a Ciência e a Tecnologia" (Portugal) for funding (UID/MULTI/00612/2013).

Conflicts of Interest: The authors have declared no conflict of interest.

Abbreviations

BAµE	bar adsorptive microextraction
BZ4	5-benzoyl-4-hydroxy-2-methoxy-benzenesulfonic acid
DAD	diode array detector
GAC	green analytical chemistry
LD	liquid desorption
MeOH	methanol
PBS	2-phenyl-5-benzemidazolesulfonic acid
PS-DVB	polystyrene-divinylbenzene
SAM	standard addition method
SBSE	stir bar sorptive extraction
SPME	solid phase microextraction

References

1. Arthur, C.L.; Pawliszyn, J. Solid phase microextraction with thermal desorption using fused silica optical fibers. *Anal. Chem.* **1990**, *62*, 2145–2148. [CrossRef]
2. Baltussen, E.; Sandra, P. Stir bar sorptive extraction (SBSE), a novel extraction technique for aqueous samples: Theory and principles. *J. Microcolumn.* **1999**, *11*, 737–747. [CrossRef]
3. Neng, N.R.; Silva, A.R.M.; Nogueira, J.M.F. Adsorptive micro-extraction techniques-novel analytical tools for trace levels of polar solutes in aqueous media. *J. Chromatogr. A* **2010**, *1217*, 7303–7310. [CrossRef] [PubMed]
4. Nogueira, J.M.F. Novel sorption-based methodologies for static microextraction analysis: A review on sbse and related techniques. *Anal. Chim. Acta* **2012**, *757*, 1–10. [CrossRef] [PubMed]
5. Ide, A.H.; Nogueira, J.M.F. New-generation bar adsorptive microextraction (BAµE) devices for a better eco-user-friendly analytical approach–Application for the determination of antidepressant pharmaceuticals in biological fluids. *J. Pharm. Biomed. Anal.* **2018**, *153*, 126–134. [CrossRef] [PubMed]
6. Kameda, Y.; Kimura, K.; Miyazaki, M. Occurrence and profiles of organic sun-blocking agents in surface waters and sediments in japanese rivers and lakes. *Environ. Pollut.* **2011**, *159*, 1570–1576. [CrossRef] [PubMed]
7. Tsui, M.M.; Leung, H.W.; Wai, T.C.; Yamashita, N.; Taniyasu, S.; Liu, W.; Lam, P.K.; Murphy, M.B. Occurrence, distribution and ecological risk assessment of multiple classes of UV filters in surface waters from different countries. *Water Res.* **2014**, *67*, 55–65. [CrossRef] [PubMed]
8. Román, I.P.; Alberto, A.C.; Canals, A. Dispersive solid-phase extraction based on oleic acid-coated magnetic nanoparticles followed by gas chromatography-mass spectrometry for UV-filter determination in water samples. *J. Chromatogr. A* **2011**, *1218*, 2467–2475. [CrossRef] [PubMed]
9. Nguyen, K.T.; Scapolla, C.; Di Carro, M.; Magi, E. Rapid and selective determination of UV filters in seawater by liquid chromatography-tandem mass spectrometry combined with stir bar sorptive extraction. *Talanta* **2011**, *85*, 2375–2384. [CrossRef] [PubMed]
10. Zhang, Y.; Lee, H.K. Ionic liquid-based ultrasound-assisted dispersive liquid-liquid microextraction followed high-performance liquid chromatography for the determination of ultraviolet filters in environmental water samples. *Anal. Chim. Acta* **2012**, *750*, 120–126. [CrossRef] [PubMed]
11. Vila, M.; Celeiro, M.; Lamas, J.P.; Dagnac, T.; Llompart, M.; Garcia-jares, C. Determination of fourteen UV filters in bathing water by headspace solid-phase microextraction and gas chromatography-tandem mass spectrometry. *Anal. Methods* **2016**, *8*, 7069–7079. [CrossRef]
12. Negreira, N.; Rodríguez, I.; Ramil, M.; Rubí, E.; Cela, R. Solid-phase extraction followed by liquid chromatography-tandem mass spectrometry for the determination of hydroxylated benzophenone UV absorbers in environmental water samples. *Anal. Chim. Acta* **2009**, *654*, 162–170. [CrossRef] [PubMed]

13. Bratkovics, S.; Sapozhnikova, Y. Determination of seven commonly used organic UV filters in fresh and saline waters by liquid chromatography-tandem mass spectrometry. *Anal. Methods* **2011**, *3*, 2943. [CrossRef]
14. Benedé, J.L.; Chisvert, A.; Giokas, D.L.; Salvador, A. Stir bar sorptive-dispersive microextraction mediated by magnetic nanoparticles-nylon 6 composite for the extraction of hydrophilic organic compounds in aqueous media. *Anal. Chim. Acta* **2016**, *926*, 63–71. [CrossRef] [PubMed]

© 2019 by the authors. Licensee MDPI, Basel, Switzerland. This article is an open access article distributed under the terms and conditions of the Creative Commons Attribution (CC BY) license (http://creativecommons.org/licenses/by/4.0/).

Article

Optimised Extraction of Trypsin Inhibitors from Defatted Gac (*Momordica cochinchinensis* Spreng) Seeds for Production of a Trypsin Inhibitor-Enriched Freeze Dried Powder

Anh V. Le [1,2,*], Sophie E. Parks [1,3], Minh H. Nguyen [1,4] and Paul D. Roach [1]

[1] School of Environmental and Life Sciences, University of Newcastle, Ourimbah, NSW 2258, Australia; sophie.parks@dpi.nsw.gov.au (S.E.P.); Minh.Nguyen@newcastle.edu.au (M.H.N.); Paul.Roach@newcastle.edu.au (P.D.R.)
[2] Faculty of Bio-Food Technology and Environment, University of Technology (HUTECH), Ho Chi Minh City 700000, Vietnam
[3] Central Coast Primary Industries Centre, NSW Department of Primary Industries, Ourimbah, NSW 2258, Australia
[4] School of Science and Health, Western Sydney University, Penrith, NSW 2751, Australia
* Correspondence: vananh.le@uon.edu.au

Received: 10 December 2018; Accepted: 28 January 2019; Published: 5 February 2019

Abstract: The seeds of the Gac fruit, *Momordica cochinchinensis* Spreng, are rich in trypsin inhibitors (TIs) but their optimal extraction and the effects of freeze drying are not established. This study aims to (1) compare aqueous solvents (DI water, 0.1 M NaCl, 0.02 M NaOH and ACN/water/FA, 25:24:1) for extracting TIs from defatted Gac seed kernel powder, (2) to optimise the extraction in terms of solvent, time and material to solvent ratio and (3) to produce a TI-enriched freeze-dried powder (FD-TIP) with good characteristics. Based on the specific TI activity (TIA), the optimal extraction was 1 h using a ratio of 2.0 g of defatted powder in 30 mL of 0.05 M NaCl. The optimisation improved the TIA and specific TIA by 8% and 13%, respectively. The FD-TIP had a high specific TIA (1.57 ± 0.17 mg trypsin/mg protein), although it also contained saponins (43.6 ± 2.3 mg AE/g) and phenolics (10.5 ± 0.3 mg GAE/g). The FD-TIP was likely stable during storage due to its very low moisture content (0.43 ± 0.08%) and water activity (0.18 ± 0.07) and its ability to be easily reconstituted in water due to its high solubility index (92.4 ± 1.5%). Therefore, the optimal conditions for the extraction of TIs from defatted Gac seed kernel powder followed by freeze drying gave a high quality powder in terms of its highly specific TIA and physical properties.

Keywords: Gac; seeds; *Momordica cochinchinensis*; extraction; trypsin inhibitors; optimisation; freeze drying; response surface methodology

1. Introduction

Trypsin inhibitors (TIs) are low molecular weight peptides which can inhibit the hydrolase activity of many kinds of serine proteases. They are commonly found in the storage organs of plants, such as seeds, roots and tubers. Three major sub-types of TIs have been reported and identified in plants [1]: the Bowman–Birk-type inhibitors, Kunitz-type inhibitors and squash family inhibitors. Their molecular weights are about 7500, 20,000 and 3500 kDa, respectively. The first two types were isolated from leguminous plants while the third was obtained from Cucurbitaceous species [1].

This study focuses on the seeds from Gac (*Momordica cochinchinensis* Spreng), a plant that belongs to the Cucurbitaceae family. The aril around the seeds of the Gac fruit is widely used as a food ingredient but the seeds are mainly discarded. However, the seeds have long been used as a traditional Chinese medicine (Mu bie zi) to treat many common diseases such as boils, pyodermatitis, mastitis, tuberculous cervical lymphadenitis, ringworm infections, freckles, sebaceous hemorrhoids and hemangiomas [2,3].

Several TIs from *Momordica cochinchinensis* (MCoTIs) have been characterised [1,3–5] and proposed to be among the most important bioactives in Gac seeds. They serve as storage proteins and may also be involved in the regulation of endogenous proteases during seed dormancy [6]. Nine MCoTIs have been isolated and sequenced from the seeds of Gac fruit [1,2,4]. Structurally, MCoTIs consist of 28–34 amino acid residues, six of which are cysteine residues that form three disulfide bonds. The Gac seed TIs have a very small molecular weight of 3–5 kDa [1,2,4] and in comparison to other TI families, they are more compact in structure and exceptionally stable [7–9]. Among these, MCoTI-I and MCoTI-II are cyclic peptides and as such, they have a very compact and stable structure [4,10]. This enables them to penetrate into cells and, therefore, they are attractive candidates for use as scaffolds for the development of novel intracellularly-targeted drugs [11,12]. Moreover, the activity of Gac seed TIs is very high, at least 50-fold more potent than those from different *Cucurbitaceous* seeds [2]. Due to their clinical potential, Gac seed TIs could be used in a variety of applications in medicine, agriculture and food technology.

As for all plant-derived natural products, extraction is the first critical step in the isolation of TIs from their sources. However, up to date, there are only a few papers dealing with the suitability of different solvents and the extraction conditions for the extraction of these valuable compounds from Gac seeds. In one study [13], a mixture of acetonitrile, water and formic acid (ACN/Water/FA) was found to be optimal for extracting cysteine knot peptides from Gac seeds, some of which are trypsin inhibitors. In our previous study [14], we showed that trypsin inhibitors were able to be effectively extracted from defatted Gac seed kernel powder using conventional solvent extraction with deionised (DI) water only. For other plant sources of TIs, water was also the optimal solvent for their extraction from Thai mung beans [15] but 0.1M NaCl was the best for their extraction from *Chenopodium quinoa* seeds [16] and 0.02 M NaOH was the best for their extraction from grass peas [17]. Clearly, aqueous media are the best solvents for the extraction of TIs.

Apart from the type of solvent used, the efficiency of extractions can be affected by other factors, such as extraction time and the ratio of sample to solvent. For the extraction of trypsin inhibitors, the extraction conditions have so far been studied by conducting one-factor-at-a-time experiments [15–17]. However, one-factor-at-a-time experiments cannot fully determine the interactions between different factors [18] and to overcome this deficiency, the response surface methodology (RSM) is used to determine the simultaneous effects of several factors on extractions.

Therefore, the present study aimed to determine the suitability of different aqueous media (DI water, 0.1 M NaCl, 0.02 M NaOH and ACN/water/FA) as extraction solvents and to further determine the optimal conditions for the extraction of TIs from defatted Gac seed kernel powder by using RSM. The optimised extraction conditions were then used to produce a TI-enriched freeze-dried powder and the physicochemical properties of the freeze-dried powder were determined.

2. Materials and Methods

2.1. Materials

2.1.1. Reagents and Chemicals

Trypsin (type I) from bovine pancreas, benzyl-DL-arginine-para-nitroanilide (BAPNA), dimethyl sulfoxide (DMSO), tris(hydroxymethyl)aminomethane, bovine serum albumin, Folin-Ciocalteu's phenol reagent, cupric sulphate, sodium carbonate, sodium tartrate and formic acid were products of Sigma-Aldrich (Castle Hills, NSW, Australia). Sodium hydroxide and calcium chloride were from Merck (Bayswater, VIC, Australia) and sodium chloride and acetic acid were from Chem-Supply (Port Adelaide, SA, Australia).

2.1.2. Gac Seeds

Gac seeds were collected from 450 kg of fresh Gac fruit from *Momordica cochinchinensis* accession VS7 as classified by Wimalasiri et al. [19]. These fruits were obtained at Gac fruit fields in Dong Nai province, Ho Chi Minh City, Vietnam (Latitude: 10.757410; Longitude: 106.673439). The seeds were separated from the seed pulp and then vacuum dried at 40 °C for 24 h to reduce their moisture and to increase the crispness of the shells to facilitate their removal. The dried seeds were de-shelled to get the kernels, which were then packaged in vacuum-sealed aluminium bags and stored at −18 °C for use within 4 years.

2.1.3. Preparation of Defatted Gac Seed Kernel Powder

The Gac seed kernels were ground using an electric grinder (100 g ST-02A Mulry Disintegrator) to a powder that could pass through a 1.4-mm sieve. The powder was then frozen with liquid nitrogen and freeze-dried using a Dynavac FD3 Freeze Dryer (Sydney, NSW, Australia) for 48 h at −45 °C under vacuum at a pressure loading of 10^{-2} mbar (1 Pa), to reduce the moisture content to $1.21 \pm 0.02\%$. The powder was then defatted using hexane (1:5 w/v, 30 min, ×3) on a magnetic stirrer at ambient temperature. The resulting slurry was suction filtered using a Buchner funnel and Whatman No. 1 filter paper (Sigma-Aldrich, Castle Hills, NSW, Australia). The residue was placed in a fume hood at ambient temperature until dry and free of hexane odour (~12 h) and stored in an air-tight jar at ambient temperature for use within a year. This defatted Gac seed kernel powder was referred to as "defatted powder" and its moisture content was measured, using a Shimazdu MOC63u moisture analyser (Gallay Medical & Scientific, Mulgrave, VIC, Australia), prior to weighing for extractions so that the results could be expressed in terms of the defatted powder's dry weight (DW).

2.2. Methods

2.2.1. Experiment Design

A summary of the experimental design for the study is shown in Scheme 1.

2.2.2. Extraction with Four Aqueous Media

Due to the known hydrophilic nature of the Gac seed TIs, aqueous solutions were investigated as solvents for their extraction. Four different aqueous media were compared for their efficiency to extract TIA from the defatted powder: (1) DI water, (2) 0.1 M NaCl, (3) 0.02 M NaOH and (4) ACN/Water/FA, (25:24:1, v/v/v).

Defatted powder (1.5 g) was suspended in 30 mL of each medium in a 100-mL conical flask and shaken for 1 h at 110 rpm at ambient temperature (17 °C) using a Citenco KQ-606 shaking water bath (Citenco Ltd., London, UK). The suspensions were filtered through a 0.45 µm syringe filter (Phenomenex, Lane Cove, NSW, Australia). The clear filtrates were collected and their TIA (mg trypsin/g defatted powder DW), protein content (mg/g defatted powder DW) as well as their specific TIA (TIA/protein content) were analysed. Each extraction was done in triplicate.

2.2.3. Optimisation of the Extraction Conditions Using RSM

The results from the extractions in Section 2.2.2 showed that the 0.1 M NaCl solution was the best media for extracting TIA from the defatted powder. Therefore, the NaCl solution was selected for further optimisation of the extraction using the response surface methodology (RSM) with three independent factors: the NaCl concentration of the extraction media, the extraction time and the ratio of defatted powder to the volume of NaCl solution. Based on the results from preliminary experiments (data not shown), three levels (Table 1) were selected for a Box–Behnken RSM design [20] to test the NaCl concentration (X_1, mol/L), the extraction time (X_2, hour) and the amount of defatted powder (X_3, g) extracted with 30 mL of the extraction NaCl solution. Therefore, 15 experimental combinations representing 12 factorial points and three central points were performed randomly in duplicate and the experimental values for TIA (mg trypsin/g powder DW), protein content (mg/g defatted powder DW) and specific TIA (mg trypsin/mg protein) were determined for each of the 15 combinations.

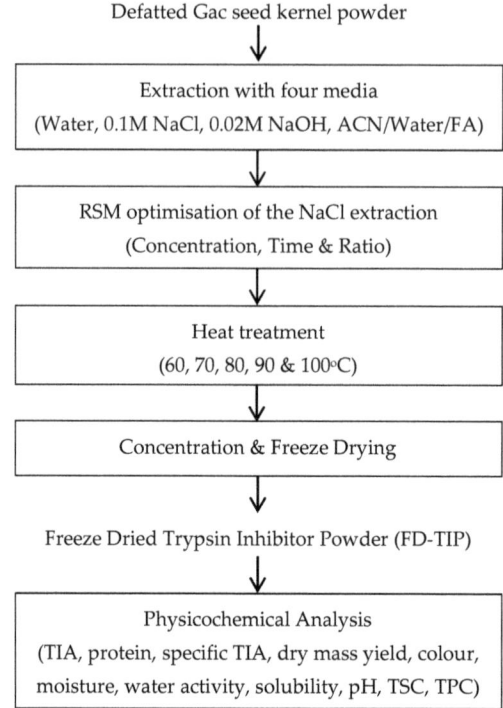

Scheme 1. Experimental design for the study. ACN: acetonitrile; FA: formic acid; TIA: trypsin inhibitor activity, TSC: total saponin content, TPC: total phenolic content.

Table 1. Independent factors and their levels for the RSM Box–Behnken design.

Levels	Independent Factors		
	X_1 NaCl Concentration (mol/L)	X_2 Extraction Time (h)	X_3 Ratio (g/30 mL)
−1	0.050	1.0	1.0
0	0.175	2.5	3.0
+1	0.300	4.0	5.0

The data obtained for the selected combinations was used to generate the second-order polynomial equation/quadratic model shown in Equation (1), which was then used to predict the optimal parameters for the extraction process [21]. To test the predicted optimal parameters, these parameters were used for a TI extraction from defatted powder done in triplicate.

$$Y = \beta_0 + \beta_1 X_1 + \beta_2 X_2 + \beta_3 X_3 + \beta_{12} X_1 X_2 + \beta_{13} X_1 X_3 + \beta_{23} X_2 X_3 + \beta_{11} X_1^2 + \beta_{22} X_2^2 + \beta_{33} X_3^2 \quad (1)$$

where Y is the response variable; X_1, X_2, X_3 are the coded independent variables for the NaCl concentration, the extraction time and the ratio of sample to solvent (Table 1), respectively. β_0, β_1, β_2, β_3 and β_{11}, β_{22}, β_{33}, β_{12}, β_{13}, β_{23} are the regression coefficients for the constant (β_0), the linear effects (β_1, β_2, β_3), the quadratic effects (β_{11}, β_{22}, β_{33}) and the interactions (β_{12}, β_{13}, β_{23}), respectively.

2.2.4. Heat Treatment of the Optimal TI Extract

The extract obtained with the optimal NaCl extraction conditions determined in Section 2.2.3. (2.0 g of defatted powder in 30 mL of 0.05 M NaCl for 1 h) was then heated at different temperatures to determine whether any unstable proteins could be removed as described by Klomklao et al. [15]. Triplicate samples (5 mL) of the extract were heated at different temperatures (60, 70, 80, 90 and 100 °C) for 10 min and then cooled with ice water. As controls, triplicate samples were kept at ambient temperature. To remove any heat coagulated debris, the extracts were filtered through a 0.45 µm syringe filter (Phenomenex, Lane Cove, NSW, Australia). The TIA (mg trypsin/g powder DW), protein content (mg/g defatted powder DW) and specific TIA (mg trypsin/mg protein) were then determined for the clear filtrates.

2.2.5. Concentrating and Freeze Drying the Extract Prepared with the Optimal Parameters

Based on the results in Section 2.2.4, which showed that heat treatment between 60 and 100 °C had no effect on the specific TIA of the extract, heat treatment was not included as a step in the production of a freeze-dried powder with high specific TIA. The optimal extraction parameters determined in Section 2.2.3. were applied to extract the defatted powder but the relative amounts were increased by 60 in order to obtain an amount of freeze-dried material sufficient for the many physicochemical analyses to be undertaken.

The defatted powder (120 g) was stirred with 1800 mL of 0.05 M NaCl for 1 h at ambient temperature and then the suspension was filtered through 3 layers of cheese cloth and finally through a No. 1 Whatman filter paper. The collected filtrate was equally transferred into three evaporating flasks (1 L) and concentrated using a rotary evaporator (Buchi Rotavapor B480, Buchi Australia, Noble Park, VIC, Australia) at 45 °C under vacuum, until around 70 mL of the concentrated extract was left in each flask (~4 h). The three concentrates were transferred into three 250 mL beakers, frozen using liquid nitrogen and freeze-dried using a Scitek BenchTop Pro freeze dryer (Lane Cove, NSW, Australia) at −60 °C and 30 mbar for 48 h.

The three beakers with the freeze dried trypsin inhibitor powder (FD-TIP) were placed in a desiccator, quickly weighed and kept at ambient temperature until analysed. The difference in weight between each beaker with powder and the empty beakers was taken to be the mass recovered in each beaker for the extract.

2.2.6. Determination of Trypsin-Inhibiting Activity (TIA)

The TIA assay was performed as described by Makkar et al. [22] except that the absorbance was measured at 385 nm, as suggested by Stauffer [23], instead of at 410 nm.

- *Reagent preparation:*

Substrate solution: A substrate solution of 92 mM BAPNA was made in 0.05 M Tris-buffer (pH 8.2) containing 0.02 M $CaCl_2$. The BAPNA was first dissolved in DMSO (40 mg/mL) and then diluted with the buffer solution pre-warmed to 37 °C (1.100 v/v). This BAPNA solution was prepared daily and kept at 37 °C while in use.

Trypsin solution: 20 mg of trypsin (type I) from bovine pancreas was dissolved in 1 mM HCl to make 1 L, stored at 4 °C for use within 2 weeks.

- *Determination of TIA*

The liquid extracts from Sections 2.2.2–2.2.4 or the freeze-dried trypsin-inhibitor powder (FD-TIP) from Section 2.2.5 was dissolved and diluted in water at a concentration to give an inhibition of Trypsin between 40 and 60%. The assay was setup as shown in Table 2 with four test tubes (a, b, c and d) prepared for each sample. All the prepared test tubes were kept in a water bath at 37 °C for 10 min to promote the formation of an enzyme-inhibitor complex; 2.5 mL of BAPNA solution pre-warmed to 37 °C was then added into each tube. The contents of the tubes were well mixed after each addition.

The tubes were then incubated in the water bath at 37 °C for another 10 min before 0.5 mL of 30% acetic acid solution was added to each tube to stop the reaction. Then, 1.0 mL of trypsin solution was added into each blank tube (Table 2). After thorough mixing, the absorbance of the reaction mixture (due to the release of p–nitroanilide) was measured at 385 nm.

- Calculations

The TIA was calculated in terms of milligrams of pure trypsin inhibited per gram on a dry-weight basis (mg/g DW) of the defatted Gac seed kernel powder or the FD-TIP (Equation (2)) [14].

$$\text{TIA} = \frac{A_I \times V \times D}{19 \times S \times (1 - m\%/100)} \quad (2)$$

where, A_I: Change in absorbance due to inhibition per 1 mL of diluted extract
$A_I = (A_b - A_a) - (A_d - A_c)$, subscripts as per Table 2.
V: Volume of undiluted extract (mL)
D: Dilution factor of the extract
S: Weight of defatted Gac seed kernel powder or FD-TIP in V mL (g)
19: Constant figure based on the absorbance given by 1 mg of pure trypsin
$m\%$: moisture content of defatted Gac seed kernel powder or freeze-dried extract powder

Table 2. Reagent composition in extracted filtrate.

Component (mL)	Reagent Blank (a)	Standard (b)	Sample Blank (c)	Sample (d)
DI water	1	1	0.5	0.5
Trypsin solution	-	1	-	1
Trypsin-inhibitor sample	-	-	0.5	0.5
Trypsin solution after reaction inactivation	1	-	1	-

2.2.7. Determination of Total Protein Yield (TPY)

Total protein yield (TPY) was defined as the amount (mg) of water-soluble protein obtained per 100 g of ground defatted seed kernel powder on dry weight (DW) basis. TPY was measured by the method of Lowry et al. [24] as described by Klomklao et al. [15], with some modifications. Briefly, 0.5 mL of diluted sample or standard was mixed with 2.5 mL of reagent A containing 2 mL of 0.5% $CuSO_4$ in 1% sodium citrate and 100 mL of 2% sodium carbonate in 0.1 N NaOH for 10 min at ambient temperature. Then, 0.25 mL of 0.5 N Folin-Ciocalteu phenol reagent was added while vortexing. After incubating for 30 min, the absorbance of the reaction mixture was measured at 750 nm using a Cary 50 Bio UV–VIS spectrophotometer (Agilent Technologies, Mulgrave, VIC, Australia). Bovine serum albumin (BSA) was used as a standard and the result was expressed as mg BSA per gram of the defatted Gac seed kernel powder or the FD-TIP (mg BSA/g DW).

2.2.8. Physicochemial Analyses on the FD-TIP

Along with the determination of TIA (Section 2.2.6) and protein content (Section 2.2.7), the following physicochemical analyses were done on each of the FD-TIP in the three beakers described in Section 2.2.5.

- Dry Mass Yield, Moisture Content, Water Activity, Water Solubility Index, pH and Colour

The dry mass yield was defined as the amount (g) of FD-TIP produced per 100 g of dried defatted Gac seed kernel powder. Equation (3) was used to calculate the dry mass yield (DM), in which FD (g) is the weight of the FD-TIP obtained after the extraction and DS (g) is the mass of the dried defatted Gac seed kernel powder used for the extraction.

$$\text{DM (g/100 g)} = \frac{\text{FD}}{\text{DS}} \times 100 \tag{3}$$

The moisture content of the FD-TIP was determined by weight difference after drying at 80 °C for 24 h in a vacuum oven drier (Thermoline Scientific, Wetherill Park, NSW, Australia). The water activity of the freeze dried powder was measured using a Pawkit water activity meter (Graintec, Toowoomba, QLD, Australia).

The water solubility index of the FD-TIP was determined according to Anderson [25] with some modifications. The FD powder (2.5 g) was dispersed in 25 mL of DI water and stirred constantly for 10 min at ambient temperature. The solution was centrifuged at 3000 rpm for 10 min in a Clements 2000 centrifuge (Clements Medical Equipment Pty Ltd., Somersby, NSW, Australia). The supernatant was vacuum oven dried at 80 °C and −70 kPas for 44 h. The water solubility index is expressed as a percentage of the dry solids obtained after drying of the extracted material compared to the original 2.5 g FD powder sample.

To determine the pH of the powders, 2.5 g of powder was dissolved in 25 mL of deionized water and the pH was measured using a labCHEM pH meter (TPS Pty Ltd., Brendale, QLD, Australia) calibrated with standard pH 4 and 7 buffers.

The colour of the powder was measured using a CR-400 chroma meter (Thermo Fisher Scientific, North Ryde, NSW, Australia) calibrated with a white standard tile. Each of the three FD-TIPs were packed into a polyethylene pouch for colour measurements, and the results for each sample were the average of five measurements expressed as the Hunter colour values for the L*, a* and b* co-ordinates as defined by the CIE (Commission Internationale de l'Eclairage). The L* value represents the lightness–darkness dimension, the a* value represents the red–green dimension and the b* value represents the yellow–blue dimension.

- *Total Saponin Content (TSC)*

The FD-TIP crude extract was dissolved in water at a concentration of 2 mg/mL and vortexed before the TSC was determined according to Le et al. [26]. Briefly, 0.25 mL of each extract was mixed with 0.25 mL of 8% (w/v) vanillin solution and 2.5 mL of 72% (v/v) sulphuric acid. The mixture was vortexed and incubated in a water bath at 60 °C for 15 min and then cooled on ice for 10 min. The absorption of the mixture was measured at 560 nm using a Cary 50 Bio UV–VIS spectrophotometer (Agilent Technologies, Mulgrave, VIC, Australia). Aecsin was used as a standard and the results are expressed as milligram aecsin equivalents (AE) per gram of the FD-TIP (mg AE/g DW).

- *Total Phenolic Content (TPC)*

The FD-TIP was dissolved in water at a concentration of 2 mg/mL and vortexed before the TPC was determined according to the method of Le et al. [26]. Briefly, 0.5 mL of each extract was mixed with 2.5 mL of 10% (v/v) Folin-Ciocalteu's phenol reagent in water and incubated at ambient temperature for 2 min to equilibrate. Then, 2 mL of 7.5% (w/v) sodium carbonate solution in water was added and the mixture was incubated at ambient temperature for 1 h. The absorption of the reaction mixture was measured at 765 nm using a Cary 50 UV–Vis spectrophotometer. Gallic acid was used as a standard and the results are expressed as milligram gallic acid equivalents (GAE) per gram of the FD-TIP (mg GAE/g DW).

2.2.9. Statistical Analyses

For designing and analysing the RSM experiment, including generating the three-dimensional (3D) surface and two-dimensional (2D) contour plots, the JMP software version 13.0 (SAS, Cary, NC, USA) was used. The adequacy of the RSM second-order polynomial model was determined based on the lack of fit and the coefficient of determination (R^2).

For extractions performed in triplicate, the means ± SD were assessed with the IBM SPSS Statistics 24 program (IBM Corp., Armonk, NY, USA) using the Student's *t*-test, when only two means were compared, or the one-way analysis of variance (ANOVA) and Tukey's Post Hoc multiple comparisons test, when more than two means were compared. Differences in means were considered statistically significant at $p < 0.05$.

3. Results and Discussion

3.1. Extraction with four Aqueous Media

The highest TIA was obtained with the ACN/water/FA extraction media (Table 3). This result was in accordance with the findings of Mahatmanto [13], who reported that the extraction mixture of ACN/water/FA yielded the highest concentration in cyclotides from Gac seeds, some of which are trypsin inhibitors. However, in the present study, a high TPY was also found for this solvent mixture, which reduced the specific TIA to the second lowest value (Table 3).

Table 3. Effect of extraction media on the extraction of Gac seed trypsin inhibitors [†].

Extraction Media	Mean ± SD		
	TIA (mg trypsin/g Defatted Powder)	TPY (mg BSA/g Defatted Powder)	Specific TIA (mg trypsin/mg Protein)
DI Water	85.22 ± 0.47 [c]	42.55 ± 3.15 [d]	2.01 ± 0.14 [a]
0.1 M NaCl	140.39 ± 8.33 [b]	80.58 ± 0.65 [c]	1.74 ± 0.11 [a]
ACN/Water/FA	168.11 ± 0.50 [a]	152.44 ± 2.08 [b]	1.10 ± 0.01 [b]
0.02 M NaOH	81.43 ± 2.02 [c]	245.19 ± 1.06 [a]	0.33 ± 0.01 [c]

[†] The defatted powder was extracted in different media at ambient temperature (17 ± 1 °C) for 1 h. Trypsin-inhibitor activity (TIA) was analysed using BAPNA as a substrate [23,24] and total protein yield (TPY) was determined using the Lowry assay [25]. Different letter superscripts for the values in the same column denote statistically significant differences ($p < 0.05$).

The solution of 0.1 M NaCl achieved the second highest TIA. However, this solvent extracted a low protein content, which therefore resulted in the highest specific TIA value. The effectiveness of 0.1 M NaCl is consistent with the extraction of TIs from *Chenopodium quinoa* seeds [16]. DI water also achieved the highest specific TIA; however, it yielded a 35% lower TIA in comparison to 0.1 M NaCl (Table 3) and, therefore, water was less effective than 0.1 M NaCl in the recovery of Gac seed trypsin inhibitors.

The 0.02 M NaOH showed the lowest capacity for extracting trypsin inhibitors from Gac seeds, as evidenced by the lowest specific activity, particularly with the highest protein concentration (Table 3). This is consistent with a study on the extractability of protein from flamboyant seeds [27], which found that their protein was most soluble in NaOH followed by NaCl and then water. The result was also in agreement with Benjakul et al. [28], who reported that the extraction of proteins from cowpea and pigeon pea was markedly increased when alkaline solution was used, compared to NaCl. The increase of protein content was observed when grass pea was solubilised with NaOH solution [17]. There are many factors involved in protein solubility and recovery including protein meal and solvent ratio, particle size of flour, temperature, length of extraction time, pH, ionic strength, type and concentration of extraction as well as the hydration properties of proteins [29]. From the results, 0.1 M NaCl was selected as the extractant for Gac seeds.

3.2. Optimisation of the Extraction Conditions Using RSM

3.2.1. Fitting the Response Surface Model

Table 4 shows the experimental values (Exp.) measured for TIA and TPY and the Exp. values calculated for specific TIA in the extracts prepared as described by the RSM Box–Behnken design in

Table 2. From the Exp. TIA values, the response surface model described in Equation (4) was generated and the values predicted (Pred.) for the specific TIA values by this model are also shown in Table 4.

$$Y = 1.557 - 0.154X_1 + 0.041X_2 - 0.043X_3 + 0.014X_1^2 + 0.234X_2^2 - 0.153X_3^2 + 0.020X_1X_2 + 0.063X_1X_3 + 0.038X_2X_3; \quad (R^2 = 0.98) \tag{4}$$

Table 4. The experimental values for trypsin-inhibiting activity (TIA), total protein yield (TPY) and specific TIA and the predicted values for specific TIA for the extracts produced using a 15 run RSM Box–Behnken design.

Run	Pattern	X_1 NaCl (mol/L)	X_2 Time (h)	X_3 Ratio (g/30 mL)	TIA Exp.	TPY Exp.	Specific TIA Exp.	Specific TIA Pred.
1	− − 0	0.05	1	3	155.04	79.85	1.94	1.94
2	− + 0	0.05	4	3	156.19	79.18	1.97	1.98
3	+ − 0	0.3	1	3	164.01	102.43	1.60	1.59
4	+ + 0	0.3	4	3	189.26	110.69	1.71	1.71
5	0 − −	0.175	1	1	152.39	93.85	1.62	1.68
6	0 − +	0.175	1	5	143.55	91.79	1.56	1.52
7	0 + −	0.175	4	1	178.60	108.97	1.64	1.68
8	0 + +	0.175	4	5	177.72	102.65	1.73	1.67
9	− 0 −	0.05	2.5	1	142.63	82.31	1.73	1.68
10	+ 0 −	0.3	2.5	1	168.29	130.12	1.29	1.24
11	− 0 +	0.05	2.5	5	115.65	81.72	1.42	1.47
12	+ 0 +	0.3	2.5	5	145.57	118.48	1.23	1.28
13	000	0.175	2.5	3	167.90	105.47	1.59	1.56
14	000	0.175	2.5	3	147.02	99.21	1.48	1.56
15	000	0.175	2.5	3	163.60	102.10	1.60	1.56

The determination coefficient (R^2 = 0.98) and the p value for the lack of fit (0.42), shown in Table 5, indicated that the model for the specific TIA of the 15 extracts (Equation (4)) had a very high and statistically significant fit; there was a 98% fit of the predicted values with the experimental values (Table 4). Therefore, the response surface model for the specific TIA (Equation (4)) was adequate and suitable for describing the effects of and the interactions between the three independent factors, the NaCl concentration of the extraction media, the extraction time and the ratio of defatted powder to extraction media and could be used to determine the optimum values for these extraction parameters.

Table 5. The regression coefficients and their significance values generated from the fitted 2nd order equation for the Box–Behnken design.

Regression Coefficient	Specific Trypsin-Inhibitor Activity (mg trypsin/mg protein)		
	Regression Coefficient Values	t Ratio	p Value
β_0	1.56	35.23	<0.0001
Linear			
β_1	−0.15	−5.68	0.002
β_2	0.04	1.52	0.188
β_3	−0.04	−1.57	0.177
Quadratic			
β_{11}	0.01	0.36	0.74
β_{22}	0.23	5.88	0.002
β_{33}	−0.15	−3.85	0.012
Interaction			
β_{12}	0.02	0.52	0.623
β_{13}	0.06	1.63	0.163
β_{23}	0.04	0.98	0.372
R^2	0.98		
p value of lack of fit	0.42		

3.2.2. Effects of the Extraction Parameters on the Specific TIA

It can be seen in Table 5 that all three of the extraction parameters had a significant effect on the specific TIA. The NaCl concentration of the extraction media (X_1) had a significant inverse linear effect, the extraction time (X_2) had a significant positive quadratic effect and the ratio of defatted powder to extraction media (X_3) had a significant inverse quandratic effect on the specific TIA of the extracts. However, there were no significant interactive effects between the three parameters. The predicted effects of the three independent extraction parameters are presented visually in the 3D surface and 2D contour plots of Figure 1A–C.

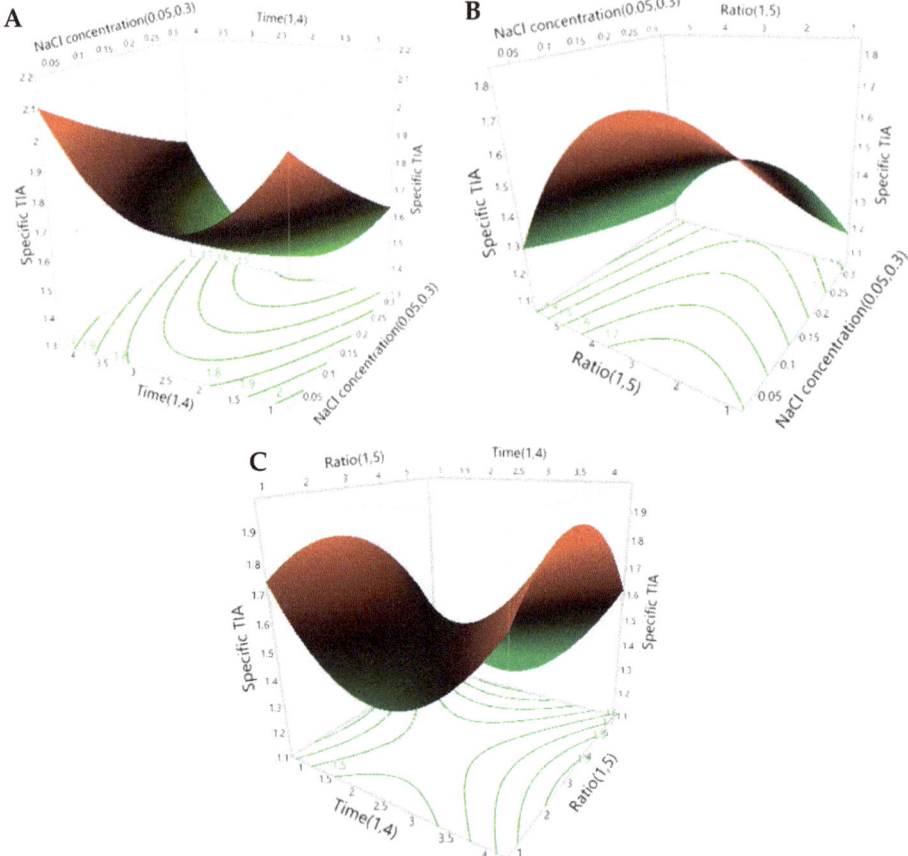

Figure 1. The 3D response surface and 2D contour plots for the specific trypsin-inhibitor activity (TIA) of the extracts. The effects of the NaCl concentration and the extraction time (**A**), the NaCl concentration and the ratio of defatted powder to solvent (**B**) and the extraction time and ratio of defatted powder to solvent (**C**) on the predicted specific TIA of the extracts are plotted according to the 2nd-order polynomial equation generated by the RSM (Equation (4)).

3.2.3. Optimal Extraction Parameters and Validation of the Model

The extraction conditions for obtaining the highest TIA, protein and specific TIA from the defatted Gac kernel powder were generated using the prediction profiler as seen in the plots presented in Figure 2. There were two sets of theoretical maximum values predicted for the extraction parameters to achieve the highest specific TIA. The first optimum set of extraction conditions was 0.05 M NaCl, an extraction time of 1 h and ratio of 2.0 g of defatted powder in 30 mL of the extraction media (Figure 2, left panels). The second optimum set of conditions was 0.05 M NaCl, an extraction time of 4 h and a ratio of 2.5 g/30 mL (Figure 2, right panels). However, for the first set, a much shorter extraction time (1 h) was predicted in comparison to the second set (4 h). Therefore, the first set was chosen as the optimal conditions for the extraction of trypsin inhibitors from the defatted Gac kernel powder (0.05 M NaCl, 1 h extraction with a ratio of 2.0 g/30 mL). Under these optimal condititions, the TIA and specific TIA improved by 8% and 13%, respectively, in comparison to the un-optimal conditions.

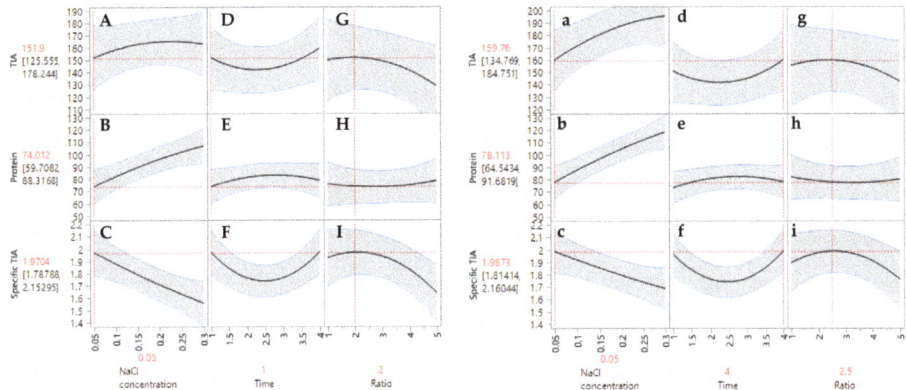

Figure 2. Prediction profiler plots for TIA, protein and specific TIA relative to NaCl concentration, time, and ratio of defatted powder to solvent. Two sets of optimal conditions were predicted as per the panels on the left (capital letters) and the panels on the right (small letters).

To validate the response model (Equation (3)), three independent extractions were conducted for 1 h using a ratio of 2.0 g of defatted powder in 30 mL of 0.05 M NaCl. There was no difference ($p = 0.45$) between the measured (1.86 ± 0.12 mg trypsin/mg protein) and predicted (1.97 ± 0.19 mg trypsin/mg protein) values for the specific TIA of the extracts. This indicated that the response model (Equation (3)) was valid and reliable, and thus, the conditions (1 h at a ratio of 2.0 g/30 mL 0.05 M NaCl) were used in subsequent studies for the extraction of trypsin inhibitors from the defatted Gac seed kernel powder.

3.3. Heat Treatment of the Optimal TI Extract

The optimal TI extract was subjected to heat treatment (Table 6) to determine whether this preparation was resistant to heat as previously observed for Gac seed TIs [13] and to determine whether its specific TIA could be increased because heat treatment has been used previously to remove protein without loss of TIA [30]. Table 6 shows that the TI extract was mostly resistant to heat treatment with only 13% of the TIA lost. However, there was also only an 11% loss of protein and, therefore, the heat treatment between 60 and 100 °C did not improve the specific TIA of the optimal TI extract (Table 6). This meant that the trypsin inhibitors as well as the other proteins in the extract were stable at these elevated temperatures and suggests that the trypsin inhibitors may be the dominant proteins in the preparation.

The heat stability of the TIs in the present study is consistent with the findings of Mahatmanto [13], who showed that boiling water could be used to successfully extract trypsin inhibitors from Gac seeds. This attribute of the Gac seed trypsin inhibitors is most likely due to their small size and their compact cyclical conformation [4]. Other TIs are also heat stable; for example, the trypsin inhibitor from barley has also been found to be heat stable when exposed to 100 °C for 15 min [31].

Table 6. Trypsin-inhibitory activity (TIA), total protein yield (TPY) and specific TIA of Gac seed extracts after heating for 10 min at diferent temperatures.

Temperature (°C)	Mean ± SD		
	TIA (mg trypsin/g DW Defatted Powder)	TPY (mg BSA/g DW Defatted Powder)	Specific TIA (mg trypsin/mg Protein)
Control	169.46 ± 4.73 [a]	91.17 ± 2.05 [a]	1.86 ± 0.09 [a]
60	155.28 ± 2.29 [a]	84.96 ± 2.64 [ab]	1.83 ± 0.07 [a]
70	149.24 ± 2.58 [b]	83.00 ± 3.09 [b]	1.80 ± 0.06 [a]
80	157.34 ± 6.84 [ab]	84.29 ± 2.15 [ab]	1.87 ± 0.13 [a]
90	147.76 ± 7.98 [b]	81.19 ± 3.34 [b]	1.82 ± 0.05 [a]
100	149.13 ± 7.31 [b]	83.25 ± 2.76 [b]	1.79 ± 0.08 [a]

The values are means ± SD of triplicate determinations for each temperature. Values in a column not sharing the same superscript letter are significantly different at $p < 0.05$.

3.4. The Physicochemical Properties of the Freeze Dried Trypsin-Inhibitor Powder (FD-TIP)

The physicochemical properties of the powder obtained by freeze-drying the optimal TI extract are shown in Table 7. The dry mass yield for the FD-TIP (16.3 ± 0.1) was higher than that obtained for extracts in previous studies using water (13.1 ± 0.1 g/100 g) [26] and ACN/water/FA (14.3 g/100 g) [13] as the extraction solvents. However, the high value in the present study is likely due to the retention, through the freeze-drying process, of the NaCl in the 0.05 M NaCl extraction media used, which could have added approximately 4.6 g/100 g to the dry mass yield.

Table 7. Physicochemical properties the trypsin-inhibitor freeze-dried powder.

Properties	Mean ± SD
Dry mass yield (g FD-TIP/100g DW defatted powder)	16.3 ± 0.1
Moisture content (%, w/w)	0.43 ± 0.08
Water activity	0.18 ± 0.07
Water solubility (%)	92.4 ± 1.5
pH	5.33 ± 0.01
Colour	
L*	96.73 ± 0.18
a*	−0.50 ± 0.03
b*	8.22 ± 0.15
Trypsin-ininbitor activity (mg trypsin/g DW FD-TIP)	695.6 ± 77.2
Protein content (mg protein/g DW FD-TIP)	444.0 ± 21.5
Specific trypsin-inhibitor activity (mg trypsin/mg protein)	1.57 ± 0.17
Total saponin content (mg AE/g DW FD-TIP)	43.6 ± 2.3
Total phenolic content (mg GAE/g DW FD-TIP)	10.5 ± 0.3

FD-TIP: freeze dried trypsin-inhibitor powder; AE: Aescin equivalents; GAE: gallic acid equivalents.

As is often seen with freeze-drying [21,32], the moisture content of the FD-TIP was very low (Table 7), which ensure the stability of the powder. Nonetheless, moisture content alone is insufficient to predict the stability and quality of dried products. For example, according to Mai et al. [33], dried Gac fruit powder with a moisture content between 15% and 18% had better physicochemical properties than powder with a moisture content of 6%. Therefore, it is important to understand whether the water in a dried product is available for microbial growth or for enzyme or chemical activity, which can all lead to degradation of the product. Water activity (a_w) of a product reflects the free water available for the growing of microbials and chemical reactions. The a_w level of 0.18 for the powder in this study can be considered as a low-moisture product with the a_w level lower than 0.70, hence having a long shelf life [34]. It is well known that reduced a_w protects against microbiological growth and degradation-causing chemical reactions [35]. Generally, the rate of deterioration can be reduced if the water activity is below 0.6 because the growth of moulds and bacteria is inhibited at those levels [36]. The powder prepared in the present study had water activity less than 0.18 (Table 7), and thus, was likely to be microbiologically stable during storage.

The FD-TIP had a high water solubility index (Table 7), which means the powder could be used as a water-based preparation or in water-based products. Its high water solubility is likely due to the aqueous nature of the 0.05M NaCl extraction media; the ionic strength of the proteins thus extracted and the presence of NaCl in the powder most likely both contributed to the powder's high water solubility [37]. It also suggests that the FD-TIP may have had a small particle size, which also helps solubility [32]. The water solubility index of the FD-TIP was higher than for bitter melon freeze-dried powder (69–79%) [21] and Gac aril spray-dried powder (37%) [38], which suggests it contained more hydrophilic components than these other powders. However, the powder was white in colour and had an acidic pH of 5.3 (Table 7), which is similar to those of the freeze-dried powders from bitter melon [21].

The results on the TIA of the FD-TIP (Table 7) showed that the recovery of trypsin inhibitors and protein were ~67% and 79% compared to the values shown in Table 6 for the control samples, respectively. The 33% and 21% losses may have been due to degradation of the trypsin inhibitors and other proteins occurring during the lengthy concentration (~4 h) and freeze-drying processes or to incomplete resolubilisation of the trypsin inhibitors and other proteins in the FD-TIP powder in water prior to the TIA and protein assays. Compared to the starting material, the defatted powder, the FD-TIP was 4.1-fold and 4.9-fold more concentrated in trypsin inhibitors and protein, respectively (Table 7 vs. Table 6). This means that the specific TIA of the FD-TIP was 16% lower (1.57 ± 0.17 vs. 1.86 ± 0.09) than for the starting material (Table 7 vs. Table 6). This specific TIA was very similar to the activity of the trypsin inhibitor from bovine pancreas (1.5 mg trypsin/mg protein), which is commercially available from Sigma-Aldrich (CAS number 9035-81-8).

The total saponin content of the FD-TIP (Table 7) was comparable to a comercial bitter melon powder (43.6 ± 2.3 vs. 40.2 ± 1.6) [21] but it was higher than for a Gac seed powder previously extracted using deionised water (43.6 ± 2.3 vs. 34.0 ± 1.4) [26]. In contrast, the total phenolic content of the FD-TIP (Table 7) was lower than for the Gac seed powder extracted with the deionised water (10.5 ± 0.3 vs. 17.8 ± 0.5) [26]. This suggests that the saline water used in the present study promoted the extraction of saponins but hindered the extraction of phenolics from the defatted Gac seed kernel powder.

4. Conclusions

Using the RSM and the Box–Behnken system, the optimal conditions for the extraction of trypsin inhibitors from defatted Gac seed kernel powder were determined to be 1 h using a ratio of 2.0 g of defatted powder in 30 mL of 0.05 M NaCl. The powder obtained by freeze-drying the extract prepared using these optimal conditions had a highly specific TIA, although it also contained some saponins and phenolics. The powder was likely to be stable during storage due to its very low moisture content and water activity and to be easily reconstituted in water due to its high water solubility. Therefore, the optimal conditions for the extraction of trypsin inhibitors from defatted Gac seed kernel powder followed by freeze drying gave a high quality trypsin inhibitor-enriched powder.

Supplementary Materials: The following are available at http://www.mdpi.com/2297-8739/6/1/8/s1.

Author Contributions: Conceptualisation, A.V.L., M.H.N. and P.D.R.; Methodology, A.V.L.; Validation, A.V.L; Formal Analysis, A.V.L.; Investigation, A.V.L.; Data Curation, A.V.L. and P.D.R.; Writing-Original Draft Preparation, A.V.L.; Writing-Review & Editing, P.D.R., M.H.N. and S.E.P.; Supervision, P.D.R., M.H.N. and S.E.P.

Funding: This research received no external funding.

Acknowledgments: AVL acknowledges the University of Newcastle and VIED for their financial support.

Conflicts of Interest: The authors declare no conflict of interest.

Abbreviations

ACN	Acetonitrile
a_w	water activity
BAPNA	Benzyl-DL-arginine-para-nitroanilide
DI	Deionised
DMSO	Dimethyl sulfoxide
DW	Dry weight
FA	Formic acid
FD-TIP	Freeze dried trypsin-inhibitor powder
MCoTI	*Momordica cochinchinensis* trypsin inhibitor
RSM	Response surface methodology
TI	Trypsin inhibitor
TIA	Trypsin-inhibitor activity
TPC	Total phenolic content
TPY	Total protein yield
TSC	Total saponin content

References

1. Wong, R.C.; Fong, W.; Ng, T. Multiple trypsin inhibitors from *Momordica cochinchinensis* seeds, the Chinese drug mubiezhi. *Peptides* **2004**, *25*, 163–169. [CrossRef] [PubMed]
2. Huang, B.; Ng, T.; Fong, W.; Wan, C.; Yeung, H. Isolation of a trypsin inhibitor with deletion of N-terminal pentapeptide from the seeds of Momordica cochinchinensis, the Chinese drug mubiezhi. *Int. J. Biochem. Cell Biol.* **1999**, *31*, 707–715. [CrossRef]
3. Chan, L.Y.; Wang, C.K.L.; Major, J.M.; Greenwood, K.P.; Lewis, R.J.; Craik, D.J.; Daly, N.L. Isolation and characterization of peptides from *Momordica cochinchinensis* seeds. *J. Nat. Prod.* **2009**, *72*, 1453–1458. [CrossRef] [PubMed]
4. Hernandez, J.F.; Gagnon, J.; Chiche, L.; Nguyen, T.M.; Andrieu, J.P.; Heitz, A.; Hong, T.T.; Pham, T.T.C.; Nguyen, D.L. Squash trypsin inhibitors from *Momordica cochinchinensis* exhibit an atypical macrocyclic structure. *Biochemistry* **2000**, *39*, 5722–5730. [CrossRef] [PubMed]
5. Felizmenio-Quimio, M.E.; Daly, N.L.; Craik, D.J. Circular Proteins in Plants: Solution structure of a novel macrocyclic trypsin inhibitor from *Momordica Cochinchinensis*. *J. Biol. Chem.* **2001**, *276*, 22875–22882. [CrossRef] [PubMed]
6. Birk, Y. Protein proteinase inhibitors in legume seeds-overview. *Arch. Latinoam. Nutr.* **1996**, *44*, 26S–30S. [PubMed]
7. Greenwood, K.P.; Daly, N.L.; Brown, D.L.; Stow, J.L.; Craik, D.J. The cyclic cystine knot miniprotein MCoTI-II is internalized into cells by macropinocytosis. *Int. J. Biochem. Cell Biol.* **2007**, *39*, 2252–2264. [CrossRef] [PubMed]
8. Contreras, J.; Elnagar, A.Y.; Hamm-Alvarez, S.F.; Camarero, J.A. Cellular uptake of cyclotide MCoTI-I follows multiple endocytic pathways. *J. Control. Release* **2011**, *155*, 134–143. [CrossRef] [PubMed]
9. Cascales, L.; Henriques, S.T.; Kerr, M.C.; Huang, Y.-H.; Sweet, M.J.; Daly, N.L.; Craik, D.J. Identification and characterization of a new family of cell-penetrating peptides cyclic cell-penetrating peptides. *J. Biol. Chem.* **2011**, *286*, 36932–36943. [CrossRef] [PubMed]
10. Mahatmanto, T.; Mylne, J.S.; Poth, A.G.; Swedberg, J.E.; Kaas, Q.; Schaefer, H.; Craik, D.J. The evolution of Momordica cyclic peptides. *Mol. Biol. Evol.* **2014**, *32*, 392–405. [CrossRef] [PubMed]
11. Craik, D.J.; Swedberg, J.E.; Mylne, J.S.; Cemazar, M. Cyclotides as a basis for drug design. *Exp. Opin. Drug Discov.* **2012**, *7*, 179–194. [CrossRef] [PubMed]
12. Wang, C.K.; Gruber, C.W.; Cemazar, M.; Siatskas, C.; Tagore, P.; Payne, N.; Sun, G.; Wang, S.; Bernard, C.C.; Craik, D.J. Molecular grafting onto a stable framework yields novel cyclic peptides for the treatment of multiple sclerosis. *ACS Chem. Biol.* **2013**, *9*, 156–163. [CrossRef] [PubMed]
13. Mahatmanto, T.; Poth, A.G.; Mylne, J.S.; Craik, D.J. A comparative study of extraction methods reveals preferred solvents for cystine knot peptide isolation from *Momordica cochinchinensis* seeds. *Fitoterapia* **2014**, *95*, 22–33. [CrossRef] [PubMed]

14. Le, A.V.; Parks, S.E.; Nguyen, M.H.; Roach, P.D. Effect of solvents and extraction methods on recovery of bioactive compounds from defatted Gac (*Momordica cochinchinensis* Spreng.) seeds. *Separations* **2018**, *5*, 39. [CrossRef]
15. Klomklao, S.; Benjakul, S.; Kishimura, H.; Chaijan, M. Extraction, purification and properties of trypsin inhibitor from Thai mung bean (*Vigna radiata* (L.) R. Wilczek). *Food Chem.* **2011**, *129*, 1348–1354. [CrossRef]
16. Pesoti, A.R.; de Oliveira, B.M.; de Oliveira, A.C.; Pompeu, D.G.; Gonçalves, D.B.; Marangoni, S.; da Silva, D.A.; Granjeiro, P.A. Extraction, purification and characterization of inhibitor of trypsin from Chenopodium quinoa seeds. *Food Sci. Technol.* **2015**, *35*, 588–597. [CrossRef]
17. Deshpande, S.S.; Campbell, C.G. Effect of different solvents on protein recovery and neurotoxin and trypsin inhibitor contents of grass pea (*Lathyrus sativus*). *J. Sci. Food Agric.* **1992**, *60*, 245–249. [CrossRef]
18. Baş, D.; Boyacı, İ.H. Modeling and optimization I: Usability of response surface methodology. *J. Food Eng.* **2007**, *78*, 836–845. [CrossRef]
19. Wimalasiri, D.; Piva, T.; Urban, S.; Huynh, T. Morphological and genetic diversity of *Momordica cochinchinenesis* (Cucurbitaceae) in Vietnam and Thailand. *Genet. Resour. Crop Evol.* **2016**, *63*, 19–33. [CrossRef]
20. Box, G.E.P.; Behnken, D.W. Some New Three Level Designs for the Study of Quantitative Variables. *Technometrics* **1960**, *2*, 455–475. [CrossRef]
21. Tan, S.P.; Vuong, Q.V.; Stathopoulos, C.E.; Parks, S.E.; Roach, P.D. Optimized aqueous extraction of saponins from bitter melon for production of a saponin-enriched bitter melon powder. *J. Food Sci.* **2014**, *79*, E1372–E1381. [CrossRef] [PubMed]
22. Makkar, H.P.; Siddhuraju, P.; Becker, K. *Trypsin Inhibitor. Plant Secondary Metabolites*; Humana Press: New York, NY, USA, 2007; pp. 1–6.
23. Stauffer, C.E. Measuring trypsin inhibitor in soy meal: Suggested improvements in the standard method. *Cereal Chem* **1990**, *67*, 296–302.
24. Lowry, O.H.; Rosebrough, N.J.; Farr, A.L.; Randall, R.J. Protein measurement with the Folin phenol reagent. *J. Biol. Chem.* **1951**, *193*, 265–275. [PubMed]
25. Anderson, R. Water absorption and solubility and amylograph characteristics of roll-cooked small grain products. *Cereal Chem.* **1982**, *59*, 265–269.
26. Le, A.; Huynh, T.; Parks, S.; Nguyen, M.; Roach, P. Bioactive Composition, Antioxidant Activity, and Anticancer Potential of Freeze-Dried Extracts from Defatted Gac (*Momordica cochinchinensis* Spreng) Seeds. *Medicines* **2018**, *5*, 104. [CrossRef] [PubMed]
27. Marfo, E.; Oke, O. Effect of sodium chloride, calcium chloride and sodium hydroxide on Denolix regia protein solubility. *Food Chem.* **1989**, *31*, 117–127. [CrossRef]
28. Benjakul, S.; Visessanguan, W.; Thummaratwasik, P. Isolation and characterization of trypsin inhibitors from some Thai legume seeds. *J. Food Biochem.* **2000**, *24*, 107–127. [CrossRef]
29. Sathe, S.; Salunkhe, D. Solubilization and electrophoretic characterization of the Great Northern bean (*Phaseolus vulgaris* L.) proteins. *J. Food Sci.* **1981**, *46*, 82–87. [CrossRef]
30. Hamato, N.; Koshiba, T.; Pham, T.-N.; Tatsumi, Y.; Nakamura, D.; Takano, R.; Hayashi, K.; Hong, Y.-M.; Hara, S. Trypsin and elastase inhibitors from bitter gourd (*Momordica charantia* LINN.) seeds: purification, amino acid sequences, and inhibitory activities of four new inhibitors. *J. Biochem.* **1995**, *117*, 432–437. [CrossRef] [PubMed]
31. Mikola, J.; Suolinna, E.M. Purification and properties of a trypsin inhibitor from barley. *Eur. J. Biochem.* **1969**, *9*, 555–560. [CrossRef] [PubMed]
32. Lee, C.-W.; Oh, H.-J.; Han, S.-H.; Lim, S.-B. Effects of hot air and freeze drying methods on physicochemical properties of citrus 'hallabong' powders. *Food Sci. Biotechnol.* **2012**, *21*, 1633–1639. [CrossRef]
33. Mai, H.C.; Truong, V.; Haut, B.; Debaste, F. Impact of limited drying on Momordica cochinchinensis Spreng. aril carotenoids content and antioxidant activity. *J. Food Eng.* **2013**, *118*, 358–364. [CrossRef]
34. Blessington, T.; Theofel, C.G.; Harris, L.J. A dry-inoculation method for nut kernels. *Food Microbiol.* **2013**, *33*, 292–297. [CrossRef] [PubMed]
35. Tuyen, C.K.; Nguyen, M.H.; Roach, P.D. Effects of pre-treatments and air drying temperatures on colour and antioxidant properties of Gac fruit powder. *Int. J. Food Eng.* **2011**, *7*. [CrossRef]
36. Fellows, P.J. *Food Processing Technology: Principles and Practice*, 4th ed.; Woodhead Publishing: Cambridge, UK, 2009; ISBN 9780081019078.

37. Vojdani, F. Solubility. In *Methods of Testing Protein Functionality*; Hall, G.M., Ed.; Blackie Academic & Professional: London, UK, 1996; pp. 11–60.
38. Tan, S.P.; Tuyen, K.C.; Parks, S.E.; Stathopoulos, C.E.; Roach, P.D. Effects of the spray-drying temperatures on the physiochemical properties of an encapsulated bitter melon aqueous extract powder. *Powder Technol.* **2015**, *281*, 65–75. [CrossRef]

© 2019 by the authors. Licensee MDPI, Basel, Switzerland. This article is an open access article distributed under the terms and conditions of the Creative Commons Attribution (CC BY) license (http://creativecommons.org/licenses/by/4.0/).

Article

Automation of µ-SPE (Smart-SPE) and Liquid-Liquid Extraction Applied for the Analysis of Chemical Warfare Agents

Marc André Althoff [1], Andreas Bertsch [2] and Manfred Metzulat [2,*]

1. Department of Chemistry, Ludwig-Maximilian University (LMU), Butenandtstrasse 5-13 (Haus D), D-81377 Munich, Germany; Marc.Althoff@cup.uni-muenchen.de
2. Chemistry Section, Science Department, Chemical Defense, Safety and Environmental Protection School, Mühlenweg 12, D-87527 Sonthofen, Germany; andreasbertsch@bundeswehr.org
* Correspondence: manfredmetzulat@bundeswehr.org

Received: 30 June 2019; Accepted: 30 August 2019; Published: 9 October 2019

Abstract: Existing autosamplers are frequently applied only for subjecting the samples to the instruments for injection. In our study, we have set up a TriPlusRSH autosampler mounted on a GC-FID-MS/MS system using the new Method Composer and Script Editor software to automatize all necessary sample preparation steps and subsequent injection of samples in the field of chemical disarmament. Those include but are not limited to: liquid-liquid extraction, drying steps, solvent exchange, and µ-SPE. Tedious and error prone off-line steps are eliminated. In particular, when investigating highly toxic substances like chemical warfare agents or anticancer drugs, automation can help to minimize health risks for lab personnel. The setup engaged features brand new prototype equipment, e.g., a centrifuge to assist in phase separation for liquid-liquid extraction. Efficiency and accuracy of the automated methods were carefully evaluated and proven to outperform the respective manual steps after optimization, e.g., the processing time is up to 60% faster and recovery rates are doubled. The developed workflows can easily be adapted to other sample preparation protocols, e.g., determination of octanol/water partition coefficients, and be used amongst different instruments and chromatography data handling systems.

Keywords: automation; sample handling; sample preparation; chemical warfare agent; SPE; LLE

---

1. Introduction

Preparing and processing of samples prior to analysis is one of the most crucial steps in analytical chemistry and at the same time is the most time consuming task [1]. Most often it is also one of the most cost-effective parts if it has to be done manually. Thus, many people are of the opinion that automation means replacing people by machines [2]. Depending on the analytes being processed automation is a desired feature to have, especially to not expose humans to the risks inhering contagious samples [3] or deadly poisonous compounds such as chemical warfare agents. The latter ones are in our main focus. Lots of methods have been developed for the general work-up of chemical warfare agent samples, their precursors, and degradation products, and are summarized in the book by Vanninen [4]. However, automation has not yet entered the field of chemical warfare agent analysis besides the use of autosamplers for the injection of samples [4]. This may be related to the fact that the number of samples in routine analysis is very low, because luckily chemical warfare agents are very rarely used. Additionally, the implementation of automation is thought to be costly and time-consuming by customers.

In our study we want to show that automation has more advantages than just saving time in laboratory processes. Furthermore, available x-y-z autosamplers [5] can be used instead of separate stand-alone instruments, like those available for solid phase extraction (SPE) from, e.g., Gilson or Agilent, or liquid-liquid extraction (LLE) from, e.g., Aurora, which are not capable of sample analysis. However, only a few studies in this research area are both publicly available and report on efficiency of the methods [6]. Häkkinen was the first to publish research on the SPE method development for different chemical warfare agents [7]. Bae et al. report on SPE of VX ([2-(Diisopropylamino)ethyl]-O-ethyl methylphosphonothioate) from various food matrices by HPLC-MS (High Performance Liquid Chromatography – Mass Spectrometry) [8]. A most recent study of the extraction from non-polar solvents was prepared by Sinha et al. [9] and Liu et al. [10] studied the recovery of sulfur mustard metabolites from urine. However, none of these studies report on the extraction of different chemical warfare agents from natural waters or compare different extraction methods with each other. This may be because of the fact that a binary yes-or-no answer is in most cases sufficient regardless of whether a chemical warfare agent was deployed or not, as the amount used is not crucial for its deadly task; c.f. reports of the OPCW (Organization for the Prohibition of Chemical Weapons) to the UNO (United Nations Organization) on the latest investigations in Syria [11].

2. Materials and Methods

2.1. Materials and Analytes

Reagents used were of *purum* grade and solvents of analytical grade. They were used without further purification and were purchased from Sigma Aldrich, Schnelldorf, Germany and VWR, Darmstadt, Germany. Chemical warfare agents VX ([2-(Diisopropylamino)ethyl]-O-ethyl methylphosphonothioate), VX-Disulfide (Bis(diisopropylaminoethyl)disulfide), HD (sulfur mustard or Bis(2-chloroethyl) sulfide), and VG (Amiton or O,O-diethyl S-[2-(diethylamino)ethyl] phosphorothioate) were either available from stock from Chemical Defense, Safety and Environmental Protection School Sonthofen, Germany or freshly synthesized by standard in-house methods [12]. Diesel fuel was obtained from a local gas station, soil and 'natural water' were collected from an area outside the laboratory building and from a puddle, respectively. All experiments were run from the same batch of 'natural water'. The specifications of the tap water were provided by the local waterworks [13].

μ-SPE cartridges (C18-EC and Silica) from ITSP Solutions were provided by CTC, Zwingen, Switzerland. A detailed description of the μ-SPE-features can be found on the webpage of ITSP Solutions Inc., Hartwell, GA, USA [14].

2.2. Hardware Equipment

Experiments were performed either manually with standard laboratory glassware or on a Thermo Fisher Scientific® GC-MS/MS instrument equipped with a TriPlusRSH autosampler partly bearing prototype equipment, e.g., a centrifuge. The respective setups for the LLE- and μ-SPE experiments are shown in Figure 1. Additional equipment used were billimex® LD/HD vials from LABC, Hennef, Germany in course of the LLE-experiments. These vials have a special design to be precisely compatible with the autosampler and were originally invented for performing dispersive liquid-liquid-microextraction on the CTC autosamplers. LD and HD stand for low density and high density of the extraction solvent, respectively.

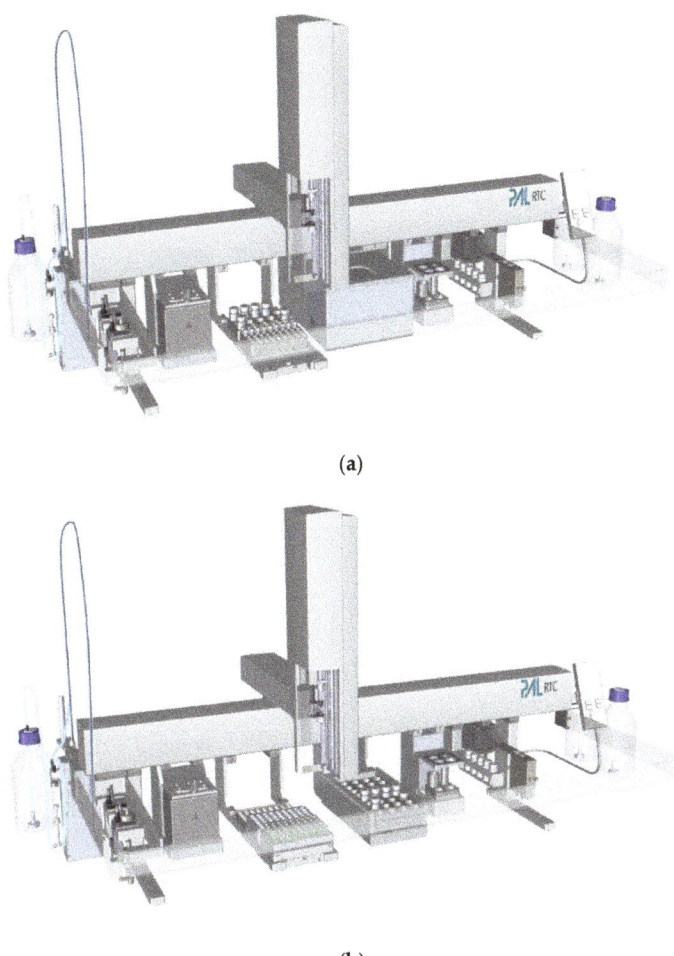

Figure 1. (a) Autosampler setup for LLE procedures: ATC with Dilutor, Agitator, Sample Rack, Centrifuge, Vortexer, Standard Wash Station and Fast Wash Station (left to right); (b) Autosampler setup for μ-SPE procedures: ATC with Dilutor, Agitator, μ-SPE Rack, Tray-Cooler, Vortexer, Standard Wash Station and Fast Wash Station (left to right).

To set-up the autosampler prototype software, applications from CTC were used. First, the experimental workflows were tested step by step by the PAL Method Composer, version 1.1.17284.927 [15]. This program allows us to drag and drop individual incremental steps of the autosampler processes to a complete workflow without the need to involve a chromatographic data software (CDS). It was recently put on the market as "TriPlus RSH Sampling Workflow Editor Software". A demo-video of this software can be found on the homepage of Thermo Fisher Scientifc [16]. Second, the PAL Script Editor Alpha, version 2.4 Alpha, was used to transfer the PAL Method Composer workflow into a compiled script of optimized processing time and compatibility with the CDS: Chromeleon, version 7.2, SR 4, from Thermo Fisher Scientific. The PAL Method Compser software uses a programming syntax comparable to the well-known C programming language. Scripts were tested offline with the PAL Script Executer, version 2.5 Beta, beforehand. All Scripts are available from the corresponding author upon request.

Instrument parameters for the analysis of the samples are given in Table 1.

Table 1. Instrument parameters for processing of the samples on GC-FID-MS/MS.

Instrument Parameter	Value
GC	Thermo Fisher Scientific Trace 1310 with PTV Injector
Data System	Chromeleon 7.2 SR 4
Analytical column	Agilent J&W GC-column (CP-Sil 8 CB Low Bleed/MS, 30 m 0.25 μm)
Oven conditions:	
Temperature program	70(1) –290 °C at 30 °C min^{-1}, post temperature 290 °C (1)
Carrier gas	Helium BIP® [17]
Injection conditions:	
Injector temperature	250 °C
Injection type	liquid injection: splitless, injection volume 1 μL
Purge flow	1.20 mL/min
Liner type	130 mm × 2.0 mm ID glass liner
MS	Thermo Fisher Scientific TSQ Duo, Triple Quadrupole scan time: 0.2 s
Transfer line temperature	280 °C
Ion source temperature	280 °C
Ionization mode	EI
FID	Thermo Fisher Scientific FID-Module temperature: 250 °C H$_2$-flow: 35 mL N$_2$-flow: 40 mL synthetic air-flow: 350 mL

2.3. *Experimental Workflows*

Experimental workflows follow recommended operating procedures according to the respective chapters of the 'blue book' [4], which are given in brief in the Appendix in a graphical depiction. Necessary adjustments of these procedures, e.g., in volume, are given in the results section, as those had to be adjusted to the equipment used. All experiments were run at least in triplicate to yield a reliable RSD.

- Liquid-liquid Extraction:

Tests of the extraction efficiency in the automated liquid-liquid extraction were performed in the range of 600–2000 rpm of the autosampler vortexer. A mixture of caffeine and theophylline in the ratio of 4:1 was used as the evaluation test mixture to find the optimum vortex speed. The mean of three replicates per speed value were determined.

Preparation of samples: stock solutions of tap water and 'natural water' were spiked with the respective analytes stock solution to reach a 10 ppm level and adjusted to pH 7 and pH 11, respectively by the addition of ammonium hydroxide solution, each as a requirement of the recommended procedures (c.f. Figures A1–A3). For preparing the soil samples the respective amount of analytes were directly spiked to the stock sample of soil. The wipe sample was a cotton bud tip onto which a 10 ppm solution of the analytes was directly loaded with a gas-tight Hamilton syringe. The analytes stock solution was prepared in dichloromethane.

- μ-SPE:

Generally the workup procedures of the 'blue book' [4] could be used to extract the analytes with the respective μ-SPE cartridges. However, adjustments and optimizations of flow rates and volumes had to be found. The flow rates for loading, elution and washing steps were optimized in the range of two to 20 μL/s. Therefor the recovery rates in terms of the peak area of the given analytes was determined. Samples prepared in tap water were processed with C18-EC μ-SPE cartridges and those dissolved in *n*-hexane were spiked with a strong hydrocarbon background of 10 mg/mL (1‰)

of diesel fuel and were processed with silica μ-SPE cartridges. Samples were spiked with 10 ppm of the respective analytes. To remove the hydrocarbon background by means of the silica μ-SPE cartridge, the following procedure was used: (i) the silica cartridge was preconditioned with 1 mL of *n*-hexane, (ii) 500 μL of sample were loaded on the cartridge, (iii) the cartridge was washed with 2 mL of *n*-hexane and (iv) the cartridge was eluted with 500 μL of methanol. Additional parameters, which were developed and optimized in course of this study, are given in the results section.

3. Results and Discussion

Generally, it can be said that the transfer and automation of the described methods on the first glance seems to be very easy. However, the transfer of a basic laboratory procedure, e.g., liquid-liquid extraction, has to be broken down to the very incremental steps. Therefore, a very careful examination of the incremental work steps is necessary.

3.1. Automated Liquid-Liquid Extraction

The given protocol for manual liquid-liquid extraction according to reference [4] was adapted to the autosampler setup with respect to the volume of the vial size suitable for the autosampler (2–20 mL). For complexity reasons the respective liquid-liquid extraction workflow for the PAL Method Composer software is only given in its very basic structure, c.f. Figure 2. For each item in the list parameters like penetration depth of the vial, penetration speed of the needle or vortex speed and duration had to be defined and optimized once. In case multiple extractions have to be prepared, the routine can be altered by adding a repeat command for entries "Dilutor Add Solvent" to the second "Clean Syringe" command. The easy to use PAL Method Composer software was very convenient in the testing of the identified incremental work steps. A brief description of the workflow is as follows: The autosampler picks up the dilutor tool to prime it (rinsing), before it adds a given amount of extraction solvent to the sample vial. Afterwards the vial is vortexed and optionally centrifuged, meaning one can decide whether centrifugation is necessary or not. In the next step the tool is changed to a regular syringe to transfer the extraction solvent layer to a second vial containing a proper amount of sodium sulfate (drying agent). After a final vortexing step, the sample is injected into the GC. Finally, the autosampler is returning to home position. The entry liquid injection contains another tool change to get the correct syringe tool with the injection syringe and not the extraction solvent transfer syringe.

For comparison with the manually performed liquid-liquid extractions the workflow was transferred into a regular script for the autosampler by PAL Script Editor software, which is fully operational with the Chromeleon CDS. First, the general extraction performance of the vortexer tool was evaluated by variation of the vortex speed and vortexing time. As can be seen from Figure 3 the mechanical limits of the vortexer are 600 to 2000 rpm. The extraction efficiency for five seconds extraction time is lower compared to ten seconds extraction time. However, at about 1200 rpm the differences can be neglected and the optimum extraction conditions are reached. Thus, for all further experiments these conditions were used. The manual procedure recommends a two-fold extraction of five minutes, meaning the automated workflow is up to 60 times faster, having the same extraction efficiency. The described workflow can easily be extended by other sample preparation tasks, e.g., derivatization steps, as we have described previously [18].

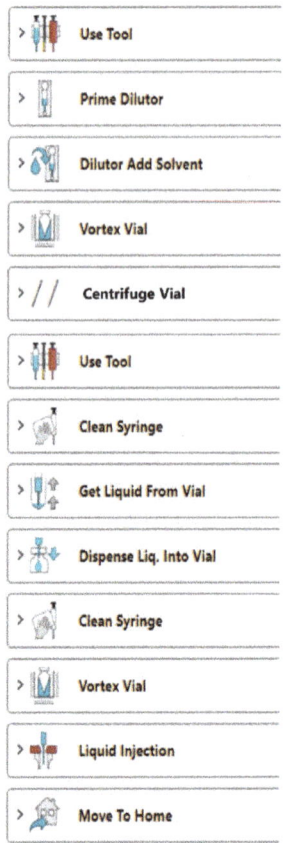

Figure 2. Scheme of the basic steps for the liquid-liquid extraction experiment represented in the PAL Method Composer software style. Each menu item consists of a set of parameters which have to be adjusted and optimized.

Consequently, we applied the developed automated workflow for the simultaneous extraction of different type of chemical warfare agents VX, VG and HD from four different matrices. To the best of our knowledge, this has not been done before. The results found (c.f. Table 2) clearly show that very high recovery rates of the analytes in the given matrix are achieved. Recovery rates vary depending on the pH-value of the matrix. The more acidic the matrix the faster the degradation process of the V-series nerve agents is and, thus, only less amount of analyte can be recovered. These findings are in very good agreement with our previous study on solid-phase microextraction of the same analytes [18]. In contrast, HD hydrolyses faster at higher pH-values. For soil and wipe samples strong matrix effects have to be taken into account and, thus, a fair amount of analytes is lost to the matrix and cannot be recovered. Overall the obtained recovery rates are excellent with respect to processing time and automation of the overall process.

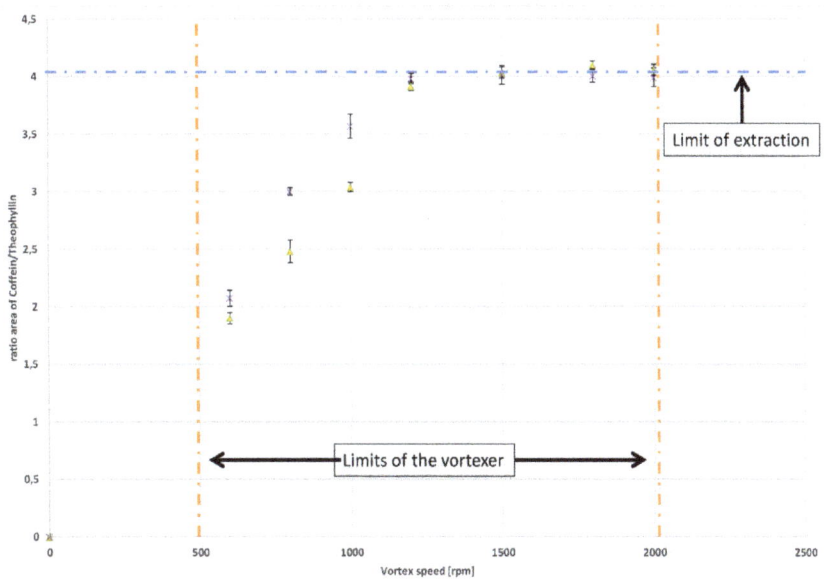

Figure 3. Results of the automated liquid-liquid extraction methodology. Noteworthy are the short extraction times of 5 to 10 s and completeness of extraction at 1200 rpm.

Table 2. Results of the automated extraction of chemical warfare agents VX, VG and HD at 5 s extraction time and 1200 rpm vortex speed.

Matrix [1]	Recovery Rate ± RSD [%]		
	VX	VG	HD
tap water (pH 7)	64.1 ± 9.6	79.8 ± 8.3	47.3 ± 3.8
tap water (pH 11)	88.0 ± 3.4	85.9 ± 5.1	22.1 ± 2.7
surface water (pH 7)	96.0 ± 3.4	99.9 ± 3.6	51.5 ± 4.0
surface water (pH 11)	98.7 ± 1.4	99.9 ± 1.8	50.7 ± 1.4
soil	66.2 ± 2.7	88.2 ± 4.0	46.8 ± 4.4
wipe	65.6 ± 3.1	58.4 ± 2.3	51.2 ± 1.4

[1] Aqueous samples were adjusted to the given pH-value prior to the extraction.

Another advantage of the automation is that comparably small amounts of sample can be handled. With our setup, we were able to easily process quantities down to 1 mL. Thus we meet all of the three key tasks of today's laboratory issues: automation, miniaturization and processing time as were described by Costa and Pawliszyn [19,20]. Best results could be achieved by using billimex® LD/HD vials, which allow an almost complete "decantation" of lower or upper phase of the extraction solvent by means of the autosampler syringe. They are fully compatible with standard autosampler equipment. A picture of the different vial types used during this study and their appearance during the extraction process is shown in Figure 4. Generally, to "decant by syringe" the exact penetration depth needed to be determined to exactly hit the interface layer of the two liquids. We are of the opinion that the miniaturization and optimization of well-established sample preparation protocols is superior to the development of new extraction methods like Liquid-Phase Microextraction or Single-Drop Microextraction, since even more and yet not existing hardware modules would be needed to be invented. Furthermore, those tools would have to be very complex, e.g., to hit a single drop on the surface of a liquid with a syringe is very hard to do manually; therefore a camera and a vial which is movable and tilt-able with respect to the syringe position would include at least a second robot arm which is more flexible than existing ones. This would make the instrumental setup more cost intensive and would also cost a lot of time in processing a sample. To compensate this shortcoming of the employed x-y-z-autosampler, freely movable industry-robot-arms, like in automobile industry could be used, but are even more complex and are not yet commercially available of the shelf for chemical analysis.

Figure 4. Autosampler vials after (i) extraction (**left**, billimex®), (ii) centrifugation (**middle**) and (iii) drying (vortexed) with Na2SO4 (**right**).

Another challenging part was to obtain a good phase separation after vortexing, especially in case of processing the soil samples. However, the prototype centrifuge we had access to during our studies proved to be the right tool to help phase separation and to force all floating particles to the bottom of the extraction vial, c.f. Figure 4 (middle). Additionally, also other common non-desirable effects, which can occur during liquid-liquid extraction, e.g., foaming, can be dealt with by using the centrifuge. Due to being prototype equipment the centrifuge commands were not available in the PAL Method Composer software. Thus, it had to be programmed using the PAL Script editor. Centrifuge parameters were not optimized during this study since they would have not affected the recovery rates. The centrifuge was not used after the extraction solvent was added to the drying agent since sodium sulfate very nicely precipitated during transportation of the vial from the vortexer to the original vial position. One possible limitation of the autosampler centrifuge could be its G-force limit of 2000 G with respect to the use of larger sampling vials (10 and 20 mL). According to the blue book [4] up to 5000 G should be applied for the phase separation process. This is only possible to achieve by using 2 mL autosampler vials.

3.2. Automated μ-SPE

The basic workflow of the μ-SPE experiment for removing the hydrocarbon background is described in the experimental part, whereas the procedure for the C18 method is given in Figure A3 of Appendix A. The main difference between standard SPE and μ-SPE is the cartridge size and sorbent volume. The μ-SPE cartridge contains only 10–45 mg of sorbent material. Flow rates are precisely controlled by the autosampler plunger motor; as low as 1 μL/s. This high precision helps to achieve high accuracy of the overall process. The four incremental steps of the experiment are: conditioning of the cartridge, loading of the sample onto the cartridge, washing off undesired compounds and elution of the analytes. The C18-method had to be adjusted regarding the applied volumes: (i) conditioning: 1 mL of methanol and 2 mL of water, (ii) loading: 500 μL of sample, (iii) elution: 1 mL of ethyl acetate. Drying of the cartridge was performed by pushing 5 mL of air through the cartridge with the autosampler syringe.

In the beginning the flow rates had to be optimized. An exemplary plot of the results for the elution flow rates is shown in Figure 5. As can be seen the elution flow rate is optimum at 12 μL/s for all analytes. The different heights of the curves for VX and VG are a matter of the response factor of the detector for the individual analyte, whereas VX-disulfide, as a degradation product of VX, was present at 1/10 of the concentration of VX. Moreover, we also found that for loading of the cartridge, a flow rate of 12 µL/s was best.

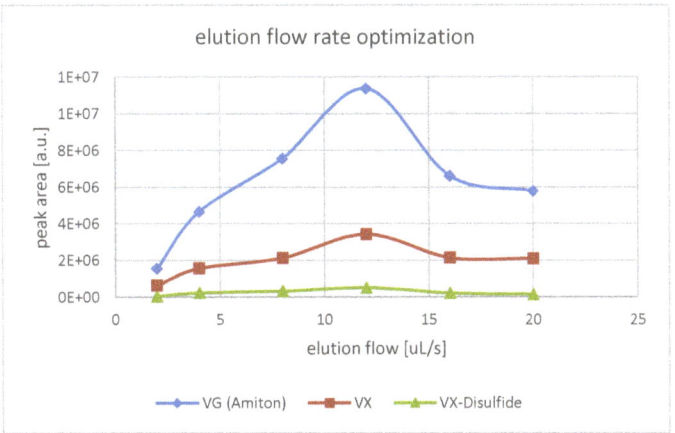

Figure 5. Representation of the elution flow rates and obtained peak areas in GC-MS for the different analytes on the silica µ-SPE cartridge.

In Figure 6 three different scans from the hydrocarbon contaminated sample show that the chosen sample preparation technique (spiking) is successful, meaning that a direct injection of the organic sample in full-scan mode (top) or SIM mode (middle) is not successful in the detection of any analyte from the sample. The last chromatogram (bottom) was obtained in SIM mode after sample processing with the developed µ-SPE methodology and very nicely shows the "hidden" analytes of the original sample in a very good response.

In Table 3 the results of the two different sample processing procedures are summarized. First, all for analytes were extracted from an aqueous sample buffered to neutral conditions as recommended by extraction with C18-EC cartridge. On average, the recovery rates range from 56 to 65% of the spiked amount. Compared to available literature resources, this is roughly twice the amount of analyte as can be obtained with the manual procedure. For VX-Disulfide and VG, no literature values have been reported yet. For sample purification in case of a strong hydrocarbon background as impurity even higher recovery rates could be obtained ranging from 76 to 98% of the originally spiked amount of analyte. Due to the non-polar properties of sulfur mustard (HD), this analyte is not efficiently retained on the silica cartridge. Moreover, it is washed off the silica cartridge together with the hydrocarbon background by the washing solvent (n-hexane). The eluted fraction containing the hydrocarbon background as well as the sulfur mustard has to be treated in another preparation step which can also be found in the blue book. However, here we report recovery rates for the cleaning of a hydrocarbon background from chemical warfare agents by silica cartridges for the first time.

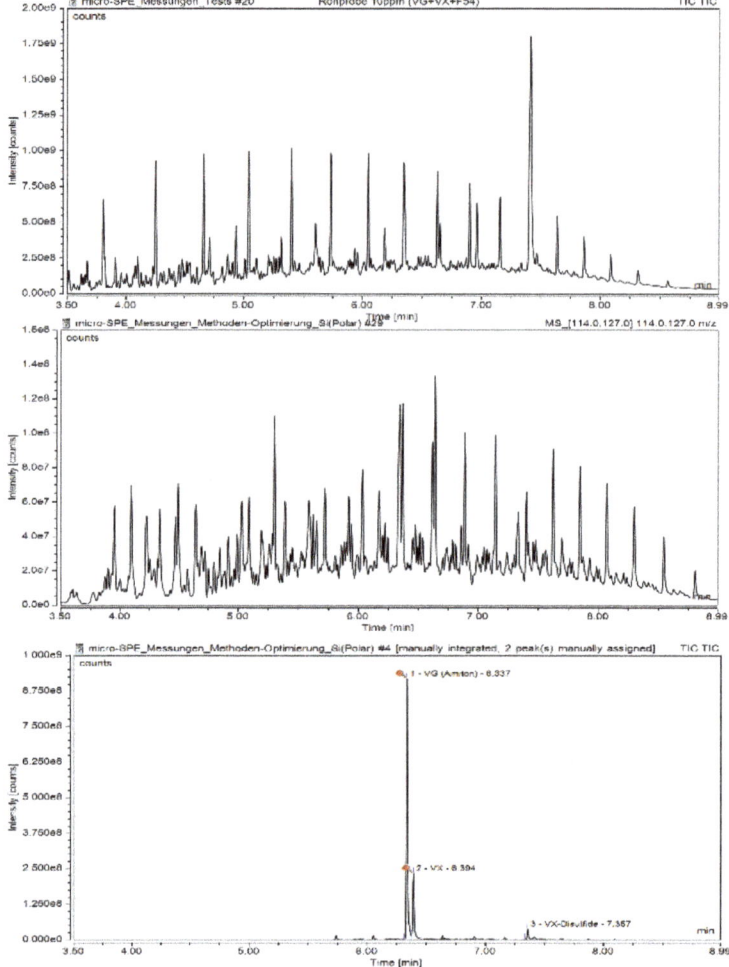

Figure 6. Chromatograms of the sample containing a strong hydrocarbon background from diesel fuel: TIC-scan of the raw sample (**top**), SIM-scan of the raw sample (**middle**) and SIM-scan after sample preparation with silica cartridge (**bottom**).

Table 3. Recovery rates of chemical warfare agents from aqueous samples and strong hydrocarbon background containing organic solvent.

Matrix (Cartridge)	Recovery ± RSD [%]			
	VX	VX-Disulfide	VG	HD
water (C18-EC)	56.5 ± 0.5	57.9 ± 2.2	65.8 ± 0.1	64.8 ± 1.9
water samples [7]	24 ± 10	−1	−1	32 ± 7
diesel fuel (silica)	76.3 ± 1.3	77.4 ± 0.2	98.2 ± 1.5	−2

[1] no values available, [2] analyte not retained by the cartridge.

The main technical differences between the well-established recommended operating procedures of the blue book and our experimental setup are given in Table 4. In sense of going green with sample preparation processes, our methodology is advantageous over existing methods: less sorbent material and less solvent volume are needed. Comparable autosamplers, e.g., from Gerstel, are (yet) not capable of processing µ-SPE cartridges. Instead they are capable of processing standard-sized SPE-cartridges. This means that in total comparably fewer samples can be processed due to the fact that more room is needed for keeping available fresh cartridges. Moreover, a high enrichment factor of up to 200 can be achieved on a 10 mg cartridge as well as the flow rate can be precisely controlled and has a highly sufficient repeatability. Additionally, the time for the overall experiment is cut down from 15 to 10 min. Furthermore, it seems to be a very convincing fact that most manually handling steps of highly toxic substances can be cut down to a minimum and thus the risk for contamination of laboratory personal is minimized.

Table 4. Comparison of µ-SPE and standard SPE.

Parameter	µ-SPE (This Work)	SPE (Gerstel MPS [21])	SPE (Laboratory)
sorbent volume	10–45 mg	0.1–2 g	0.1 g – bulk
solvent volume	125 µL	≤10 mL	3 mL
processing time	10 min	unknown	15 min
flow rate	1–100 µL/s	10–250 µL/s	ca. 1 drop/s
enrichment factor	≤200	5–10	Indefinite

4. Conclusions and Future Perspective

Although many steps of the described procedures could be automated and are working very well, some experimental limitations do remain unresolved – so far. To the best of our knowledge, there is no device available to be mounted on the kind of autosampler we used to measure and adjust the pH value meaning that one initial step of all the blue book procedures remains manual work. As miniaturization of equipment proceeds, less solvent and auxiliary materials are needed and very reliable methods can be branded as green technologies afterwards.

Another auxiliary step, not explicitly mentioned in the procedures itself but sometimes crucial in sample processing, is filtration. Real-life samples can possibly contain floating components or sediment material, which would disturb the described method by blocking syringes or µ-SPE cartridges. This would be easily noticed during manual processing of the sample but not by automatic procedures. Eventually CTC informed us, that the autosampler can be also adapted with tools from Brechbühler company for processing Thomson filter vials [22], which would easily allow filtering sample fractions of 1 mL in an automated fashion. For the described method of LLE of µ-SPE, this would mean our script would have to be extended by a few basic operations of the autosampler and this issue could be addressed. However, we were not able to get access to such a setup and could not extend our study in this direction.

The PAL Method Composer software proved to be easily suitable to put complex workflows by means of drag-and-drop together and be tested without the need of blocking the analytical instrument. The PAL Script Editor allowed a complex programming of the instrument based on the preliminary results obtained from the PAL Method Composer software and full integration of the analytical instrument and its CDS. Moreover, the workflows can be transferred with slight alternations to other CDS form different instrument manufacturers, as was done with some other routines, e.g., those described in our previous work [18], in cooperation with one of our partner laboratories of the German Federal Police.

The herein developed workflows are running very stable and have found their way into daily laboratory routines (SOP's) in the processing of chemical warfare agents and other analytes. They are faster, have a higher precision, use less material and chemicals, and pose less risk to the laboratory personnel as manual handling time of chemical warfare agents is absolutely minimized. Moreover, we can imagine that the liquid-liquid extraction procedure in particular can be easily adapted for other experimental tasks, e.g., determination of octanol/water partition coefficients or be employed in the analysis of pesticide residues or in the field of anticancer drug research where similar procedures are daily routine.

Finally, we are of the opinion that by our efforts in automation (c.f. [18]) and miniaturization reported in this work have entered a new area in the processing of chemical warfare agents.

Author Contributions: Conceptualization, M.A.A.; Data curation, M.A.A.; Formal analysis, M.A.A.; Investigation, A.B.; Methodology, M.A.A.; Project administration, M.M.; Resources, M.M.; Software, M.A.A. and A.B.; Supervision, M.M.; Validation, M.A.A.; Visualization, A.B.; Writing–original draft, M.A.A.

Funding: This research received no external funding.

Acknowledgments: We acknowledge the collaboration between the German Armed Forces (Bundeswehr) and the Ludwig-Maximilian University (LMU) according to the official collaboration agreement between the two institutions. The authors of this study are indebted to CTC Company, Zwingen, Switzerland for giving access to prototype equipment, providing the µ-SPE cartridges from ITSP™ solutions Inc. and support in programming and creation of the autosampler sketches. billimex® LD/HD-AS vials were provided by courtesy of LABC-Labortechnik Zillger KG, Hennef, Germany.

Conflicts of Interest: The authors declare no conflict of interest. The companies providing material for this study had no role in the design of the study; in the collection, analyses, or interpretation of data; in the writing of the manuscript, or in the decision to publish the results.

Appendix A

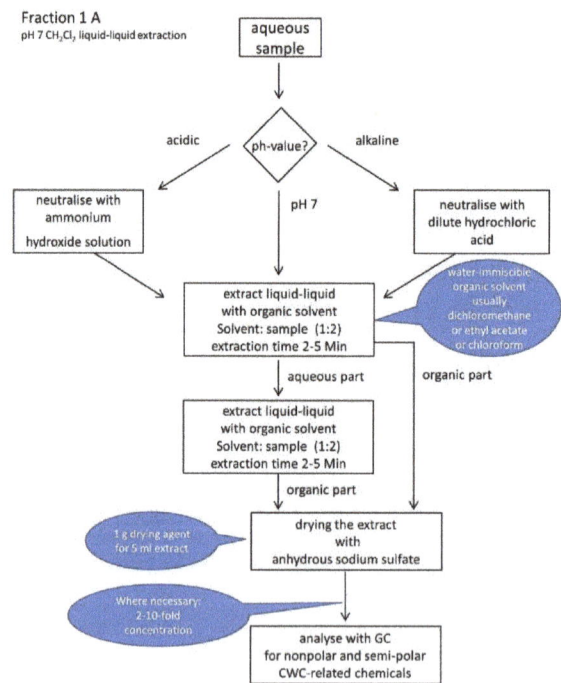

Figure A1. Recommended operating procedure according to the blue book for the processing of liquid-liquid extraction at neutral conditions.

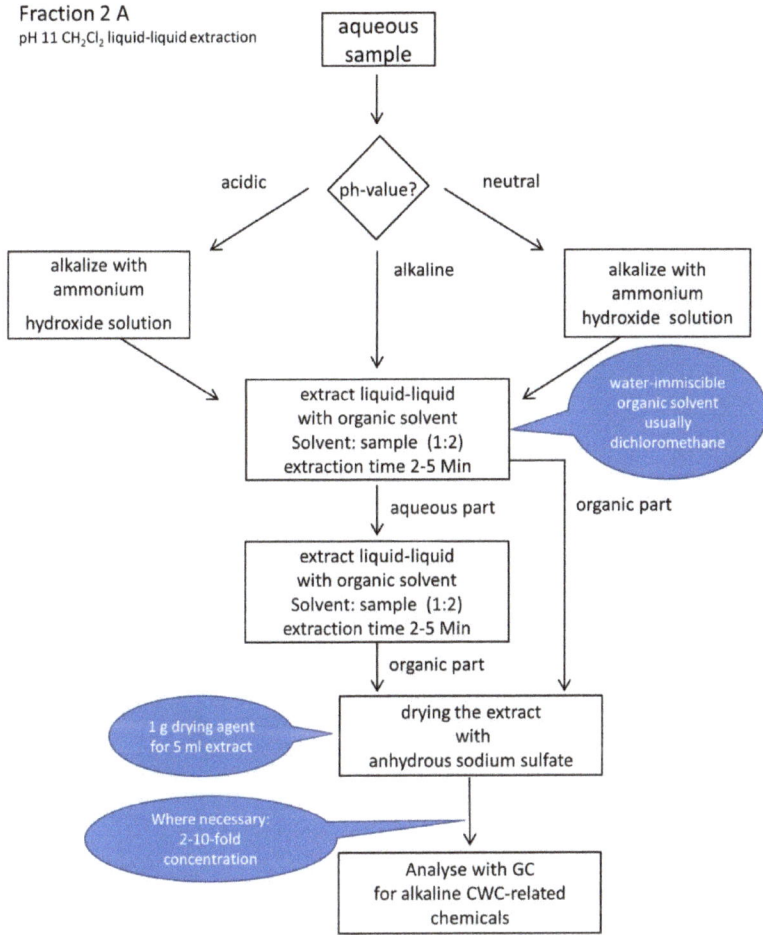

Note! Do not concentrate the organic extract to dryness since certain volatile CWC-related chemicals such as sarin are lost from the sample and, on the other hand, certain less volatile CWC- related chemicals such as VX in residues of organic extracts are firmly adsorbed to glass surfaces.

Figure A2. Recommended operating procedure according to the blue book for the processing of liquid-liquid extraction at alkaline conditions.

Sample Preparation

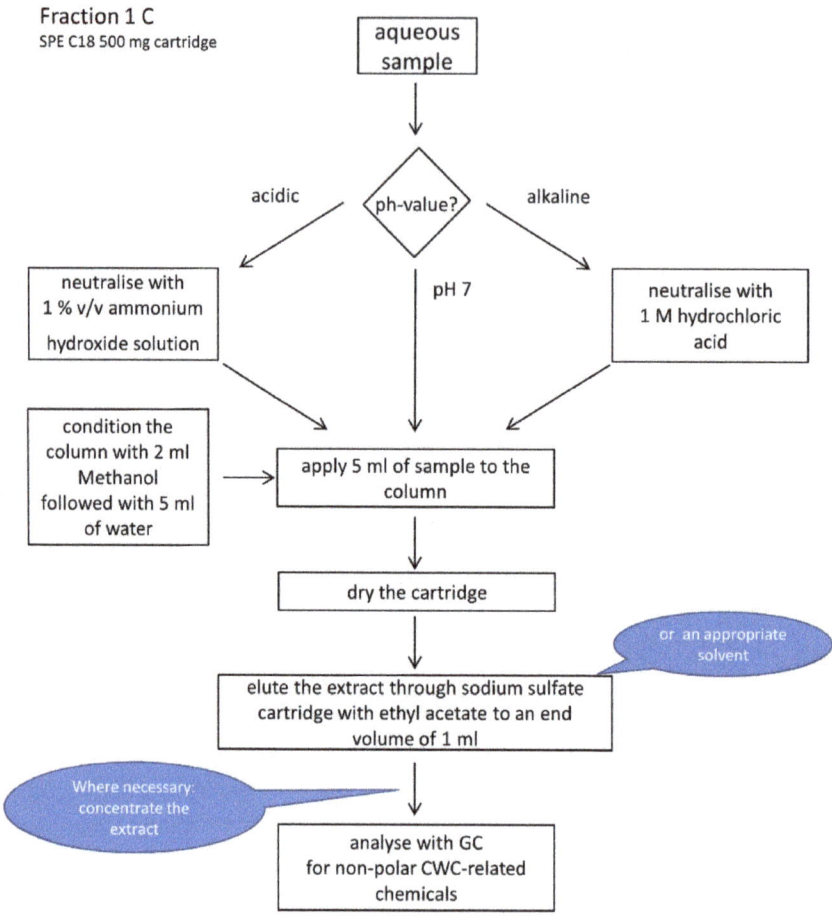

Note! Do not concentrate the organic extract to dryness since certain volatile CWC-related chemicals such as sarin are lost from the sample and, on the other hand, certain less volatile CWC- related chemicals such as VX in residues of organic extracts are firmly adsorbed to glass surfaces.

Figure A3. Recommended operating procedure according to the blue book for the processing of solid-phase extraction by C18-sorbent at neutral conditions.

References

1. Pawliszyn, J.; Lord, H.L. *Handbook of Sample Preparation*; Wiley: Hoboken, NJ, USA, 2012.
2. Chui, M.; Manyika, J.; Miremadi, M. Where Machines Could Replace Humans and Where They Can't (Yet)? Available online: http://www.oregon4biz.com/assets/e-lib/Workforce/MachReplaceHumans.pdf (accessed on 25 September 2019).

3. Armbruster, D.A.; Overcash, D.R.; Reyes, J. Clinical Chemistry Laboratory Automation in the 21st Century—Amat Victoria curam (Victory loves careful preparation). *Clin. Biochem. Rev.* **2014**, *35*, 143–153.
4. Vanninen, P. *Recommended Operating Procedures for Analysis in the Verification of Chemical Disarmament*; University of Helsinki: Helsinki, Finnland, 2017.
5. Wittsiepe, J.; Nestola, M.; Kohne, M.; Zinn, P.; Wilhelm, M. Determination of polychlorinated biphenyls and organochlorine pesticides in small volumes of human blood by high-throughput on-line SPE-LVI-GC-HRMS. *J. Chromatogr. B* **2014**, *945–946*, 217–224. [CrossRef] [PubMed]
6. Kuitunen, M.L. Sample Preparation for Analysis of Chemicals Related to the Chemical Weapons Convention in an Off-site Laboratory. In *Encyclopedia of Analytical Chemistry*; Meyers, R.A., Ed.; Wiley: Hoboken, NJ, USA, 2010. [CrossRef]
7. Häkkinen, V.M.A. Analysis of chemical warfare agents in water by solid phase extraction and two-channel capillary gas chromatography. *J. Sep. Sci.* **1991**, *14*, 811–815. [CrossRef]
8. Bae, S.Y.; Winemiller, M.D. *Quantification of VX Nerve Agent in Various Food Matrices by Solid-Phase Extraction Ultra-Performance Liquid Chromatography–Time-of-Flight Mass Spectrometry*; Edgewood Chemical Biological Center: Edgewood, WA, USA, 2016; p. 26.
9. Sinha Roy, K.; Goud, D.R.; Chandra, B.; Dubey, D.K. Efficient Extraction of Sulfur and Nitrogen Mustards from Nonpolar Matrix and an Investigation on Their Sorption Behavior on Silica. *Anal. Chem.* **2018**, *90*, 8295–8299. [CrossRef] [PubMed]
10. Liu, C.C.; Liu, S.L.; Xi, H.L.; Yu, H.L.; Zhou, S.K.; Huang, G.L.; Liang, L.H.; Liu, J.Q. Simultaneous quantification of four metabolites of sulfur mustard in urine samples by ultra-high performance liquid chromatography-tandem mass spectrometry after solid phase extraction. *J. Chromatogr. A* **2017**, *1492*, 41–48. [CrossRef] [PubMed]
11. Sellström, A. Report of the United Nations Mission to Investigate Allegations of the Use of Chemical Weapons in the Syrian Arab Republic on the Alleged Use of Chemical Weapons in the Ghouta Area of Damascus on 21 August 2013. Available online: https://www.un.org/zh/focus/northafrica/cwinvestigation.pdf (accessed on 26 September 2019).
12. Althoff, M.A.; Bertsch, A.; Metzulat, M.; Kalthoff, O.; Karaghiosoff, K. New Aspects of the Detection and Analysis of Organo(thio)phosphates related to the Chemical Weapons Convention. *Phosphorus Sulfur* **2016**, *192*, 149–156. [CrossRef]
13. *Trinkwasseranlayse der Stadt Sonthofen*; Stadt Sonthofen: Sonthofen, Germany, 2018; Volume 2018, pp. 1–2.
14. ITSP Solutions Inc. Available online: https://www.itspsolutions.com/ (accessed on 22 May 2019).
15. PAL Method Composer. Available online: https://www.palsystem.com/index.php?id=850 (accessed on 27 July 2019).
16. TriPlus™ RSH™ Sampling Workflow Editor Software. Available online: https://www.thermofisher.com/order/catalog/product/1R77010-1200 (accessed on 27 July 2019).
17. BIP Gasreinigungstechnologie. Available online: http://www.tig.de/produkte/bip-gasreinigungs-technologie.html (accessed on 25 September 2019).
18. Althoff, M.A.; Bertsch, A.; Metzulat, M.; Klapötke, T.M.; Karaghiosoff, K.L. Application of Headspace and Direct Immersion Solid-Phase Microextraction in the Analysis of Organothiophosphates related to the Chemical Weapons Convention from Water and Complex Matrices. *Talanta* **2017**, *174*, 295–300. [CrossRef] [PubMed]
19. Costa, R. Newly Introduced Sample Preparation Techniques: Towards Miniaturization. *Crit. Rev. Anal. Chem.* **2014**, *44*, 299–310. [CrossRef] [PubMed]
20. Pawliszyn, J. *Handbook of Solid Phase Microextraction*; Elsevier: Hoboken, NJ, USA, 2011.
21. Solid Phase Extraction Specifications. Available online: http://www.gerstel.com/pdf/SPE_Spec_en.pdf (accessed on 26 September 2019).
22. Jaffuel, A.; Huteau, A.; Moreau, S. Direct Liquid Chromatography Tandem Mass Spectrometry Analysis of Glyphosate, AMPA, Glufosinate, and MPPA in Water Without Derivatization. *Column* **2018**, *14*, 36–40.

© 2019 by the authors. Licensee MDPI, Basel, Switzerland. This article is an open access article distributed under the terms and conditions of the Creative Commons Attribution (CC BY) license (http://creativecommons.org/licenses/by/4.0/).

MDPI
St. Alban-Anlage 66
4052 Basel
Switzerland
Tel. +41 61 683 77 34
Fax +41 61 302 89 18
www.mdpi.com

Separations Editorial Office
E-mail: separations@mdpi.com
www.mdpi.com/journal/separations